D0850527

THE
DESCRIPTION AND CLASSIFICATION
OF VEGETATION

BIOLOGY SERIES

General Editor: R. Phillips Dales
Professor of Zoology in the University of London at Bedford College

U.S. Editor: Arthur W. Martin
Professor of Physiology, Department of Zoology, University of Washington

---

Structure and Habit in Vertebrate Evolution
G. S. Carter

The Biology of Estuarine Animals
J. Green

Animal Mechanics
R. McNeill Alexander

Marine Biology: An Introduction to Its Problems and Results
H. Friedrich
Translated by G. Vevers

Practical Invertebrate Zoology
Compiled by R. P. Dales

Molecular Biology and the Origin of Species
Clyde Manwell and C. M. Ann Baker

The
Description and Classification
of Vegetation

by

DAVID W. SHIMWELL

Lecturer in Biogeography and Conservation
Department of Geography, University of Manchester

(*Quot homines tot sententiae*)

UNIVERSITY OF WASHINGTON PRESS
SEATTLE

First published in Great Britain 1971

Copyright © 1971 by David W. Shimwell

Published in the United States by
the University of Washington Press 1972

Library of Congress Catalog Card Number 75-180238

Printed in Great Britain

To my parents

CONTENTS

LIST OF FIGURES

PLATE

PREFACE

The subject of vegetation description and classification is extremely diverse and complex, and in no other branch of biology or indeed science does the Latin maxim *quot homines tot sententiae* apply more aptly than to the diversity of opinions expressed on the subject. This diversity is such that a declaration of my intention to tackle the subject as a single text invoked reactions from several eminent ecologists to the effect that a comprehensive review is only possible by a team of writers each covering his own line of approach. Having completed the manuscript I would in part endorse these sentiments, mainly from the point of view that there is not scope in a single text to include data on the background to statistical and computer methods or to include details of the vegetation types recognized by a particular school of formal classification. But what I feel can be and has been readily achieved is a review of the basic characterisitcs of vegetation and the ways in which they can be and have been used to describe and classify vegetation. As such, I would like to think that the text provides a basic manual for the ecologist and phytosociologist of the various methods and their relative popularity and value for depicting the background to a particular ecological or phytogeographical problem.

Inevitably, the text must have numerous shortcomings simply because of the diversity of material available for review. Thus, there is more emphasis placed upon certain features of vegetation and methods than others. This applies particularly to the sections on the development of a functional life-form system classification and the methods of the Zurich–Montpellier School of phytosociology. I make no apologies for this detail since the greater part of plant geographical literature involves the use of physiognomic and functional characteristics for vegetation description especially in the New World and the tropics, while the Z-M system of phytosociology is the most universal of all systems.

The book is divided into eight chapters. The first section comprising Chapters 1 and 2 provides an introduction firstly to the nature of vegetation via quantitative studies on species abundance, density of individuals, and correlations between species occurrence; secondly, Chapter 2 attempts to review the major trends and traditions in vegetation classification and the use and abuse of the term association and formation.

The next three chapters are devoted to physiognomic, functional and structural features of vegetation which are useful for descriptive purposes and is aimed at the wider plant geographical platform of description. More detailed studies on structure in time are considered in Chapter 4, while Chapter 5 illustrates the major methods of vegetation analysis in terms of these structural and physiognomic features.

Chapter 6 is devoted to methods of analysis which are primarily floristic and the two chapters contain details of the formal methods used by the Zurich–Montpellier, Scandinavian and Danish Schools of phytosociology.

The belief that vegetation is a continuum rather than a series of classifiable units forms the theme of Chapter 7 via a study of the various methods of direct and indirect gradient analysis. Finally, Chapter 8 assesses the present state and future needs of vegetation description and classification and reviews the applications of the subject via education and conservation.

Each chapter is thus an entity in itself with a minimum of cross-reference to other chapters and a minimum of author's *sententiae* in an attempt not to fall foul of the Latin sub-title! The most eminent British ecologist, Sir Arthur George Tansley, said in 1922 (*J. Ecol.*, 10, p. 245): 'The increasing recognition of the genetic principle on the Continent is a great gain on the one hand, and increased attention in England and America to the Continental methods of analysing communities will also contribute in an important degree, to mutual understanding and eventual accord.' This text is offered as a contribution to the achievement of such understanding and accord.

In the compilation of the book I have had valuable assistance from a number of persons to whom I wish to offer my gratitude; to Dr D. J. Bellamy for his suggestion that I should undertake the project and for his initial enthusiasm and direction; to Professor Reinhold Tüxen for his hospitality during my stay at Todenmann and for his unfailing patience in expounding some of the basic phytosociological principles; to my wife and Mr Andrew R. Hall for drawing the text figures; to Messrs T. C. D. Dargie and J. S. Ledwood for permission to include unpublished data; to the Geography Department, University of Hull for permission to reproduce Plate I; to Blackwell Scientific Publications Ltd. and C. V. Engelhard & Co. for permission to include figures; and finally to Miss Eileen M. Sharpe for her skill and patience in the typing of the manuscript.

Botany Department
University of Hull
September 1971

INTRODUCTION

Most people are familiar with the meaning of the word *vegetation* either in the form of its definition from the *Oxford English Dictionary*— 'plants collectively' or in the form of a more complex definition supplied by the ecologist, for example that of Tansley (1939): "Plants are gregarious beings, because they are mostly fixed in the soil and propagate themselves largely in social masses, either from broadcast seeds (or spores), or vegetatively by means of rhizomes, runners, tubers, bulbs, or corms; sometimes by new shoots ('suckers') arising from the roots. In this way they produce vegetation, as plant growth in the mass is conveniently called, which is actually differentiated into distinguishable units or plant communities." But to elaborate let us begin with the descending definition spiral typical of many sciences. Simply, on a regional basis, vegetation may be regarded as being composed of all the different types of plant communities within the region. The *plant community* in turn is not merely a random aggregation of plants but an organized complex with a typical floristic composition and morphological structure which have resulted from the interaction of species populations through time. The plant community may be studied by reference to a *stand* of that community which is a concrete example of the community in question located in the field. On the other hand, the *community-type* is an abstract concept based upon the knowledge of the nature of several similar community stands.

The term *species population* mentioned above is the fundamental vegetation characteristic and it is the continuous flux of species populations which makes the vegetation and the community so variable. The distribution of each species population is affected by both interspecific and intraspecific factors which are directly related to the genotypic adaptability or phenotypic plasticity of the species. Each species population has a potential optimum size which is affected by internal and external interference so that the population size and space is invariably modified according to the competitive, reproductive and tolerance capacities of the individuals, relative to the magnitude and type of interference. This is the first level of sociological organization. Secondly, every species population has certain essential requirements of its physical environment or habitat. Several species have pronounced

1

climatic or edaphic requirements and will only grow under such conditions. Every species has a characteristic requirement for a range of external physical environmental conditions for growth, i.e. has a certain *ecological amplitude*. Different species have different ecological amplitudes and this phenomenon results in variations in the specific composition of vegetation—the second level of sociological organization. Adding the two levels of sociology together, it may be stated that each species has its own *ecological niche*. The sum total of ecological niches of both plants and animals plus their environment may be referred to as the *ecosystem* (Tansley, 1935) or the *biogeocenose* (Sukachev, 1947).

The different levels of social organization within the plant community and the ecosystem present several characteristics to the observer which appear to be useful for community description. Each observer sees the problem in a different light and tends to place greater or lesser weight on a particular character for description. *Quot homines tot sententiae.* Whereas the students of vegetation in former years were mainly plant geographers who combined vegetation description with geographical exploration, the modern student has become more orientated towards study of vegetation at the intensive level. Here the quantitative aspects of species populations and their pattern, and species interrelationships become of importance and are often the most rewarding to the investigator. Quantitative characteristics such as population density, cover, frequency and biomass are all of extreme interest on an intensive scale. On a more extensive scale these characteristics become increasingly more difficult to record due to the time and volume of work involved and consequently approximations must be made or new characteristics sought. Qualitative characters of vegetation and communities rather than species populations assume greater importance. Floristic composition, layering, structure, physiognomy, function, periodicity and interspecific correlations are some of the more useful characteristics for vegetation description at an extensive level.

With the accumulation of data from vegetation descriptions there develops an increasing desire for a reference scheme or classification for the data and hence the vegetation. Again the dilemma of the decision on the characters useful for classification raises its ugly head and diverse approaches are developed. *Quot homines tot sententiae.* And so we arrive at the imbroglio of present-day *phytosociology* or *phytocenology*, for as Kuchler (1967) points out there is not even universal agreement on which is the more etymologically and philosophically correct name for the subject. Clearly there is a need to begin with some of the basic premises of vegetation description and classification. The next two chapters are therefore devoted to this end. Chapter 1 reviews the methods of analysis on an intensive scale, recording such species population characteristics as density, frequency and cover, and then proceeds to review some simple statistical methods for the detection of non-randomness and the study of pattern within the populations. Chapter 2 is aimed at a similar basic introductory level, and by means of 'genealogical trees' attempts to place the views of the major investigators in relation to one another to give some understanding of the development of different concepts. Once these basic background reviews are comprehended the broader studies on vegetation structure, function and floristics, structural, functional and floristic classifications and gradient analysis of subsequent sections may be readily interpreted in relation to this introduction.

Chapter 1

THE QUANTITATIVE DESCRIPTION OF
SPECIES POPULATIONS

Practical methods

The approaches to vegetation description can be conveniently divided
into four main analytic methods involving the use of (a) quadrats,
(b) transects, (c) isonome studies, and (d) plotless samples. The first
two are widely applied in primary surveys of vegetation on a large
scale, whilst all are easily used at the intensive level of study.

Quadrats

The *quadrat* is conventionally a rigid square or rectangular structure
of variable size, but usually a metre square. Alternatively, a circle of
0.01 m² has been used by Raunkiaer (1934). Clapham (1932) adopts a
strip shape whilst Levy and Madden (1933) were probably responsible
for the introduction of the point quadrat. Perhaps the most common
type in use in modern phytosociological investigations at an intensive
level is the multiple quadrat of the type outlined by Archibald (1949).
He used a light, portable metal frame, one metre square, subdivided by
flexible cross-strings to give units of smaller size, which varied according
to the community type under study. Floristically diverse communities
were analysed with a multiple quadrat providing squares of 25, 625,
2,500 cm² and 1 m², whilst floristically poor communities permitted
the use of larger squares of 4, 16, 64 m². Additional types in use today
are often divided into smaller units each 10 cm² to give progressive
increases in area of 100, 2,500 and 10,000 cm², respectively. Archibald
considered it necessary to understand the relationship between the
average species number per unit area relative to a progressive increase
in unit area if the 'specific character' of the community is to be estimated,
and his quadrat studies were aimed at this end. By using quadrats of
increasing size data on the species number, density, cover and frequency can be obtained as a basis for statistical analysis aimed at a
definition of the 'specific character' of the communities in quantitative
terms. In each quadrat (a) the number of species can be recorded to

show increase with area; (b) the use of a knitting needle or skewer thrown at random as a point quadrat will score hits and give a value for percentage cover; (c) the numbers of individuals of each species counted at each quadrat size will give an indication of density distributions; (d) the percentage frequency of the species in the whole quadrat can be calculated by recording species presence or absence in each of the smaller unit areas.

Taking each of these points in turn a deeper understanding of the nature of vegetation may be obtained. For the time being the first aspect of quadrat study is best postponed to the section on minimal area and homogeneity where its value becomes more obvious. The second aspect, *cover*, may be defined as the area of ground occupied by a perpendicular projection on to it of the foliage and stems of individuals of a particular species. Cover values are usually expressed as percentage figures and may be either estimated or measured. Estimations are in common usage in broad surveys and usually appear in the form of cover-abundance scales as numbers or letters (see Chapter 3). Such methods of cover categorization are subject to personal errors, each observer recording a slightly different conception of the actual amount of cover. Measurement of cover is conveniently recorded using the point quadrat method, which involves the recording of species hit by pins or knitting needles suspended in a frame which can be adjusted to the height of the vegetation. The pins may be lowered individually and all the species hit by the pin recorded. For each species, the total number of hits from a series of sample frames is expressed as a percentage of the total number of pin 'shots'. The total percentage cover will nearly always exceed 100% except in open vegetation types, because of the layering and overlapping of species at several points.

Percentage cover measured in this manner is a good method for the calculation of species abundance particularly in grassland vegetation, a type where large-scale layering is minimal and where a detailed analysis of herbage composition is of applied value. There are three main drawbacks to the universal use of this method. First, the inevitably slow progress of sampling can be tedious and it is obviously impracticable to apply its use to large-scale surveys. Secondly, its use is more or less restricted to low, unlayered vegetation types since in rank tall-herb communities or communities with pronounced layering it is often difficult to observe the point hit of the needle. Thirdly, there is often an exaggeration of the estimate of percentage cover when a large-diameter pin is used (Goodall, 1952). In general, there is a progressive increase in cover percentage with the increase in point size (Table 1). Beginning with a pin of 'infinite diameter' (an ordinary household pin mounted in a stick or an optical cross-wire sighting tube

TABLE 1

Pin diameter and cover estimates

Species	Number of points	Estimated cover (*pin diameters in mm*)				
				(1·84G)		(4·75G)
		0	1	2	3	5
1. Ammophila arenaria	100	46·0	63·0	65·5	66·0	78·0
2. Ammophila arenaria (Goodall)	200	39·0	–	66·5	–	71·0
3. Ammophila arenaria (Goodall)	200	60·5	–	74·0	–	82·0
4. Festuca rubra	100	67·0	66·5	76·0	79·0	82·0
5. Glaux maritima	100	18·0	34·5	37·5	39·0	42·5
6. No contact	200	57·5	50·5	40·5	37·5	36·0
7. Fumaria officinalis (Goodall)	200	20·5	–	31·5	–	30·0
8. Meum athamanticum	200	25·0	32·5	48·0	46·5	40·5
9. Ranunculus acris	200	16·0	24·0	32·5	33·5	30·5
10. Anthoxanthum odoratum	200	64·0	66·0	74·0	72·0	68·0
11. Thuidium tamariscinum	200	13·5	17·0	16·0	17·0	21·5
12. Hylocomium splendens	200	9·0	17·5	18·0	22·0	24·0

Source of data: 1, 4, 5, 6, dune grassland, foreshore and dune slacks, Spurn Head, Yorkshire; 2, 3, 7, from Goodall (1952); 8, 9, 10, 11, 12, hill farmhouse pastures, Mid-Garraries, New Galloway.
Pin-type: 0, household pin; 1, frame pin; 2, match-stick; 3, No. 9 crochet hook; 5, garden cane; G. Goodall.

of the type used by Winkworth and Goodall, 1962), progressing through a quadrat frame pin or knitting needle (1 mm), a match (2 mm) to a slender garden cane (5 mm) the progressive increase may be noted. The examples of *Ammophila arenaria* and *Festuca rubra* illustrate this trend neatly. On the other hand, the data for species such as *Meum athamanticum* and *Ranunculus acris* indicate a peak cover value at 2 mm (Table 1). This is probably directly related to the habit or life form of the plant, *Ammophila* and *Festuca* being tussock and stoloniferous species, *Meum* and *Ranunculus* single-stalked, non-vegetative reproducing plants. Non-stoloniferous grasses also seem to reach a peak value at 2 mm although the data for *Anthoxanthum* is somewhat inconclusive. Mosses and herbaceous species in open communities show an increase in cover value with pin size. All these statements are rather broad generalizations and there appears to be little comparative data apart from that of Goodall, which will serve to either uphold or negate the suggested relationships between peak cover values, pin size and life form. However, this is not a project to inspire even the most dogged student! Finally, in this discussion, the term *cover repetition* needs to be explained. Instead of simply recording species hits with a point pin, the number of times a species encounters the pin is recorded and forms the basis of a measurement of performance. Data on this topic is confusing and seems to be subject to a large amount of sampling and personal error. Greig-Smith (1964) remarks that cover repetition values will increase with pin diameter and that values are also likely to be increased for most species if the pins are inclined. On the other hand, Winkworth (1955) has shown that this is not necessarily the case and that the opposite effect of decrease in value may be caused. Pointing out that the main object of point cover repetition studies is some representation of the foliage area of a species per unit area of ground, Wilson (1959a, 1959b) indicates that inclined contacts will never give as wide a range of foliage cover values as vertical contacts which will theoretically range between 0 and 100% for vertical and horizontal foliage, respectively. He also indicates that the angle at which the foliage is held will affect cover repetition determination with either vertical or inclined points, and after detailed formulation concludes that an angle of pin inclination of 32·5° with the horizontal is least affected by foliage angle and will give the most representative estimate of foliage area. The foliage index area is calculated by multiplying the number of contacts per quadrat by the factor 1·1 which will give a value within ±10% accuracy.

Foliage area at different heights above ground is also of interest especially in agriculture, since it will indicate the proportion of sward which will be removed by grazing or mowing. If the point quadrat is marked at appropriate intervals, records of foliage area at each centi-

TABLE 2

Estimated cover and leaf area for *Cynosurus cristatus* in a hay meadow

Sample	Mean number hits per 100 point quadrats	Leaf area index
(a) Total crop (vertical)	8·78	
(inclined)	22·4	24·64
(b) Lower layer (v)	6·2	
(to 10 cm) (i)	11·3	12·43
(c) Upper layer (v)	3·3	
(i)	13·6	14·96
(d) Stubble (v)	7·1	
(i)	12·1	13·31

metre layer may be assessed. When a quadrat is inclined, however, more than 1 cm length occurs in each horizontal 1 cm layer, calculated as cosec 32·5° − 1·86 cm. The data in Table 2, besides providing a comparison of sampling with vertical and inclined pins also demonstrate the type of result to be expected from leaf area index investigations on a typical herbage grass of a hay meadow, *Cynosurus cristatus*. Detailed stratal sampling of the two layers above and below 10 cm (a level chosen as roughly equivalent to the cutting height of a mowing machine) produce an even better estimate of leaf area, whilst the sampling of the stubble after mowing gives a more accurate measure also, due to the comparative ease of observation on a short sward. The figure for the upper layer provides a measurement of *Cynosurus* performance, and data on other species will enable an understanding of the composition of the hay crop.

The term *density* refers to the number of individuals per unit area usually calculated by a simple count of plants in a series of random quadrats and the expression of this as an individual: quadrat size ratio figure for the total sample. The method can be extremely time consuming but is also extremely accurate allowing direct comparisons of quadrats in different areas. The density distribution varies greatly from individual to individual mainly as a function of reproduction type (sexual or vegetative), dispersal type and population age. The differences in density of three species, *Senecio jacobaea*, *Geranium robertianum* and *Hornungia petraea* are shown in the sample area in Figure 1. *Senecio jacobaea* has an irregular distribution associated with its light wind-blown propagule; *Geranium robertianum* has a markedly clumped pattern again correlated with the length of seed throw in its sling dispersal mechanism; *Hornungia petraea* shows a high density of individuals around a single presumed parent plant with no specialized dispersal mechanism. If a multiple quadrat of increasing area 100 cm², 2,500 cm², etc., is used to analyse the sample area the three density patterns show fairly diagnostic trends. The closely aggregated *Hornungia* reach optimum density at a smaller size of quadrat as is also the case with young clonal species such as *Hieracium pilosella*. But by far the largest number of species fall into the second category exemplified here by *Geranium* and *Senecio* where density increases to a median quadrat size for optimal density.

The *frequency* of a given species refers to the chance of recording it in any single quadrat throw. If a species has a frequency of 100% per 1 m² it will be present in each 10 cm² quadrat unit. Comparison of its occurrence in a number of quadrats in the same vegetation type gives rise to a knowledge of the species' overall frequency. Greig-Smith (1957) has defined and used two types of frequency—*shoot frequency* obtained by recording as present all foliage overlapping into a quadrat;

rooted frequency records a species present only when it is actually rooted in the quadrat. Frequency determinations are subject to several sources of error and depend upon three major factors enumerated by Kershaw (1964):

(*a*) *Quadrat size* is one obvious factor affecting frequency figures and is one which needs no further elaboration except to emphasize that quadrat size used should be stated;

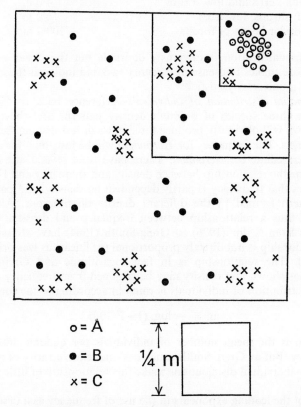

Figure 1. Densities of three species populations on a scree slope in Biggin Dale, Derbyshire. A, *Hornungia petraea*; B, *Senecio jacobaea*; C, *Geranium robertianum*.

(*b*) *Individual plant size*—again a factor which needs little explanation. Individual plants of *Carex paniculata* reach a clump size of up to 100 cm diameter with a shoot diameter three times this figure. *C. disticha* is a single-stemmed rhizomatous species and calculations of either rooted or shoot frequency on quadrats in mixed fen communities where this and the latter species occur give markedly different frequency values for the two species. Because of the obvious relationship of species to area several investigators have attempted to standardize the size of quadrat used for different layers of vegetation where certain species predominate. Cain (1932) has suggested the following sizes:

1. Soil layer, cryptogam-dominated layers   0·01 or 0·1 m²
2. Herbaceous layers   1·0   or 2·0 m²
3. Rank herbs and low shrubs   4·0 m²
4. Tall shrubs, low trees   16·0 m²
5. Superior layers of forests   100·0 m²

Some standardization is obviously desirable but the *Carex* example cited above makes nonsense of category two and indicates the impossibility of a universal yardstick.

(*c*) *Spatial distribution of individuals*—reference back to Figure 1 with the three species of different density patterns and the quadrat size shown will obviously produce three more or less distinct frequency percentages. The frequency for *Hornungia* will be low, high for *Senecio* and intermediate for *Geranium*. There have been several attempts to determine the relationship between density and frequency and Figure 1 illustrates that frequency is partly dependent on density, and partly on the pattern formed by the different density distributions. Although there is thus a relationship between frequency and density, various workers from Kylin (1926) to Greig-Smith (1964) have stressed that the relationship is not directly proportional (or linear) as was originally assumed. The relationship is in fact logarithmic and Greig-Smith (1964) has shown that density may be obtained from frequency counts, if the distribution of individuals is random according to the formula

$$m = - \log_e (1 - F/100)$$

where m is the mean number of individuals per quadrat and F the frequency. But as Greig-Smith concludes, the relative rarity of random species—individual distributions make this relationship of little importance.

One of the leading exponents in the use of frequency as a descriptive tool was Raunkiaer (1918, 1934) who grouped species frequencies into five equal classes A to E as follows:

Class A	0—20%	53% total
B	21—40%	14%
C	41—60%	9%
D	61—80%	8%
E	81—100%	16%

His *law of frequency* which was based on an extremely detailed survey of 8,087 frequency claculations, stated that

$$A > B > C \gtrless D < E$$

or 53 / 14 / 9 / 8 / 16 expressed as percentages of all frequencies, which produced a characteristic reversed J-type of histogram (Figure 2). Kenoyer (1927) was able to confirm this overall shape and on the basis of 1,425 frequency percentages from American vegetation types produced figures to the effect—A 69%, B 12%, C 6%, D 4%, E 9%. Raunkiaer (1934, p. 397) viewed the distribution pattern as one deter-

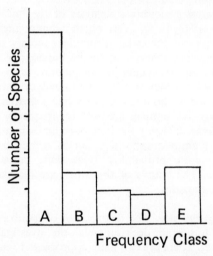

Figure 2. Raunkiaer's J-type of frequency distribution.

mined by the dominant species (dominance in terms of physiognomy) which were best fitted to exist under the environmental conditions thus restricting other species by competition, allowing them to exist in small numbers. This accounted for the low values of classes B, C and D and high A. Kenoyer offered a similar explanation and concluded that the differences between the class percentages derived from American examples were due to the greater richness of the American flora compared with that of Scandinavia, and the fact that American workers assessed rooted frequency whilst the Scandinavians recorded bud (shoot) frequency.

There have been several reviews of Raunkiaer's law of frequency. Gleason (1929) wrote explaining the law in the simple terms that in any community there are more species with few individuals than with many and that the law only applies to a certain size of quadrat, being obscured or lost if the quadrat size is too large or too small. Cain (1934), following up this criticism, prepared frequency distribution patterns from data collected in nine different-sized quadrats from 1 m² to 100 m² ignoring Raunkiaer's size of 1/10 m². In four of the smaller quadrat sizes (1, 3, 4 and 9) he recorded a class E percentage higher than D, and showed also the decrease in A, increase in E with increase in quadrat size—a neat demonstration of Gleason's suggestions. Other workers have gone so far as to suggest that the correct size of sample quadrat is that one which produces a J-distribution, and if the quadrat analysis does not give rise to such a histogram then the quadrat should be accordingly adjusted (Lüdi, 1928).

Perhaps the most penetrating analysis of the law of frequency is that of Preston (1948) in an essay on the commonness and rarity of species. He points out that since Raunkiaer mainly used twenty-five quadrats for frequency determination the boundaries as often quoted are incorrect. If a species occurs in five quadrats it has a frequency of 20%, if in six the frequency is 24%. The boundary should therefore be 22% and so on for all boundaries. This has a greater meaning when more than twenty-five samples are used for frequency determinations. He further discusses critically the boundary limits in a series of groups of species rarity/commonness which can be related to the lognormal distribution curve—a probability (Gaussian) curve drawn on a logarithmic base. The details of this work are discussed later in the section on basic statistics.

Quadrat siting and size. The distribution of quadrats in an area of vegetation is an important decision which the investigator must decide upon at the outset. There are two main approaches to quadrat sampling —the single plot method which consists of closely examining a single quadrat in each community type or each stand of a community and the multiple plot method where the sample consists of a series of quadrats scattered through a stand type and their overall properties are taken as the representative stand assessment. Within these two types the location of the quadrats may be decided by five methods: (*a*) the subjective assessment of a sample which gives the impression of being fairly representative of the stand or community type; (*b*) a series of partial random samples; (*c*) a series of random samples; (*d*) a series of regular spaced samples in a checkerboard arrangement; (*e*) a series of contagious samples in the form of a belt transect.

The first method involves the observer's own intuitive concepts of how a representative sample or samples should appear and is closely tied up with what is known as homogeneity. If a stand appears to be heterogeneous it is common practice to divide the area up into a convenient number of even-sized sub-areas which are more homogeneous than the whole, and to sample at random within these sub-units—a partial random sampling technique. With true random sampling, the idea is to so distribute a set of quadrats that the location of each quadrat is independent of all the other quadrats and also of any prominent features of the stand. For example, if there is a note-worthy rare species in a stand or a clump of individuals of an obvious growth form, subconsciously the observer feels that these ought to be represented in at least one quadrat. Consequently, if a random sample is being obtained by the tossing of a stick over the shoulder with eyes closed there is inevitably a desire to throw the stick so as to include one of the prominent features observed. Greig-Smith (1964) has shown that this action of selecting random samples is an extremely instructive exercise and that if the landing position of the stick is plotted and the positions tested for randomness by a grid analysis method, in most cases the samples will not be found to be random. He considers that random samples are most readily obtained by laying down two lines at right angles as axes and then using random numbers (obtained from standard statistical tables such as those of Fisher and (Yates 1943) for co-ordinates to position the samples. Cain and Castro (1959) commenting that tables of random numbers are not always available, suggest a method of obtaining a random sample based on the use of common playing cards. The two axes are laid out and the units along them designated in groups of ten as spades, hearts, clubs, diamonds. Face cards are discarded and the remaining deck shuffled, cut and the fifth card dealt used to locate the coordinate on the x axis. The pro-cedure is repeated to give a position on the y axis. The drawn cards are returned to the pack and the process repeated to give a series of random samples.

Two types of sampling remain—regularly spaced samples in which the quadrats are scattered regularly throughout the stand, and contiguous quadrats. Regular samples seldom have an advantage in information gained over random samples, but there are a number of exceptions. Where there is an obvious stripe pattern or ridge and furrow as is often found in upland regions due to periglacial sorting of soil com-ponents or simply in wet meadows drained by parallel regular land drains, then it is clearly advantageous to space the quadrats so as to hit either a ridge or a furrow, since the associated species and vegetation nearly always vary. Alternate samples of ridge and furrow would eventually lead to the recognition of two community types whereas random sampling would always include quadrats which were half ridge, half furrow and inevitably heterogeneous. Similarly, the con-tiguous method of quadrat sampling is also likely to include hetero-geneous quadrats which occur across a readily observable boundary. The main form of contiguous sampling is the belt transect (q.v.), but the related method of accumulated rectangular forest plot sampling has been used widely in the study of tropical forest vegetation. A base line is laid out and subplots 10×20 m either side the line sampled until data on a rectangular plot has been accumulated.

With respect to quadrat size, the two concepts of *homogeneity* and *minimal area* have invoked a voluminous literature. Homogeneity is regarded by most phytosociological schools, but especially the Scandinavian workers, as the first important feature to be taken into consideration in the choice of a sample quadrat for description. A neat definition is provided by Dahl and Hadac (1949):

'A plant species is said to be homogeneously distributed in a certain area, if the possibility to catch an individual of a plant species within the test area of given size is the same in all parts of the area. A plant community is said to be homogeneous if the individuals of the plant species which we use for the characterization of the community are homogeneously distributed.'

This approach can be conceived as the underlying approach to the selection of a stand of vegetation by most of the traditional schools of phytosociology. Modifications of the basic definition are rife but the simple subjective assessment of homogeneity by eye (*Pflanzenso-ziologieblick*) prevails in European phytosociology. Goodall (1954a) attempting an objective assessment of vegetation units argues that if the unit really exists it should show at least some degree of homogeneity and that this should be greater within the stand than between two stands. This difference in itself should be sufficient to allow recognition and definition of stands. Lambert and Dale (1964), however, suggest that if the stand concept is abandoned in favour of more objective sampling, the tests of homogeneity are not necessary and the more satisfactory situation of homogeneity emerging from analysis, rather than its imposition at the outset, obtains. These three views subjective, objective and rejection are taken as tacit examples of the numerous views on this subject, the debating chamber of phytosociology. One of the most recent theoretical approaches to the problem is revealed in van Leeuwen's 'Relation Theory' (1966) where complicated relationships in vegetation and ecosystems are reduced to four simple elements, equality and inequality in space and time. Recourse to the derivations of this theory leads van Leeuwen to conclude that homogeneity and stability of vegetation cannot occur together in the same situation (see Chapter 4).

Perhaps the most important concept associated with homogeneity is that of *minimal area*. Again there are various terms and definitions involved with this concept but all have a common basis in the idea that minimal area is the smallest area which provides enough environmental space (environmental and habitat features) for a particular community type or stand to develop its true characteristics of species complement and structure. Three methods of the determination of the minimal area of a community have been described, one based on species composition, a second on species frequency and a third on homogeneity of composition. The first in its species-area relation curve is the most widely discussed method in the literature. Such curves are produced when sample area is plotted on the x axis of a graph and species number or percentage of total species on the y axis. Two sampling procedures have been followed to determine this species-area curve (Figure 3). An initial plot obviously smaller than the minimal area is chosen and the species present listed. The plot area is then doubled and additional species appended to the list, but keeping them separate from the species in the first plot. This process is continued as long as it is deemed necessary in a manner variously figured by Poore (1955) and Cain and Castro (1959) (Figure 3, A) as a series of 'nested plots'. Sizes for discontinuation vary. Temperate vegetation types seldom require continuation beyond 100 or 128 m^2, but tropical types may require up to 256 m^2 to produce the appropriate curve. The second method is the accumulated rectangular forest-plot method used by Cain and Castro for tropical rain forests and described previously in connection with sampling (Figure 3, B). In this case, the succession of sub-plots will build up a species-area curve. Such curves all have a similar form, rising sharply at first and curving gradually to almost parallel the x axis. The region at which it begins to flatten off has been accepted by various plant sociologists as the minimal area and the minimum-sized quadrat necessary for stand description (Figure 3, C). Any stands of less than the minimal area are regarded as fragmentary (Braun-Blanquet, 1951). A detailed review of the use of the minimal area methods by Hopkins (1955, 1957) led him to the conclusion that the minimal area could not in fact be determined objectively from the species-area curve. This conclusion is echoed in part by Cain and Castro (1959) who decided that if the species-area relationship is to be a useful tool, the selection process of the minimal area should be more accurate, mechanical and not merely visual. They suggest that any range of percentile rise should be detectable on the curve and propose a simple method involving the use of a triangle and a ruler. For example, if it is desirable to determine which point on the curve represents a co-ordinated 20% increase of total species number for a 20% increase in sample area, the triangle edge is placed at the base of the chart so that it passes through the intersection of x and y and through the point which represents the co-ordinate of the two 20% scores. A line parallel to the edge of the triangle and at a tangent to the curve marks the area where rate of increase of species and rate of increase of area

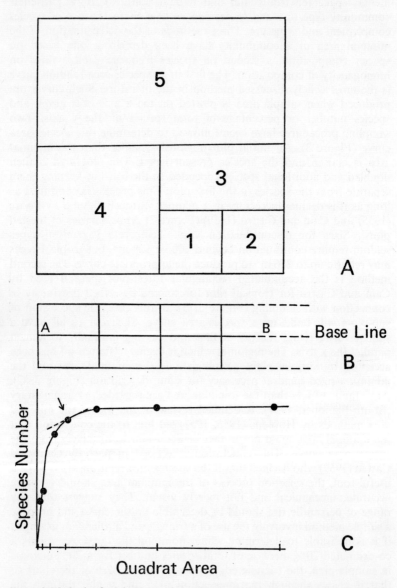

Figure 3. Methods of minimal-area determination: A, sample area doubling method; B, accumulated rectangular forest-plot method; C, typical sample species-area number curve.

are equal. A ruler placed along the right-hand side of the triangle may be slid upwards until its end describes this tangent. The mid-point of the tangent represents the quadrat size to be chosen and the investigator is free to choose any ratio between species increase and area increase that he desires.

The use of frequency in minimal area determinations probably began with the work of Raunkiaer (1909, 1918) who considered that the basic community unit—the formation—was a 'sensibly homogeneous' area of vegetation dominated by certain species of high frequency. Du Rietz (1921) decided that if frequency is estimated using an increasing quadrat size, and if the number of species with a 90% frequency or greater is plotted against quadrat size, the curve will become parallel to the x axis of the graph. Such species are known as constants and the minimal area comes to be defined as that area in which the full number of constants occurs and remains the same for a further increase in area. Finally, Archibald (1949) suggested that the minimal area for a particular community is that quadrat size which has 95% frequency for at least one species.

The third approach to minimal area is that suggested by Goodall (1954b) who based his determination on the homogeneity of a series of values of quantitative data on species in multiple samples chosen at random. He provided the definition of minimal area as 'the smallest sample area for which the expected differences in composition between replicates are independent of their distance apart', or in other words are due to simple random variation. This approach is more of a direct discussion of the pattern scale of each species rather than a direct attempt at minimal area determination and this aspect is dealt with more thoroughly in the section on pattern. His work also indicates that because of the existence of pattern in seemingly homogeneous vegetation any calculation of minimal area will never be more than an approximation, subjectively based on whether the sample area is large enough to represent the characteristic structure and floristics of a plant community.

Transects and isonomes

The *transect* approach to sampling is usually followed where there is a readily observable gradation in vegetation in relation to a marked environmental gradient. A base line is laid down and quadrats sampled at known intervals along the line either in a contiguous transect or a regular interrupted transect form. The transect may be in the form of a belt or strip with several quadrats at each sampling point or simply a series of point quadrats along a tape measure. All the usual measures of density, cover and frequency can be employed in the description of a transect and the data obtained is conveniently presented in the form of a

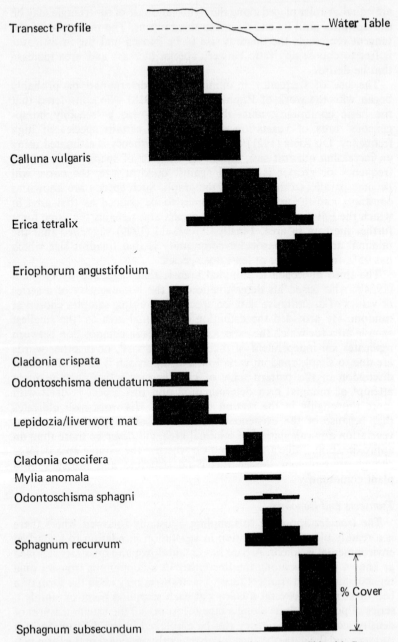

Figure 4. Transect across a dry heath–wet heath transition zone at Skipwith Common, East Yorkshire. (Figures are for percentage cover assessed subjectively.)

graph or histogram plotting species measures against position on the transect. The best examples of transect representation of vegetation zonation are to be found in the work of the Scandinavian schools of vegetation description in regions where there is often a clear-cut zonation of vegetation types due to such environmental features as altitude, frost heave and late snow-beds. The method of representation is exemplified with data from an environmental transition of dry heath to wet heath from Skipwith Common, Yorkshire (Figure 4). The zones are fairly clear-cut in relation to the height of the water table and each zone can be seen to have a series of characteristic species which are more or less restricted to that particular zone, e.g. *Odontoschisma denudatum* to the dry heath, *Sphagnum subsecundum* in the wet heath and *Cladonia coccifera* to the transition zone. Similarly, the increase of species abundance can be picked out at a glance, e.g. the increase of *Erica tetralix* towards the wet heath.

The variation in the distribution and abundance of species may also be illustrated by the *isonome* method introduced by Ashby and Pidgeon (1942). A grid of contiguous quadrats is laid down at random in a selected sample area and each species is recorded for abundance. For each species the density distribution is recorded on squared paper and squares with closely similar values joined by a series of lines or isonomes to give a contoured representation of species abundance. On to this may be superimposed environmental data such as pH values or variations in the microtopography. The approach is neatly illustrated in Figure 5 using data from a single metre square quadrat chosen at random and divided into one hundred 10 cm squares each of which is divided into quarters. The density for four species is calculated and plotted against microtopography and pH based on readings in each of the 10 cm squares. The two mosses *Sphagnum papillosum* and *Campylium stellatum* and the other species pair *Carex demissa/Narthecium ossifragum* are picked out by the two environmental parameters; *Campylium* and *Carex* having their highest densities in the lower flushed areas of higher pH, *Sphagnum* and *Narthecium* being more or less restricted to the higher hummock topography and more acid pH values. The method exemplifies the danger of a random quadrat thrown in heterogeneous vegetation or in an area where the minimal area for the community is small. The environmental gradients can be seen to be too sharp to enable an accurate assessment of the floristic relationship of the communities with other wet flush types. In some phytosociological traditions there is therefore a trend towards altering the rigid square, rectangular or circular quadrat and taking an area which looks 'sensibly homogeneous' (*sensu* Raunkiaer) where the micro-environmental gradients may be assessed to be small or minimal and where broad environmental gradients are avoided. Thus, the

B

sample area in Figure 5 would be divided into two irregular areas which correspond roughly with the *Sphagnum-Narthecium* hummock and *Carex-Camphylium* flush, described and compared with quadrat analyses from more extensive communities of the same two types and

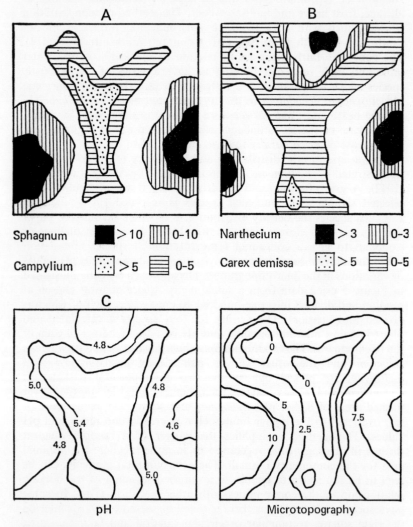

Figure 5. Isonome studies on a quadrat complex: A, density distribution of *Sphagnum papillosum* (1) and *Campylium stellatum* (2); B, density distribution of *Narthecium ossifragum* (3) and *Carex demissa* (4); C, pH distribution; D, microtopography (in centimetres).

if they contained the characteristic floristic elements maintained as a valid quadrat analysis. If the quadrat area proved to be so small as to lack the characteristic elements, the stand would be designated fragmentary.

Plotless sampling

In theory the point quadrat is a method of plotless sampling although it is usually used in association with a certain square quadrat or line transect. Several workers have paid attention to methods of plotless sampling, particularly in forest vegetation where minimal area is difficult to calculate and certain practical difficulties are met with in sampling relatively large areas for tree abundance. The main use of plotless sampling is to be found in the Wisconsin school of phytosociology and appropriate references on the subject are those of Cottam and Curtis (1949, 1955, 1956) who use four different procedures beginning with the selection of random points by the tossing of a quadrat stick (Figure 6).

(*a*) *Closest individual method*—a measurement is taken from the sampling point to the nearest individual, and procedure repeated.

(*b*) *Nearest neighbour method*—the distance from the first individual in method (*a*) to its nearest neighbour. By these methods the density of trees or the reciprocal of density, *mean area* (the mean distance between individuals) can be determined. Theoretically, the mean value calculated by these methods is equivalent to half the square root of the mean area. This is true for the first method but not the second and Cottam and Curtis (1956) found the correction factor for the latter method to be 1·67 instead of the theoretical 2. This factor was calculated empirically from a random population and the discrepancy arises because of the general lack of random populations in Nature.

(*c*) *Random pairs method*—extending from method (*a*) the nearest individual to the sample point is used and a 90° exclusion angle erected on either side of it. The distance from this individual to the nearest one situated outside the exclusion angle is measured.

(*d*) *Point-centred quarter method*—from the sampling point, lines are erected at right angles to give four quarters in which the distance from the sample point to the nearest individual is measured giving four measures at each sample point. In this case the mean of all distances has been shown to be equal to the square root of the mean area (Cottam *et al.*, 1953).

The use of any of these methods enables the calculation of a mean area and thus the density of all individuals of different species. In addition to distance measurements the basal area of each indivdual encountered is recorded and the two properties combined to give an *importance value* (Curtis and McIntosh, 1951; Curtis, 1959). Here the relative density (the proportion of the density of a species relative to that of the stand as a whole) and relative dominance (basal area of a

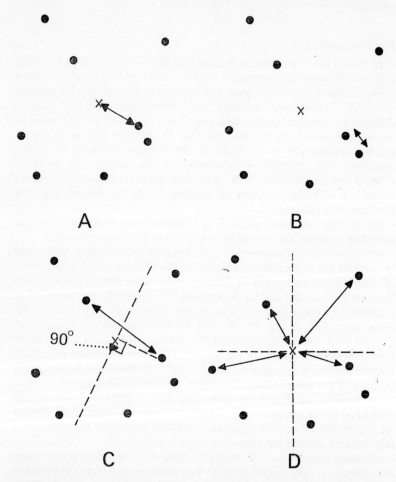

Figure 6. Plotless sampling methods of the Wisconsin School: A, closest individual method; B, nearest neighbour method; C, random pairs method; D, point-centred quarter method.

species to stand area) are summed to give each species an importance value index. Cottam and Curtis (1956) have shown that the accuracy of the estimate of density and hence importance value increase with method in the order enumerated above. The last method, the point-centred quarter method, is the most efficient in spite of the fact that a greater number of samples are taken based on few sample points. Its use is recommended in preference to the random pairs method because of the greater amount of information obtained and the relatively short time involved in fieldwork.

More recently, Yarranton (1966) has described the use of a 1 in. mesh herring net for the sampling of saxicolous bryophyte and lichen vegetation. The sampling interval depends upon the size of the individuals to be sampled and in the case of Yarranton's work samples were taken at every fourth mesh intersection. The point samples are taken where two touching species can be recorded. Where the species hit by the sample point does not contact a second species, the sample is recorded as no contact. The data collected is then used to calculate χ^2 interspecific association values and correlation coefficients (q.v.). The major objection to this contact sampling method and its use in the fashion adopted by Yarranton is that the sampling point may lie at the margin of two microhabitats each occupied by a single species. The net nature of the sample means that the whole of a boulder in all its exposures may be sampled simultaneously which will inevitably supply a large amount of data from microhabitats and major environmental gradients will be obscured rather than clarified. Nevertheless, the herring net method does have certain possibilities in the fairly rapid collection of point quadrat data from relatively homogeneous vegetation.

Some statistical methods and their applications

Greig-Smith (1964) has stressed the importance of sampling a plant community so as to obtain the maximum amount of information and since the collection of quantitative data is extremely time-consuming it is important to approach the subject with some major aim in view. As Kershaw (1964) states the approaches to these major aims are largely a matter of common sense, but often, if the sampling procedure is not correct, then the conclusions inferred may be worthless. Frequently, the procedure may be devised by a careful review of the problems beforehand with only an elementary background of statistical knowledge. But the sampling methods and techniques of comparison and correlation can be complex and advice from a statistician before the exercise is started may prove to cut down both fieldwork and calculation time. Here it is assumed that the reader has a basic knowledge of statistics; that a population (or universe) is the total number of individuals defined by some characteristic of those individuals; that a random sample is a sample from a given population each individual of which has an equal chance of being encountered; that samples will have a variable property which may be represented as a frequency distribution (or variate); that the variate will be distributed over the population in a relatively small number of ways or standard distributions which may be sampled theoretically to give an expected or ideal sample; that these standard distributions will give the probability of locating a given amount of the variate in any part of the frequency distribution; and that the difference between the observed and expected distributions can be studied by the use of a significance test in order to assess the reality of the results and whether this difference is simply an error attached to the sampling of the data. These are perhaps the basic ingredients of the type of statistical knowledge required by phytosociologists and if this knowledge is lacking the elementary books by Goodman (1957) and Moroney (1951) provide excellent introductions to the subject.

The detection of non-randomness

One of the main applications of statistical methods to data obtained from vegetation analyses has been aimed at the detection of non-randomness. The history of the subject stretches back to the work of Gleason (1920) and Svedberg (1922) and the approaches used by these two investigators have become accepted as the standard procedures for detecting a non-random distribution of individuals in a quadrat. The method consists of relating the observed number of individuals of a particular species per quadrat to the expected number derived from the Poisson series which is written:

$$e^{-m} \left(1, m, \frac{m^2}{2!}, \frac{m^3}{3!}, \frac{m^4}{4!} \, \ldots \right)$$

If the distribution of a species is random, then the proportions of quadrats containing 0, 1, 2, 3, 4 ... individuals will fall on the above Poisson series, where m is the mean density of individuals per quadrat and e the base of natural logarithms. From this series the expected number of quadrats falling into each of the classes can be easily calculated. Several workers have tested data collected from random quadrats for a fit to the Poisson expectation and in some cases species have been found to have an almost random distribution (Svedberg, 1922; Blackman, 1935). Svedberg also indicated that several species exhibited the phenomena of either 'overdispersion' or 'underdispersion' in the sense of their relations to the distribution curve of the data not the actual pattern of their individuals and Greig-Smith has advocated the use of the terms 'contagious' and 'regular' in place of these terms in the former context (1964, p. 61). Greig-Smith also discusses in detail the appropriateness of the Poisson series as the basic of random expectation, and further study of the biological aspects of the type of distribution expected is left with this reference.

Several tests and measures of departure from randomness have been suggested. Greig-Smith (1964) lists and critically discusses eleven such measures, two of which are described below, the χ^2 *test of goodness of fit* and the *variance : mean ratio* (coefficient of dispersion). The application of these tests are neatly illustrated by Kershaw (1964) who shows the type of results to be expected from random and contagious populations. In Nature, random populations are quite hard to detect but examples which seem to give the best results are those of the two composites *Taraxacum officinale* and *Senecio jacobaea* which in the author's experience can give a 50 : 50 random : contagious result in population analyses. The observed numbers of individuals containing 0, 1, 2 and 3 individuals from a pasture population of *Senecio* were as follows:

Individuals per quadrat	0	1	2	3
Frequency in 100 quadrats (f)	46	36	15	3

χ^2 *goodness of fit*

Mean population density,

$$m = \frac{Saf}{100} = \frac{36+30+9}{100} = 0 \cdot 75.$$

From the Poisson series $e^{-m}, me^{-m}, m^2/2!e^{-m}, m^3/3!e^{-m}$ the expected number of quadrats containing 0, 1, 2, 3 individuals can be calculated. The first value $e^{-m} = e^{-0 \cdot 75}$ is looked up in tables of Poisson distribution readily available in a shortened form in most statistical books and also in Greig-Smith (1964), Appendix B, Table 3.* Thus $e^{-m} = e^{-0 \cdot 75} = 0 \cdot 4736$.

$$me^{-m} = 0 \cdot 4736 \times 0 \cdot 75 = 0 \cdot 3552$$
$$\frac{m^2}{2!}e^{-m} = 0 \cdot 4736 \times \frac{0 \cdot 56}{2} = 0 \cdot 1326$$
$$\frac{m^3}{3!}e^{-m} = 0 \cdot 4736 \times \frac{0 \cdot 422}{6} = 0 \cdot 0317$$

The expected distribution is therefore:

$$0 \cdot 4736 \times 100, \ 0 \cdot 3552 \times 100, \ 0 \cdot 1326 \times 100, \ 0 \cdot 0317 \times 100$$

hence the data

Number of individuals per quadrat	0	1	2	3
Expected frequency	47·4	35·5	13·3	3·2
Observed frequency	46	36	15	3
Difference	1·4	0·5	1·7	0·2

* The Methuen *Biological Laboratory Data* by L. J. Hale is adequate for most situations in this context.

The χ^2 goodness of fit test is now applied by the use of the formula

$$\chi^2 = \sum \frac{(\text{difference})^2}{\text{expected}}$$

i.e. $\dfrac{(1\cdot4)^2}{47\cdot4} + \dfrac{(0\cdot5)^2}{35\cdot5} + \dfrac{(1\cdot7)^2}{13\cdot3} + \dfrac{(0\cdot2)^2}{3\cdot2}$

$= 0\cdot0413 + \cdot0007 + 0\cdot2173 + 0\cdot0125$

$= 0\cdot2781.$

The obtained figure is looked up in a table of χ^2 with two degrees of freedom—two less than the number of calculations used to obtain the χ^2 total—which gives a nearest value of $0\cdot211$ when $p = 0\cdot9$. The observed data may thus be regarded as an extremely good fit with the expected series and the sample population of random distribution. When figures of $0\cdot05$ (a 5% level of significance) chance of a difference between two sets of figures arise, it is normally considered that another hypothesis is necessary to explain the level of odds. The difference can then be regarded as real or related to non-random populations rather than accidental as with random populations.

Variance : mean ratio. From the data on numbers of individuals per quadrat as above, the average density may be obtained by adding the numbers and dividing by the number of quadrats in each plot as before:

$$\bar{x} = \frac{S(x)}{n} = \frac{75}{100} = 0\cdot75.$$

The variation in observed and expected values may be expressed statistically as the *variance* of the sample:

$$\text{Variance} = \frac{S(x-\bar{x})^2}{n-1} \text{ or algebraically as } \frac{S(x^2) - \dfrac{(Sx)^2}{n}}{n-1}$$

where $n-1$ is the degrees of freedom used for the sample mean rather than the true population mean (n).

From this formula and the data

$$S(x) = 75, \quad n = 100, \quad \bar{x} = 0\cdot75$$
$$S(x^2) = 36(1)^2 + 15(2)^2 + 3(3)^2 = 123$$
$$(Sx)^2 = (75)^2 = 5625.$$

Therefore $\dfrac{(Sx)^2}{n} = 56\cdot25.$

Therefore the variance of the population is

$$\frac{123 - 56\cdot25}{99} = \frac{66\cdot75}{99} = 0\cdot6742.$$

Thus the variance : mean ratio $= \dfrac{0.6742}{0.75} = 0.9$.

This difference from the expected ratio of 1 now needs to be tested by a t test, e.g. $t = \dfrac{\text{Observed} - \text{Expected}}{\text{Standard Error}}$

Because the variance of a Poisson series is equal to its mean, the measure of departure from expectation can be derived from the ratio of the variance to the mean. The departure of this ratio from unity has a standard error of

$$\sqrt{\frac{2}{n-1}} = \sqrt{\frac{2}{99}} = 0.1421.$$

Thus the value of t becomes $0.1/0.1421 = 0.7037$ which on reference to the values for deviation in a normal distribution (not t tables because of the sample size) gives a probability of >0.5. This difference could easily arise by chance and the population distribution may be regarded as random. It is interesting to note that the example quoted here using actual data gives values which are almost reversed for χ^2 and variance : mean ratio calculations. This, however, has little significance other than the fact that its comparison with Kershaw's data gives some idea of the type of range of result to be expected for randomly distributed populations.

Most species populations show a marked contagious distribution which is normally manifest by a highly significant difference between observed and expected numbers of occurrences and the chance of this difference arising accidentally are often in excess of 1,000 : 1. Consider the following data collected from 100 sample quadrats on the frequency of *Geranium robertianum*.

Individuals per quadrat	0	1	2	3	4	5	$\geqslant 6$
Frequency in 100 quadrats	36	9	5	10	14	11	15

χ^2 *method*. Mean population density $= 2.50$.

The Poisson series for this mean value is derived from the following probabilities:

$$e^{-m} = e^{-2.5} = 0.0821$$
$$me^{-m} = 0.2053$$
$$\frac{m^2}{2!}e^{-m} = \frac{2.5^2}{2} \; 0.0821 = 0.2566$$
$$\frac{m^3}{3!}e^{-m} = \frac{2.5^3}{6} \; 0.0821 = 0.2138$$
$$\frac{m^4}{4!}e^{-m} = \frac{2.5^4}{24} \; 0.0821 = 0.1336$$

$$\frac{m^5}{5!}e^{-m} = \frac{2 \cdot 5^5}{120} \ 0 \cdot 0821 = 0 \cdot 0668$$

$$\frac{m^6}{6!}e^{-m} = \frac{2 \cdot 5^6}{720} \ 0 \cdot 0821 = 0 \cdot 0278$$

Individuals per quadrat	0	1	2	3	4	5	$\geqslant 6$
Expected frequency	8·2	20·5	25·7	21·4	13·4	6·7	2·8
Observed frequency	36	9	5	10	14	11	15
Difference	27·8	11·5	20·7	11·4	0·6	4·3	12·2

$$\chi^2 = \frac{(27 \cdot 8)^2}{8 \cdot 2} + \frac{(11 \cdot 5)^2}{20 \cdot 5} + \frac{(20 \cdot 7)^2}{25 \cdot 7} + \frac{(11 \cdot 4)^2}{21 \cdot 4} + \frac{(0 \cdot 6)^2}{13 \cdot 4} + \frac{(4 \cdot 3)^2}{6 \cdot 7} + \frac{(12 \cdot 2)^2}{2 \cdot 8}$$

$$= 179 \cdot 390 \qquad p = \ < 0 \cdot 001$$

Variance : mean ratio method

$$n = 100, \quad m = \bar{x} = 2 \cdot 5$$
$$S(x)^2 = 1158$$
$$(Sx)^2 = 62500$$
$$\frac{(Sx)^2}{n} = 625 \cdot 00$$

$$\frac{S(x)^2 - \dfrac{(Sx)^2}{n}}{n-1} = \frac{1158 - 625}{99} = 5 \cdot 384$$

Therefore variance : mean ratio = 5·384/2·5 = 2·154.
Standard error = 0·1421.

Therefore t = 2·154/0·1421 = 15·158 with 99 degrees of freedom, $p = \ < 0 \cdot 001$. Thus the probability of the difference arising by chance as calculated by either method is much less than 1,000 : 1.

With the contagious distribution patterns developing from the examination of individual distribution and their obvious departure from the Poisson series, several other mathematical series have been postulated for the fit of field data. Archibald (1948) investigated the possibility of the application of Neyman's contagious distribution and found a fairly good fit for several species while Barnes and Stanbury (1952) obtained agreeable results in the application of both Neyman's and Thomas's Double Poisson distributions. Inevitably most fits and tests for goodness of fit to the various distributions have been subjected to review and criticism. For example, Thomson (1952) showed that only one in three species tested fitted either Neyman's or Thomas's distribution. Of the two major goodness of fit tests the variance : mean ratio test has been subject to the greater criticisms, mainly because it is derived from a single aspect of departure from a Poisson expectation, e.g. the mere occurrence of high or low variance. Evans (1952) cites a hypothetical example which appears non-random but the data is such that the variance is equal to the mean and the result of a variance : mean ratio test thus gives an answer of randomness. A χ^2 test on the other hand produces a non-random answer to population distribution and the major point emerging is that either of these two methods may detect non-randomness when the other fails to do so, but that χ^2 tests are more reliable than the variance : mean ratio test (Greig-Smith 1952).

Preston's approach to commonness and rarity

In a paper aimed at the deduction of how abundance and rarity are distributed among species Preston (1948) examines data on moth species caught in a light trap and remarks that the numbers of individuals per species trapped may be formally related to a Poisson distribution. He further remarks that the ordinary Gaussian curve (or the 'normal curve of errors') is also satisfactory as a model, provided the curve is curtailed at the veil line as discussed later. Preston advances the theory that species fall into a series of groups of rarity/commonness which simply as a matter of convenience may be plotted on a logarithmic base. He shows that species fall into a natural series of groups of commonness thus:

Species group	A	B	C	D	E	F	G	H etc.
Number of individuals observed of that species	1	2	4	8	16	32	64	128

The groups form a sequence of *octaves* of frequency—the most natural grouping possible according to Preston, in that it is a logarithmic series, apparently a simple Gaussian curve plotted geometrically, i.e. a lognormal curve (Figure 7) represented by the general equation

$$n = n_0 e - (aR)^2$$

where n_0 is the number of species in the modal octave, n is the number in a particular octave distant R octaves from the mode and a is a constant calculated from empirical data (usually around 0·2). Species represented by 9–15 individuals fall into octave D, all having roughly the same degree of commonness; 17–31 fall into the fifth octave, etc. Species numbers falling on an octave boundary are divided so as to give half to each octave. Preston assumes as an example that the number of species falling in each octave of the series to be

$$n - 100 \, e^{-(0·20R)^2}$$

where the modal octave has 100 species and R is the number of octaves to the left or right of the mode. The curve and distribution of species are shown in Figure 7.

He then proceeds to explore the relationships between the octave series and Raunkiaer's Law of Frequency, the index of which is merely a measure of probability that a species will occur in any one sample quadrat. In addition, he suggests that the Poisson distribution series may be applied accurately to most species in the quadrat. Thus, in an attempt to define the boundaries of the octave series categories more precisely he calculates the probability that a species is not unrepresented while comparing the results with the perfect representation of the Poisson series, the octaves of the above series and Raunkiaer's

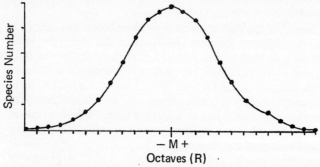

$$R = \pm 100, 96.1, 85.2, 69.8, 52.7, 36.8, 23.7, 14.1, 7.7, 3.9,$$
$$1.8, 0.8, 0.3.$$

(b) Their relationship to Raunkiaer's frequency classes

Octave name	ζ	ϵ	δ	γ	β	α	A	B	C
Poisson representation	3/128	3/64	3/32	3/16	3/8	3/4	1½	3	6
Probability (%) that a species is not unrepresented	2.4	4.6	8.9	17.0	31.3	52.7	77.7	95	99.8
Raunkiaer Class	A				B	C	D	E	

Figure 7. Preston's octaves of commonness and rarity.

frequency classes. The results are shown in Figure 7 where the main points of interest here are the octave : frequency class relationships as:

Raunkiaer Class A = ζ, ϵ, δ, and 90% of γ
B = β plus 10% of γ
C = 85% of α
D = 15% of α plus 70% A
E = 30% A plus all octaves to the right of A.

The lower octaves can be seen to cover frequency values which fall less than one. The continuation of the curve is acceptable in theory, but in practice species cannot be represented by a fraction of an individual and in consequence a veil line will be drawn across part of the distribution curve. Doubling the size of the sample will double the number of individuals which is equivalent to moving each species one octave to the right. One more octave is thus withdrawn from under the veil and so as the sample size is progressively doubled the curve becomes more like the theoretical distribution pattern. In his discussion of the boundaries of Raunkiaer's frequency classes Preston assumes a veil line between octaves $+3$ and $+4$. This means that category A will contain:

$$
\begin{aligned}
85\cdot2+96\cdot1+100\cdot0+(0\cdot9\times96\cdot1) &= 368\cdot0 \\
B\ (0\cdot1\times96\cdot1)+85\cdot2 &= 94\cdot8 \\
C\ (0\cdot85\times69\cdot8) &= 59\cdot2 \\
D\ (0\cdot15\times69\cdot8)+(0\cdot70\times52\cdot7) &= 47\cdot3 \\
E\ (0\cdot30\times52\cdot7)+89\cdot1\ (\text{octaves to right}) &= 104\cdot9 \\
\hline
&\ \ 674\cdot2
\end{aligned}
$$

This total species number is equivalent to Raunkiaer's 100% in his percentage-based frequency calculations. If the above data is expressed as percentage figures, the distribution shown in Figure 7 results. These figures are so remarkably similar to those obtained by Raunkiaer from experimental data that it may be correctly said that the Gaussian principles are appropriately applicable and that Preston placed his veil line in the correct position to emulate Raunkiaer's work. If the veil line is placed in a different position the frequency percentages will change but the change will be in the manner of the Gaussian curve.

This problem of quadrat size is thus one of the basic debating points in descriptive ecology at the intensive level. The work of Cain (1934) and Preston (1948) has already been discussed with reference to this problem. Moreover, a major criticism of the variance : mean ratio and goodness of fit tests as evidence of non-randomness has been aimed at their dependence on quadrat size (Skellam, 1952). Using a series of artificial communities composed of coloured discs Greig-Smith (1952) has shown an apparent change in the distribution of individuals in a population as the size of the sample is increased. At smaller quadrat sizes (10 to 25 cm) he obtains an indication of randomness with the variance : mean ratio approaching unity, but at 30 cm and above the ratio increases rapidly to a significant level which indicates a certain amount of contagion in the population. Perhaps the major significance of this work lies in the fact that some grasp of the scale of contagion or contagion levels of a particular species can be determined by sampling an area several times with different-sized quadrats. Such studies are aimed at the detection of *pattern* in species and the relationships of pattern to environment.

Pattern

Kershaw (1964) recognizes three types of pattern exhibited by species—environmental, sociological and morphological. The first two are often readily observed in the field without recourse to detailed sampling techniques. Environmental patterns are often related to obvious micro-topographical changes whose associated variation in soil depth, nutrients, pH and drainage produce a marked spatial pattern in the distribution of species. Sociological pattern can 'be defined as the product of the interaction of a species on another species so that the distribution of their populations will reflect certain patterns which may or may not be related to micro-environmental factors. The mysterious concept of physiological dominance is involved here and evidence can be drawn from the shading effects of a dense canopy. *Fagus sylvatica* by virtue of the low light intensity beneath its canopy excludes virtually all species with the possible exception of *Allium ursinum* which in woodland with a closed herb layer will often form a marked circular pattern associated with the lower light intensities of the canopy. Conversely, young saplings of *Fraxinus* will produce dense stands in clearings where the light intensity is greater. *Taxus*, the yew, by virtue of its dense canopy and its extremely acidic leaf litter excludes all plants from beneath its canopy.

Morphological pattern, however, is often not easily detectable in the field, since it depends upon the actual reproductive ability of the species and the population, both sexual by spores and vegetative by rhizomes and stolons. Greig-Smith (1952) pioneered the development of the techniques of pattern analysis and later Kershaw (1957, 1958, 1960) used Greig-Smith's artificial examples to illustrate a wide range of pattern scales in natural vegetation. Other notable work on the subject is that of Greig-Smith (1961a, b), Anderson (1967a, b), Greig-Smith and Chadwick (1965), Austin (1968) and Usher (1969). The method of pattern analysis is based upon the sampling of a grid of contiguous quadrats instead of throwing a range of quadrat sizes over an area. The increase in quadrat size is effected by the grouping or 'blocking' of adjacent quadrats into pairs, fours, eights, sixteens, etc. According to Greig-Smith the original method of analysis used density as the measure of species abundance. Kershaw (1957) modified and extended the approach by using frequency and percentage cover in analyses of grassland communities. Instead of counting individuals in each quadrat to give a density value, the subdivided quadrat may be used to record presence or absence in each sub-unit of the basic grid quadrat and a percentage frequency calculated for each unit in turn. A change of recording is required for use of cover data and Kershaw (1958) introduced a double 'Perspex' (Lacite) frame to enable point cover readings at 1 cm intervals along a transect, by the lowering of a pin through the two frames. Five adjacent sets of readings are grouped and expressed as a percentage cover figure, and the groups of five readings blocked together as before. Kershaw also suggested the use of a number of parallel lines of quadrats instead of a grid and this sampling approach has become standard procedure. For example, in his thesis on pattern within a *Zerna erecta* dominated chalk grassland community, Austin (1968) collects data from eight contiguous transects, each transect being divided into 128 basic units each 5×5 cm in size, giving 1,024 units for the total study. Each block thus has a maximum size of 40 cm wide by 640 cm long and is regarded as two analyses each consisting

TABLE 3

Mean square/block size data for two species of a limestone grassland community

	Block size (Ns)	Sum of squares	Degrees of freedom	Mean square (variance)
Sesleria caerulea	1	465·92	128	3·64
	2	526·72	64	8·23
	4	314·56	32	9·83
	8	99·36	16	6·21
	16	77·12	8	9·64
	32	65·44	4	16·36
	64	49·18	2	24·59
	128	3·62	1	3·62
Carex pulicaris	1	57·6	128	0·45
	2	31·36	64	0·49
	4	30·72	32	0·96
	8	7·68	16	0·48
	16	13·92	8	1·74
	32	2·24	4	0·56
	64	1·36	2	0·68
	128	0·35	1	0·35

Position of variance peaks for density measures on different species

	Ns	1	2	4	8	16	32	64	128
Poterium sanguisorba		x				x		x	
Helianthemum chamaecistus			x		x		x		
Carex flacca		x				x			x
Antennaria dioica			x		x				
Epipactis atrorubens		x		x					

of four transects or replicates. The counts from each line transect are then grouped into blocks of increasing size.

Having obtained the data it is then subjected to an analysis of variance in which the total variance between grid units can be assigned to differences between blocks of 2, 4, 8, 16, etc. grid units, tested against the residual variance within the smallest block size. After the calculation of the sum of squares $[S(x^2) - (Sx)^2/n]$ and dividing this figure by the number of degrees of freedom $(n-1)$ the variance (mean square) is calculated and plotted against the block size (Ns). Table 3 indicates the type of results obtained for several species in a closed limestone grassland sward and Figure 8 shows their Pattern on a graph plot. The manner of sampling is that of Austin (1968) using two 5×5 cm unit transects 128 units in length.

In the graphs produced the different scales of pattern are detectable as peaks at various block sizes corresponding to 'the mean area of clump'—the clump being an area where a particular species occurs at a higher or lower density in surrounding areas where no visual pattern is readily observed. In most sets of data three scales of pattern are apparent, but not necessarily three scales within the same species. According to Kershaw (1957) these three scales are: (a) small scale pattern due to size of individuals; (b) secondary level—possibly due to the morphology of the individual; (c) larger scales of pattern due to sociological or environmental factors. These produce the characteristic peaks which Austin terms primary (Ns 1, 2, 4), secondary (8, 16, (32)) and tertiary (32, 64, 128). With this background information it is interesting to look at the results in Table 3 and Figure 8.

(a) *Primary peaks* Ns 1, 2 peaks equivalent to the size of a single morphological unit such as a group of rosettes in *Poterium sanguisorba*, tillers in *Carex flacca* and an individual in *Epipactis*. Austin suggests that this peak in *C. flacca* represents the length of rhizome-bearing live tillers. The chamaephyte *Helianthemum* and the rosette hemicryptophyte *Antennaria* produce an initial block size peak of 2. In common with many dominant coarse grasses *Sesleria* produces its primary peak at Ns 4—the size of a tussock.

(b) *Secondary peaks* Ns 8, 16, (32) thought by Kershaw (1963) to be a result of competition between species, i.e. sociological pattern—present in most species.

(c) *Tertiary peaks* Ns (32), 64, 128 considered to reflect the size of the clone or clump of the species or due to other more obscure ecological and historical factors.

Greig-Smith (1964), on the subject of peak heights, concludes that the higher the peak the greater the intensity of pattern, that is it will correspond to a measure of non-randomness found in random samples. Alternatively, if the arrangement of a particular species is random over the area sampled, the mean square for all block sizes should be the same. Contagion produces variance peaks at a block size equivalent to the size of the clump and if the clumps are themselves random or contagious the mean square value will remain the same with increasing block size. Kershaw (1957) has indicated that if the number of grid

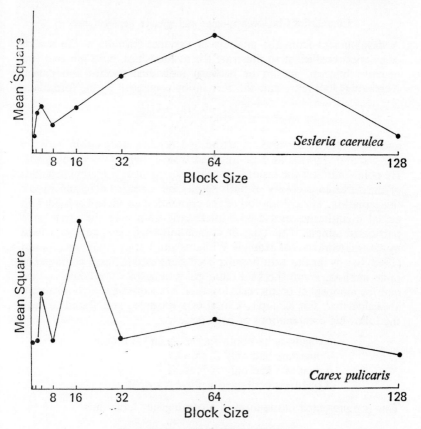

Figure 8. Mean square/block size plots.

units sampled is inadequate, the peaks tend to drift one block size to the right, especially in cover estimations where sensitivity falls with cover degree. This drift can be combated by an increase in sample size. Recently, Usher (1969) has described a peak drift to the left associated with the starting point for sampling. For instance, if the series of units starts near the centre of a clump there is a strong possibility that the mean square peak will be one block size smaller than the actual size. The correction of the drift to the left can be readily made by collecting a larger series of samples and performing several analyses. Apart from these two possible drift interferences the methods of analysis provide an important tool in the study of vegetation composition and structure.

Correlations between species and species performance

A development from the analysis of variance formula is the use of correlation coefficient to illustrate the relationships, both positive and negative between species or between individual species abundance. Kershaw (1964) represents the correlation coefficient by the formula:

$$r = \frac{S[(x_1 - \bar{x}_1)(x_2 - \bar{x}_2)]}{\sqrt{[S(x_1 - \bar{x}_1)^2 \, S(x_2 - \bar{x}_2)^2]}}$$

and illustrates its use in demonstrating a negative correlation between *Festuca* and *Agrostis* in data collected from North Downs grasslands. He points out that the result is not unexpected since a preliminary plot of the percentage covers of both species on a scatter diagram reveals this situation. The calculation of the correlation coefficient provides an actual quantitative measure of interrelationship over the range of a particular sample. This type of correlation coefficient has been used by several workers, for example Williams and Varley (1967) and Austin (1968) but by far the most popular coefficient used is the χ^2 contingency table method. From the field data, pairs of species are taken in turn and the numbers of occurrences together, by themselves and both absent are calculated. For example, a total of x quadrats were found to have the following combinations of two species:

  a quadrats had both species A and B present
  b quadrats had only B present
  c quadrats had only A present
  d quadrats contained neither A nor B

This is represented in the form of a contingency table thus

SPECIES A

SPECIES B		+		−		
+	a	64	b	36	a+b	100
−	c	58	d	42	c+d	100
	a+c		b+d		n 200 (a+b+c+d)	
	122		78			

If actual data is supplied from 100 quadrats the expected occurrences in the four cells of the contingency table may be calculated, e.g. 61 in cells a and c and 39 in b and d. As it has been stated in a previous section, the χ^2 departure from expectation test may be calculated as the sum of $\dfrac{\text{deviation}^2}{\text{expected}}$ in each of the four cells. For a contingency table this is best derived from the direct formula:

$$\chi^2 = \frac{(ad-bc)^2 n}{(a+b)\,(c+d)\,(a+c)\,(b+d)}$$

and the result referred to a table of χ^2 with one degree of freedom.

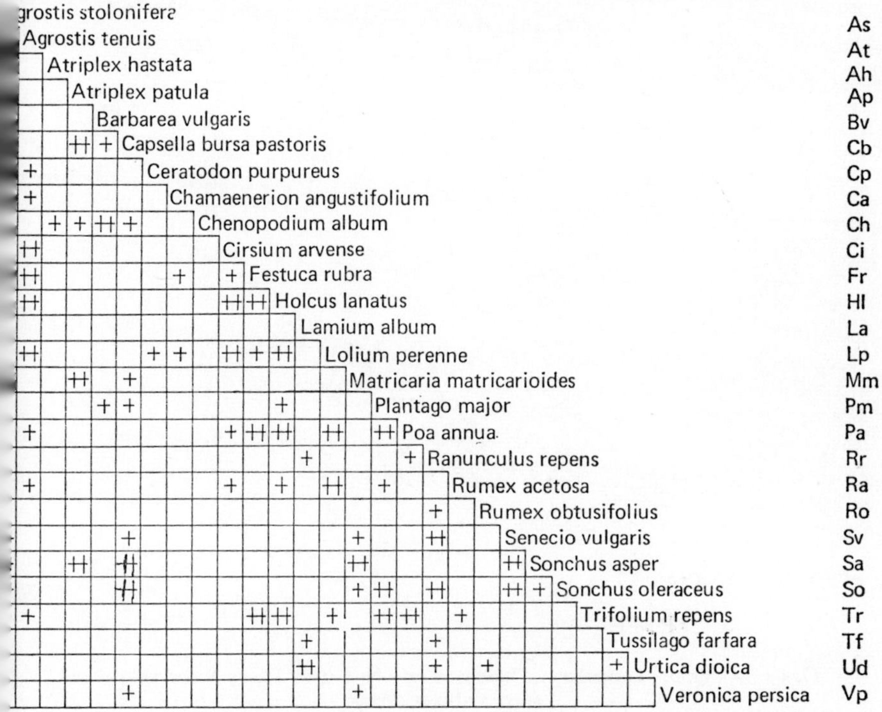

Species	As	At	Ah	Ap	Bv	Cb	Cp	Ca	Ch	Ci	Fr	Hl	La	Lp	Mm	Pm	Pa	Rr	Ra	Ro	Sv	Sa	So	Tr	Tf	Ud	Abbr
Agrostis stolonifera																											As
Agrostis tenuis																											At
Atriplex hastata																											Ah
Atriplex patula																											Ap
Barbarea vulgaris																											Bv
Capsella bursa pastoris			++	+																							Cb
Ceratodon purpureus	+																										Cp
Chamaenerion angustifolium	+																										Ca
Chenopodium album		+	+	++	+																						Ch
Cirsium arvense	++																										Ci
Festuca rubra	++						+	+																			Fr
Holcus lanatus	++								++	++																	Hl
Lamium album	++																										La
Lolium perenne	++						+	+			++	+	++														Lp
Matricaria matricarioides		++		+																							Mm
Plantago major			+	+								+															Pm
Poa annua	+							+	++	++			++	++													Pa
Ranunculus repens										+							+										Rr
Rumex acetosa	+							+	+			++		+													Ra
Rumex obtusifolius													+				++										Ro
Senecio vulgaris							+							+			++										Sv
Sonchus asper		++		++										++													Sa
Sonchus oleraceus				++									+	++			++					++					So
Trifolium repens	+								++	++				+			++	++			+						Tr
Tussilago farfara												+					+										Tf
Urtica dioica												++					+		+				+				Ud
Veronica persica				+									+														Vp

Figure 9. χ^2 matrix for 34 quadrats showing positive correlations between species (+ = 5%>P>1%; ++ = P<1%).

When the number of samples is small—as is often the case—a certain inaccuracy is introduced which may be compensated for by the use of Yates's correction. This involves the subtraction of 0·5 from each of the two values greater than expectation and the addition of 0·5 to those less than expectation. Fisher and Yates (1943) consider this modification necessary when any expected value is less that 500. Thus, from the contingency table data quoted:

$$\chi^2 = \frac{(63\cdot5 \times 41\cdot5 - 36\cdot5 \times 58\cdot5)^2\ 200}{(122 \times 78 \times 100 \times 100)} = 0\cdot525$$

The χ^2 value of $0\cdot525$ gives a probability of $<0\cdot5$ of this correlation occurring and in this case is fairly strong positive correlation between the two species.

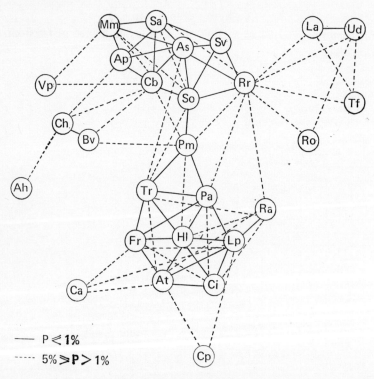

Figure 10. Species constellation based on χ^2 values (see *Figure 9* for values and species abbreviations).

Having obtained data on observed in relation to expected joint occurrences of species pairs in a series of communities the values may then be entered in a χ^2 matrix to show positive and negative species relationships (Agnew, 1961) in the manner of Figure 9. Here the data is collected in the form of a belt transect of contiguous quadrats on waste land on the university campus. For each quadrat presence and absence of species is recorded and in the final χ^2 calculations species with only one or two occurrences are omitted. The correlations obtained are represented in the form of a two-dimensional table (Figure 9) where + + refers to a probability of positive correlation of $\leqslant 1\%; +5\% \geqslant P > 1\%$. From the table the species may be represented as a two-dimensional constellation where the distances between species are calculated as the reciprocal of χ^2. This means that two highly positively correlated species with a high χ^2 value will be positioned close together, for example *Lolium perenne* and *Agrostis stolonifera* in Figure 10. The strength of the correlation is further emphasized by a bold connection for a probability value of $P \leqslant 1\%$. From the Figure, two major groups of positively correlated species are immediately obvious:

(a) A group comprising *Agrostis stolonifera*, *Senecio vulgaris*, *Sonchus asper*, *Atriplex patula*, *Capsella bursa-pastoris*, *Matricaria matricarioides*, *Ranunculus repens* and *Sonchus oleraceus*.

(b) A group comprising *Agrostis tenuis*, *Festuca rubra*, *Holcus lanatus*, *Lolium perenne*, *Plantago major*, *Trifolium repens* and *Rumex acetosa*.

An immediate explanation is available for the distinction of the two groups. Group (a) occurred on areas of recolonization where the turf had previously been cut; group (b) is undisturbed turf. This is an example of an objective verification of a readily observable sociological and environmental phenomenon. Apart from the work of Agnew (1961) other classic examples of this approach are those of Welch (1960) and De Vries (1953) using χ^2 correlation and Williams and Varley (1967) using the variance coefficient.

A parallel approach to the analysis of variation in the data is to calculate the equation which best fits the data, i.e. to calculate the regression of a factor y on another x. A regression analysis can be much more informative in that it represents the relationship of each sample to a median expected value and pictorially represents its departure from the expected. Regression analyses are of wide use in ecological investigation, to exhibit correlations between species, between species and environmental factors and between species abundance and performance in two or more localities. The manner of calculation of regression can be obtained from most standard statistical works. In this text, its use is illustrated in Figure 11 which shows regression analyses of the shrub *Hippophaë rhamnoides*, age against girth in two localities on the Spurn Peninsula, Yorkshire. Plot (b) is a result of sampling an exposed dune crest in what is probably the most exposed section of the peninsula, assessed from the fact that the last breach by the sea occurred at this point. Here the mean annual girth increase for the population is calculated at 1·4 mm per annum. Plot (a) shows samples collected at the more sheltered point dunes area where the

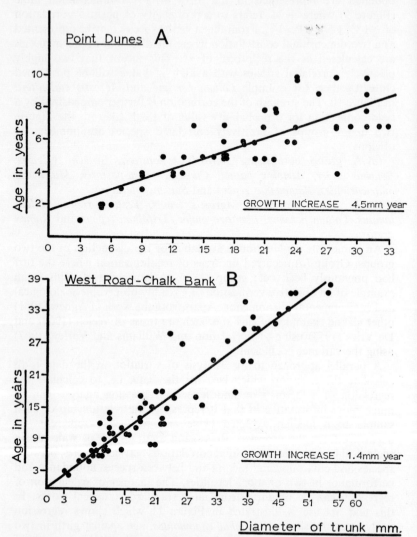

Figure 11. Regression analysis of age/girth relationships in *Hippophaë rhamnoides* populations in sheltered (*a*) and exposed (*b*) localities of the Spurn Peninsula, Yorkshire (data of J. S. Ledwood unpublished).

peninsula is continuously advancing into the Humber due to the deposition by longshore drift. The growth rate per annum can be seen to be more than three times that of the exposed locality. The regression analysis thus gives a meaningful representation of an otherwise obscure mass of performance data and having ascertained the marked difference the observer is then able to seek the local climatological and ecological explanation for the difference.

CHAPTER 2

ASSOCIATION, FORMATION, CLASSIFICATION
AND CONFUSION

The taxonomic nomenclature of vegetation classification and methodological synthesis is rife with confusing pleonastic verbosity.

Like this opening statement, the classification of vegetation suffers greatly from overstatement, some ambiguity and, inevitably, misinterpretation. The history of vegetation classification is chaotic, being partly shrouded by the difficulties of language barriers, partly by the unavoidable comparison of methods with accepted species taxonomic methods, and partly by an aura of academic mystique. In no other branch of science does the *quot homines tot sententiae* maxim apply more aptly. Fortunately, a great deal of light has been thrown on the subject by Whittaker (1962) whose detailed analysis of the history and development of the main traditions in the classification of natural communities must rank as one of the most important works on the subject. But even with reference to this work, the quantity and diversity of interpretations discussed are still overwhelming. In the subsequent sections of this chapter the four major trends are reviewed in as terse and concise a manner as possible with six text figures presented as types of genealogical and chronological 'family trees' to aid interpretation. Even so, no more than two-thirds of the total methods of classification will be represented! But before the trends are considered, two related points need to be elaborated upon—(i) why classify? (ii) the approach to classification.

Why classify?

For the human race, classification is a natural and inherent, intuitive process; to create some semblance of order from an otherwise disorderly matrix by the pigeon-holing and categorization of the matrix entities. But when it comes to vegetation and natural communities, can this process work in a similar manner to the classification of colours or species? Webb (1954), the eminent taxonomist views the scene with a certain amount of exasperation and contends that until problems of unit diagnosis and delimitation have been solved all attempts at precise classificatory procedures are useless. Like the ordinary taxonomist the plant sociologist must select his characters for description and delimitation of the units. Whether the plant taxonomist chooses morphological or genetical characters and the plant sociologist floristic or structural characters the problems are the same. *Asplenium trichomanes*, until the work of Lovis (1964), was generally regarded as a single homogeneous species until it was shown that there were two ploidy levels within the species and that the only qualitative morphological manifestation was spore size. One morphological species or two genetical (biological) species? Similarly, all heath communities where *Calluna vulgaris* is the physiognomic dominant and often the only phanerogam in the community are classed together as *'Callunetum'*. Coppins and Shimwell (1971) have shown a wide variation in cryptogams beneath a *Calluna* canopy with distinct species associated with ages of the *Calluna* and burning patterns. One physiognomic community or several floristic communities?

But comparisons between classical taxonomy and vegetation taxonomy can easily be misinterpreted in their interrelationships. Cowles (1899) introduced the view that the units of vegetation systematics are analogous with the species and genus of classical taxonomy and this view has been widely subscribed to, especially in English-speaking countries. Huxley (1940) states that the fundamental problem of systematics is the detection of evolution at work. Thus the quasi-organismal concepts of vegetation of Clements (1916) and Tansley (1920) were compared with this natural classification, whilst the interpretations of these authors of other schools of thought, were inevitably biased toward such an interpretation when, in fact, they bore no semblance of a natural classification.

The basic approaches to vegetation classification

In the initial observation of a particular landscape the phytosociologist perceives a certain amount of pattern and a repetition of types of landscape reflected largely by physiographic and vegetational features. These types are mentally grouped by the observer before he has begun to take stand samples. At this stage, an assessment of similarity and dissimilarity of types tends to be based on a subjective conception of pattern and overall character. This basis for the provisional recognition of types is governed by his belief in the significance of certain type properties as the best basis for classification, and in essence, this is the inherent intuitive process with some visual conditioning. On the one hand, he may perceive a distinct alteration of yellow–blue–yellow; *Geum urbanum–G. rivale–G. urbanum*; and woodland–scrub–woodland or a gradation from yellow through green to blue; introgression from *G. urbanum* to a hybrid and introgression through to parent *G. rivale*; and a continuous gradient of woodland to scrub back to woodland. The recognition of one of these two situations depends upon (*a*) the scale of anthropogenic effects on the system, (*b*) the choice of initial study sites, (*c*) the influences of the total natural environment and (*d*) the amount and source of the investigator's background knowledge. Any representation will thus be based largely on personal choice, intuition, subjectivity, cultural influence and initial experience. If the investigator perceives what appears to be a continuous gradient in characteristics he will describe a continuum. If a more or less distinct alternation or unit progression is observed the investigation will result in classification.

The units of classification

Much of the confusion and chaos in vegetation classification revolves round the different interpretations placed upon the vegetation units known as the *association* and the *formation*. For the purpose of analysis of the various lines of development of these concepts, the interpretations may be divided into three historical periods—(*a*) the period up to 1910 when, at the Third International Botanical Congress, Flahault and Schröter produced a basic definition of the association; (*b*) the period 1910–30 when definite trends of interpretation developed and became localized in schools of thought; and when there was intense criticism of each and every approach; (*c*) the period commencing after the Sixth International Botanical Congress in 1935 when the standing committee elected at the previous Congress in 1930 'to investigate the possibility of attaining general agreement on some of the more important concepts and terms in plant sociology', Brooks and Chipp (1931, p. 17), produced and deliberated its recommendations (Du Rietz, 1936).

The period up to 1910

The concept of the plant *association* is one of the oldest in plant geography, even pre-dating usage of the term 'ecology'. It is generally first attributed to Humboldt (1805) or earlier, whilst the first use of the suffix '-etum' to denote communities of association rank dates from the work of Schouw (1823). The suffix was added to the stem of the generic name, e.g. 'Quercetum' for a woodland whose chief constituent is the oak. No apparent definition of the term is to be found in the work of Schouw, but his usage implies a physiognomic interpretation of the concept. One of the earliest relevant definitions is the oft-quoted one from Grisebach (1838) who is credited with the introduction of the term *formation*. Clements (1905) translates this original definition as 'a group of plants which bears a definite physiognomic character . . . characterized by a single social species, by a complex of dominant species belonging to one family, or, finally, it may show an aggregate of species, which, though of various taxonomic character, have a common (physiognomic) peculiarity'. In his earlier writings Grisebach tended to regard the formation as a vegetation type co-existent with a definite edaphic habitat. Later, in his classical work, *Die Vegetation der Erde*, he uses the term to denote broad climatic vegetation types such as tundra and the forests of well-defined geographical regions. The next few decades to the turn of the century witnessed a rapid proliferation of phytogeographical studies where communities, stands and vegetation types were described indiscriminately as either associations or formations related to edaphic or climatic features. These two controversial and misinterpreted points, association or formation, ecological plant geography or floristic plant geography, gave rise to numerous semantic quibbles and a chaos of phytogeographical literature. Thus in the early years of this century there were already several voices condemning the chaos, notably Flahault (1901) who remarked that several botanists had already lost their way in the confusion and readily acknowledged having employed the terms without thinking of their definitions. If this was the position in 1900 what chances have we seventy years later of understanding the origins?

The variety of early interpretations is reviewed by Moss (1910) and Clements (1916) although these reviews provide little help in the way of an indication of trends. There are, however, three distinguishable trends in vegetation classification which are here called the Zurich–Montpellier (or Southern) Tradition, the Northern Tradition (cf. Whittaker, 1962) and the English Tradition, relating to British and American trends. As a starting point for all three traditions, the work of Grisebach (1872) in his concept of the formation is used along with the ill-defined association as used by the predecessors of Flahault (1893).

The Zurich–Montpellier Tradition (Figure 12) is compounded of two schools of thought; that of Schröter (1894) who uses the term formation for certain '*associations* des plantes' in Valais, using physiognomic criteria for distinction of units; that of Flahault (1893, 1901) employed the term association as the fundamental vegetation unit characterized by physiognomic dominants. At the 1900 Botanical Congress he advocated its adoption as such (Flahault, 1901) and met with some success since in the following year Schröter and Kirchner (1902) also suggested the use of association for lower-level units of vegetation. Whilst the two schools of thought accepted dominance as the criterion for defining associations, Brockmann-Jerosch (1907) stressed constancy as a basis for characterization. Constant species were those occurring in over 50 per cent of the samples of an association or 'stand-type'. Constants of wide-distribution were termed *formation-ubiquitists* whilst more critical species were known as *character-plants*. These floristic views found an exponent in Gradmann (1909) and it is in his work that the later developments of this tradition are highlighted particularly with reference to the relationship between the formation and the association. He states that after fairly intensive field recording it soon becomes apparent that certain associations have much more in common with each other than with all the others. These

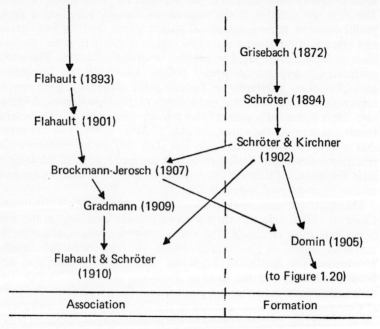

Figure 12. The origins of the Zurich–Montpellier Tradition.

can be united into groups or formations of closely allied associations. Formations thus comprise all the species of all the allied associations and because all these can seldom if ever be recognized in a single locality, the formations appear as abstractions.

It was perhaps this work more than anything else which influenced the development of the Zurich–Montpellier School and inspired the definition of the association supplied by Flahault of Montpellier and Schröter of Zurich at the 1910 International Botanical Congress:

'An association is a plant community of definite floristic composition, presenting a uniform physiognomy, and growing in uniform habitat conditions.'

Finally, in this section, the early work of Domin (1905) needs to be mentioned. In his studies of the vegetation of the central Bohemian mountains he defines his formations according to edaphic-physiognomic properties, e.g. the steppe formation, heath formation and pine-forest formation along the same lines as the early Zurich schools of Schröter. Within these formations he distinguishes *Hauptfacies* (= associations) such as *Stipa-* steppe or *Carex humilis* steppe. Species are also referred

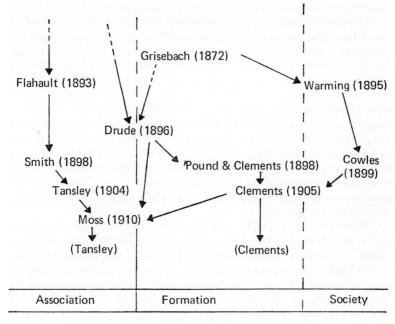

Figure 13. The origins of the English Tradition.

to as characteristic or strange to a particular formation as in the manner of Brockmann-Jerosch (1907). However, after this period he shows a pronounced trend towards the northern tradition (q.v.).

The English Tradition (Figure 13), being a tradition uniting British and American, has a number of origins, all in continental Europe. In the period up to 1910, the trends in the two continents were quite separate but after this date there is a marked convergence of ideas, accounting for the amalgam of the present section. The other feature they have in common is the attention paid to the succession of associations, especially on the succession of associations within the same formation.

Drude (1896) provided a major influence in both trends. He emphasized that in the plant cover of a region there is an alternation of the principal plant associations with habitat. He provided an enlightening definition of a formation as: (*a*) any principal association which has found its natural termination in itself; (*b*) which consists of biologically related life forms; (*c*) which is confined to similar substrata; assuming that no actual replacement of association would occur on the site of such a principal association, without external changes. This association is '*abgeschlossen*'—has reached its climax development.

Pound and Clements (1898) more or less stuck to this definition of the formation. Meanwhile, Warming (1895), being confused with the complexity of interpretation which surrounded the formation had recommended its disuse and Cowles (1899) followed this lead to use the term *society*—roughly the English equivalent of Warming's '*Plantesamfund*'. The plant society was defined as a group of plants living together in a common habitat under similar life conditions. This term and the general dynamic approach was incorporated in a formal system of classification developed by Clements (1905) and expanded and well expounded by him in 1916 (q.v.). The major features of the Clementsian system are reviewed in the sections on the next chronological period, but it is worth concluding this discussion with a note that Clements' association is basically equivalent to Drude's *abgeschlossen* principal association (formation).

The British trend which finds its earlier conclusions in the review of Moss (1910) and *Types of British Vegetation* also owes much to Drude's definition of the formation. This, with the additional extras of Flahault's physiognomic concepts of the association, brought to Britain by one of his pupils Robert Smith (1898) and accepted by Tansley (1904), greatly influenced the thinking of Moss (1907, 1910) and this in turn greatly influenced early British ecological thought. A statement taken from his 1910 review illustrates his opinions.

'A plant formation comprises the progressive associations which culminate in one or more stable or *chief associations* and the retrogressive associations which result from the decay of the chief associations, so long as these changes occur on the same habitat.'

The formation is thus related to and determined by habitat conditions, the succession of vegetation on this habitat type proceeds through subordinate progressive associations to a stable chief association (analogous to the formation or *abgeschlossen* principal association of Drude), and then degenerates from the chief association through subordinate retrogressive associations. The chief associations always represent the highest limit of development that can be achieved in the particular formation in which they occur and this limit is determined by the general life conditions of the formation.

Moss's views on the three levels of community rank were as follows (from his 1913 work): a plant formation—'the whole of the vegetation which occurs in a definite and essentially uniform habitat'; a plant association—'of lower rank than a formation and characterized by minor differences within the generally uniform habitat'; a plant society —'of lower rank than an association and marked by still less fundamental differences of habitat.'

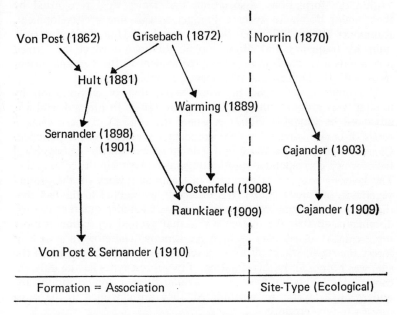

Figure 14. The origins of the Northern Tradition.

The Northern Tradition (Figure 14) with its central Swedish or Uppsala School has its origins in the vegetation analyses of von Post up to 1862 where individual stands were placed in *vegetationslokaler*— approximately the equivalent to later interpretations of the association in this school. Hult (1881) followed the concept of the formation from Grisebach describing numerous 'formations' from the vegetation of northern Finland, for example, a *Betula nana* formation and a *Poa annua* formation. These were later recognized as merely associations or only societies in the generally accepted physiognomic sense of the words and the concepts were criticized for narrowness by, amongst other people, Warming (1909) and Moss (1910). Hult grouped stands by their affinities of structural layering and the further development of this line of approach by Sernander (1898, 1901) and von Post and Sernander (1910) led to the foundation of the Uppsala School of plant sociology.

In many ways the early work of Warming (1889) in his description of the vegetation of Greenland in terms of formations defined by physiognomy and ecology is a compromise between the small units of Hult and the edaphic formations of Grisebach. Ostenfeld (1905, 1908) follows Warming's methods in Iceland and the Faeroe Isles in combining habitat and physiognomic characters for characterization of formations. Within the formations, associations and facies were recognized by their major dominant species. Parallel to this line of development, Raunkiaer (1909), following the basic trends of Hult, defined narrow units by frequency and physiognomic dominance which he termed formations and which were roughly equivalent to the Uppsala formations, with the background importance of habitat stressed.

The other major trend in Scandinavia, that of classification by habitat, was probably initiated by Norrlin (1870) in Finland, and was advanced by Cajander (1903) in which the concept of the *ecological series* of communities along environmental gradients was expounded. Cajander defined associations by dominant species and also distinguished facies which corresponded to other vegetation layers in the community. The layering aspect was extended in Cajander's work on the ground vegetation of forests (1909). His observations pointed to the fact that the same type of ground vegetation occurred under different canopy dominants, so that the inference was that ground vegetation is more representative of site factors than the canopy. Therefore, a system of forest *site-types*, where all lower layer communities are related to the site, was proposed. In this situation, if two forest types having different canopy dominants but the same or similar floristic composition in the undergrowth layers are encountered, they are classified in the same forest site-type group.

Between the Congresses: 1910–35 (Figure 15)

In these interim years the *Zurich–Montpellier Tradition* developed in three major trends each of which will be discussed in turn.

The Braun-Blanquet Association trend began with the analysis of snow-bed communities in the Alps by this author in 1913. Braun-Blanquet placed paramount importance upon the floristic composition of the plant community, pushing dominance and physiognomy into the background in a method which probably represents an extreme interpretation of the views of Brockmann-Jerosch (1907) and Gradmann (1909). The concept of the character species developed by these latter authors and Rübel (1912) as species with a greater constancy of occurrence in a particular vegetation type was given an eminent position in the definition of associations, which were grouped together into higher units, not by physiognomy but by floristic comparison. In 1915 these higher units or *Assoziationsgruppe* were given the name *Verbände* or Alliances each with their own character species (Braun-

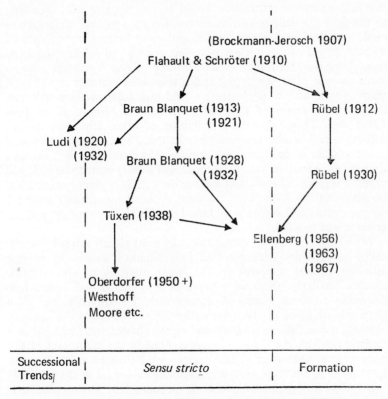

Figure 15. Divergence within the Zurich–Montpellier Tradition.

Blanquet, 1915). By 1921 the major aspects of the system had already developed with all the tools of description and characterization such as frequency, sociability, cover, abundance, fidelity, constancy, etc., in use. The fundamental concepts of the system as it then was were outlined by Braun-Blanquet and Pavillard (1922) in their well-known *Vocabulaire de sociologie végétale*. The only main characteristic not included in this vocabulary was the *differential species*, used to dis-

C

tinguish vegetation-units lower than the level of the association and this was apparently incorporated by Braun-Blanquet and Jenny (1926). The term formation had by this time dropped out of use in this trend.

The central concept of the Association was its abstract nature, i.e. the field observer never saw an Association in the field; it was only a stand just as a herbarium only contains a specimen of a species. If the stands after detailed recording are collated on overall floristic similarity into tables then the abstraction would become apparent. On returning to the field the validity of this abstraction as a natural phenomenon could be tested and trends toward and away from a central core of well-developed stands picked out. The concept of fidelity which so many later workers found so hard to accept was a concept with its roots deep in plant geography and a term used amongst experienced plant geographers whose extensive fieldwork had led them to know every square metre of their beloved Bernina Mountains or Bas-Languedoc. They knew the distribution of practically every species within the region and they could predict that a species would be found in a particular association. Furthermore, they could label their associations by these species. When the concept of fidelity came to be reviewed in other regions, by ecologists lacking the extensive plant geographical traditions of the Z–M School, species declared *Gesellschaft treue* for an association in a region were found outside this community and the concept was rejected.

Above the level of the association, the most closely related (floristically) associations were grouped into alliances where the floristic affinities became apparent in the larger numbers of characteristic species. Similarly, these were grouped into orders and orders into classes in a formal system of classification. To emphasize the plant geographical bases of the system, the highest unit was the circle of vegetation as a definite geographical region characterized by a peculiar floristic element. All these units are defined in the major text of this period, Braun-Blanquet's *Pflanzensoziologie* (1928) later (1932) translated into English. This work was no doubt influential in the establishment of the Station Internationale de Géobotanique Mediterranéene et Alpine (SIGMA) at Montpellier in 1930, from where the traditions of the school spread throughout the Mediterranean and Slavonic regions of Europe in the 1930s.

The second major trend within the Zurich–Montpellier tradition may be conveniently termed the *Rübel Formation* trend which culminated in the monograph of Rübel (1930), *Pflanzengesellschaften der Erde*. Rübel's methods may be traced back to Brockmann-Jerosch (1907) and his early monograph on the vegetation of the Bernina Montains reflects this approach. By 1930 he had developed a system of classification with the association as the basic unit defined by constancy and character species. These were grouped into alliances on overall floristic relationships and alliances were grouped into formations on physiognomic affinities. The formation was conceived as a community-type defined by physiognomic dominance and co-dominance in the upper stratum with life forms which show a positive indication of habitat factors, e.g. Mediterranean maquis dominated by broad scerophyllous species. These were then grouped into formation-groups of related physiognomic types in different geographical regions and the formation-groups united into formation-classes which corresponded roughly with the major climatic vegetation types of the world, e.g. Pluviilignosa included all tropical and sub-tropical rain forests (see Appendix II).

A third, though lesser known, trend is seen in the *Lüdi Successional* trend (Lüdi, 1920, 1932). Pavillard (1920) in his analysis of the species of plant communities into a series of causative, dynamogenetic groups, i.e. constructive species, conserving, consolidating, neutral and destructive species probably gave this trend much of its initial impetus. Lüdi adopted this greater emphasis on succession and regarded the association as a semi-stabilized stage in a successional series with emphasis on habitat relationships. His use of the character species was essentially the same as that of Brockmann-Jerosch both in its definition and its restricted use in association definition (Lüdi, 1928).

The work and thoughts of one man, F. E. Clements, dominated phytosociological methods in the *English Tradition* in the period 1910–35. In his monograph on the succession of plant communities (1916, 1928), Clements stresses the developmental aspect of vegetation. Unlike the Zurich–Montpellier system he conceived the formation not as an abstraction but as an organic entity covering a finite land area which was marked by a climatic climax. His basic ideas discussed in detail in Chapter 4 were that vegetation develops from a pioneer stage, through higher development stages to a climax which is primarily determined by climate. This development was called seral development. No matter what the pioneer stages of the sere were, how xerophytic or hydrophytic, there was always convergence to a single climatic climax at the end of the sere—a monoclimax. A complex of nomenclature was devised to describe the different communities in the formation. Formations were subdivided into *associations* which could be characterized by their co-dominant species. Within the association, if a single species was the physiognomic dominant then the term *consociation* was used, while these were further subdivided into *societies* each characterized by subordinate species. However, these terms were restricted to the considered climax vegetation type. All other developmental units leading up to a climax, all seral stages were given parallel terms such as *associes*, *consocies* and *socies*.

A second important contribution to American ecology came from Cooper (1916, 1926) who like Clements stressed the dynamic approach to vegetation but, as such, found little time for formal classification rather in the same manner as most British ecologists of the period. Like Nichols (1917, 1923) and Moss (1910), Tansley (1920), he rejected the monoclimax theory and favoured *polyclimax* or the definition of the formation units by habitat. Associations were also characterized by

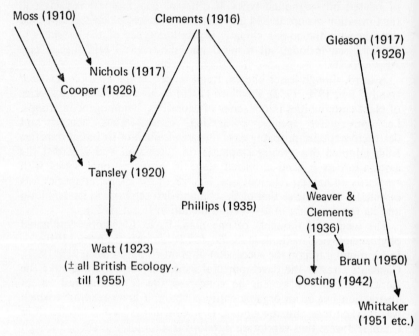

Figure 16. The English Tradition, 1910 onwards.

uniformity of edaphic factors and Nichols (1917) looked upon the formation as an *association–complex* related to distinct edaphic conditions. Similarly, Tansley (1920) derived many of his basic attitudes to the nature of vegetation from Moss (1910) and like Clements he regarded the units of vegetation as organic entities or quasi-organisms. He accepted the definition of the association from the Brussels Congress of 1910 and within this unit he recognized the Clementsian consociation and society. He also accepted Clements doctrine that the term association should be limited to mature units which are in stable equilibrium with their environment and that the transitory communities of similar rank (associes, etc.) were useful in indicating the distinction between developmental and mature vegetation units. But unlike the Clementsian formation Tansley's became to be more or less synonymous with edaphic conditions, i.e. as a unit caused by the habitat and expressed by distinctive life forms. Formal classification of units in British ecology is notably lacking as is any amount of criticism of continental and other classification systems and the levels of classification. British ecologists with this strong 'developmental-edaphic' background have always tended to place more emphasis on the ecological factors which produce or underlie vegetation patterns, succession and unit typification,

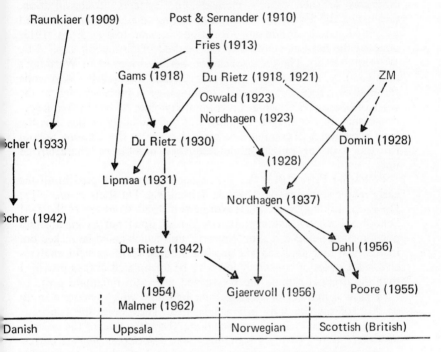

Figure 17. The major trends within the Northern Tradition, 1910 onwards.

probably one reason for the wealth of multi-variate analyses in later British ecological literature.

Within the English tradition, one other important piece of work needs to be mentioned here, namely the controversial 'individualistic concept of the plant association' Gleason (1917, 1926). He considered that vegetation was not an organic entity and that the association was merely a combination of species individuals whose overall unequal distribution produced differences in vegetation which suggested continuity. Independently in Russia, Ramensky (1924) produced a similar vegetation continuity-species individuality hypothesis, in which he suggested the representation of plant communities in ecological series with communities corresponding to the cover abundance curve peaks of the component species. The individualistic concept of the association has found a considerable following in recent years in America.

During the 1910–35 period the *Northern Tradition*, mainly through the Uppsala School proliferated at a fantastic rate and developed a great rivalry with the Zurich–Montpellier Tradition via numerous criticisms of each other's methods and systems of classification. Gradually, the Sernander tradition of calling small units characterized by physiognomic dominants formations became lost as Fries (1913) adopted the term association in the method of distinction applied by Warming (1909). The term formation persisted longest in Warming's own country, Denmark, in the work of Raunkiaer (1918, 1928) while the first convergence to Uppsala terminology is probably seen in the work of Böcher (1933) where his community types defined by frequency-dominants were called associations. But unfortunately by this time the term sociation was coming into general usage in the Uppsala School for this type of physiognomic-frequency dominance characterized unit.

Following Fries (1913) the term association was adopted in all the early major works of the Uppsala School, e.g. Du Rietz *et al.*, 1918; Du Rietz, 1921; Osvald, 1923; Nordhagen, 1920. Contrary to the Z-M School beliefs, the association was looked upon not as an abstract phenomenon but as a real or 'concrete' unit which could be picked out in the field and analysed via quadrat studies. Copious field analyses enabled a series of constancy levels to be formulated always provided that the original sample plots exceeded a certain minimum area. In such plots a number of species could be found which occurred in all or 90% of the quadrats in a single area. These were the constants and were used to define the association. Above the level of the association there were two main groupings—the *association-complex* corresponding to topographic, dynamic and ecological relationship trends; and the *formation* comprising a number of association-complexes related by physiognomy.

Not all the criticisms of each other's schools of thought were destructive. Several commentaries were aimed at some shape and form of reconciliation between Uppsala and Z-M, whilst others took a single small aspect of one or the other systems and brought it to the forefront of importance as a third alternative method of approach. Perhaps the most important in this respect was the work of Rübel (1925) who pointed out the 'micro' nature of the associations of the Uppsala School when compared to the larger 'macro associations' of the Z-M tradition. Rübel (1927) proposed the term *sociation* for such small associations and this was accepted in the 1928–30 period by the Uppsala School (Du Rietz, 1930). Above this level a series of groupings were built up—*consociation* approximating to Clements' consociation but also units with several dominants; *association*—more or less the same level as the Southern Tradition but in no way equivalent to that of the English Tradition; *federation*—equal to the alliance of Braun-Blanquet; *subformation*—used to describe a distinct geographical facies of a formation and considered by Du Rietz to be equivalent to the association of the English Tradition; *formation*—corresponding to Clements's usage as well as Tansley's polyclimax concept. The approximate equivalents of these units are summarized in Figure 18.

The lowest units, the sociations, were characterized by homogeneity

Uppsala 1921	Uppsala 1928-30	Gams	Rübel	Braun Blanquet	English
–	Socion	Synusia	Synusia	–	Society
Association	Sociation		Sociation	Facies	–
–	Consociation		Consociation	(Sub-Ass)	Consociation
Ass-Complex	Association	Phytocoenoses	Association	Association	–
–	Federation		Alliance	Alliance	Association?
Formation	Sub-Formation		Formation	Order	Association?
	Formation			Class	Formation

Figure 18. A comparative chart of some community units and classificatory levels based on Du Rietz (1930).

in all layers, consociations homogeneous in one layer only and higher units not at all. This led Du Rietz to conclude that most units of higher rank than sociations were natural units in one layer only while the other layers were composed of alternating *synusiae* with little or no interrelationships. This term *synusia* had been introduced by Gams (1918) as a unit composed of species of similar life forms with similar ecological requirements. In Gams's techniques, the association (sociation) was a topographical unit composed of ecological units or synusiae in different combinations. The use of stratal units or synusiae became a main aspect of the Uppsala School (Du Rietz, 1930). Du Rietz had argued that the sociations could be grouped differently into the higher units if one used different layers for characterization. He cited an example of seventy-nine sociations which comprised the arctic subformation of the boreal ericaceous heath formation. If only the field layer was used as the basis for classification, the sociations could be grouped into sixteen consociations, forming seven associations and two federations. But the component sociations of a consociation had extremely diverse bottom (cryptogam) layers and some authors would prefer to classify using homogeneity of bottom layers as the basis. In this fashion the sociations could be placed into eighteen consociations and these into eleven associations, four federations and three formations, with variable field layers. This was clearly an inappropriate representation. Du Rietz (1930) therefore adopted Gams's synusial approach, dividing each sociation into its elementary single-layered units and grouping these the synusiae of each layer independently of those of other layers into synusiae of higher rank. The sociations are thus divided into elementary socions and these latter grouped into consocions, associons, etc., in each layer. The units are shown in Figure 19 and the relationships to idiobiological units illustrated.

Idiobiological Unit	Synusiae	Plant Community
Class	Panformion	Panformation
Order	Formion	Formation
Family	Subformion	Subformation
Tribe	Federion	Federation
Genus	Associon	Association
Section	Consocion	Consociation
Species	Socion	Sociation

Figure 19. The synusiae and community units of the Northern Tradition (after Du Rietz, 1930).

A similar representation is to be found in the Estonian School at that time, mainly in the work of Lipmaa (1931, 1935). Here the elementary units of vegetation were considered to be *unistratal-associations* or *unions* dominated by closely related life forms and confined to a single layer. Lipmaa considered that these unions were different from the synusia of Gams, but one often finds use of the two terms as alternative names. Moreover, the unions were not considered equivalent to either the socion or consocion because they (unions) appeared as broader units of varied floristic composition. One further point of difference is that Lipmaa stressed the habitat for the definition of units, whilst Du Rietz always maintained a classification of vegetation by properties of the vegetation itself.

The Sixth Botanical Congress of 1935 was, like its 1910 predecessor, thus faced with a complex of nomenclature and interpretations which seemed to have accelerated in the previous five years. The Congress produced three major resolutions (Du Rietz, 1936): (i) to use the term *sociation* in the sense of the Scandinavian phytosociologists for vegetation units characterized mainly by dominance in the component layers; (ii) to use the term *association* in the sense of the Zurich–Montpellier School as units recognized mainly by characteristic and differential species; or to use it for units of the same order of sociological value (*sensu* Uppsala); (iii) to unite sociations and associations into *alliances* in the sense of Z-M and the alliances into higher units.

The Modern Period, 1936 onwards

The modern period has seen a remarkable spread of the Zurich–Montpellier Tradition through north-central Europe, largely due to the energies of Tüxen and his pupils at Bundesanstalt für Vegetationskartierung at Stolzenau in Germany and latterly Todenmann über Rinteln—probably now the major centre of the Zurich–Montpellier Tradition. The aims and outlooks of the school have been stated by Tüxen (1937) as falling into three component parts: (i) the preparation of an inventory of all plant communities presented as well-ordered Association tables;* (ii) the cartographic representation of the units so delimited; (iii) the study of the life-conditions and developmental possibilities of the vegetation units—developed particularly by the Stolzenau Institute with reference to landscape architecture, potential natural vegetation and the German *Lebensraum* concept.

The publication of the work of Braun-Blanquet and Tüxen (1943) gave the first comprehensive view of the application of the classification system to central European vegetation, the production of various issues of *Prodrome des Groupements Végétaux* and the series of papers by Braun-Blanquet under the general title *Die Pflanzengesellschaften Rätiens* in the new phytosociological journal *Vegetatio* (1948 *et seq.*) added further impetus to the snowballing tradition. At the 1950 International Botanical Congress in Stockholm the Northern and Zurich–Montpellier Traditions once again clashed head-on over the units of classification. After 'une discussion très animée' and 'longues et parfois fatiguantes explications théoriques' (Braun-Blanquet, 1968) the Zurich–Montpellier system of classification was generally accepted. A growing awareness of the development of the Z-M tradition led to several explanations and evaluations of the step-by-step procedure of describing Associations; Ellenberg (1956), Tomaselli (1956), Poore (1955), Becking (1957) and Küchler (1967).

The annual symposia of the International Society for Plant Geography

---

* This definition of the association is hereinafter written Association to distinguish it from other interpretations.

and Ecology held at Stolzenau and Rinteln have provided a platform for the development of a generally accepted classificatory system and the colloquia held prior to the symposia have produced trends towards a unified system of sociological nomenclature in north-west Europe (Lohmeyer *et al.*, 1962). The symposia have also provided an opportunity for the exposition and legislation of syntaxonomic method on a voluminous amount of literature. The naming of the Association is thus now governed by five rules (from Moravec, 1964, summarized by Braun-Blanquet, 1968): (i) the validity of a unit should be supported by the existence of a complete Association table, delimiting the group of character species and species of high constancy, or by a constancy table; (ii) the name of the first author to name the Association without a valid description should be preserved in brackets and followed by the first author to publish a valid Association or constancy table; (iii) generally, only published works are considered; (iv) for the denomination of the unit, the name conforming with the rules of international nomenclature established by the Botanical Congresses must be used; (v) if an association is later found to contain two or more Associations, the first name is included in the synonymy. It is recommended that the original name in an amended form should be used for one of the Associations thus defined.

These rules have been greatly aided by the publication of a series of *Bibliographia Phytosociologica Systematica* relating to countries or specific vegetation types and published in Series B of *Excerpta Botanica*. All these features make the north-west European branch of the Z-M Tradition the most positive phytosociological school at the present date.

The post-1935 development of the Northern Tradition has been a trend towards the methods of classification of the Z-M Tradition. The third resolution of the 1935 Congress—to unite sociations and associations into Alliances—has been followed by Nordhagen (1937, 1942) in Norway who grouped his sociations into Alliances and also used character species for their definition. Du Rietz (1942) also adopted the concepts of *Scheidearten* and *Leitarten* which are roughly equivalent to the concepts of the differential and character species. The *Scheidearten*, however, are probably only ecological indicator species and unlike the practice in the Z-M Traditions are used to define formations which are equivalent to Z-M Classes. Recent works in the Uppsala trend have shown some lack of interest in classification. Malmer (1962) in his excellent work on mire vegetation of southern Sweden uses the term '*small-association*' for his basic units and unites these into series which are basically topographical and ecological but show marked floristic affinities. Later, Malmer (1968) discusses the classification of mire vegetation according to three major ecological gradients and compares the ways in which different authors treat the gradients. Finally, he erects a classification system comparable with the Z-M system, the units of which are defined by constancy.

Domin (1933) introduced an 11-category scale of cover-abundance which was adopted by Dahl and Hadac (1941) in their study of Ostøy Island in Norway. They used associations in the sense of Nordhagen (1937) and united these into alliances and orders in the same fashion. Their work was probably extremely important in strengthening the unification between the Z-M and Uppsala Traditions. Later, Dahl (1956) developed this approach erecting tables of floristic similarity between alliances using character, preferential, differential and relatively dominant species to indicate relationships. At the same time Poore (1955) criticized the methods of Braun-Blanquet and used a similar method to Dahl of defining associations by constant species. Poore also introduced the term *nodum* as a neutral term for a plant community of any rank, analogous with the term *taxon* in plant taxonomy. McVean who had worked with Dahl in the field along with Poore (1957) stated the new approach to Scottish vegetation which was essentially a mixture of Dahl and Nordhagen's approaches. This was later pursued in the excellent monograph of McVean and Ratcliffe (1962), *Plant Communities of the Scottish Highlands*. The Poore-McVean method is now a basic in all British autecological studies.

The English Tradition in Britain more or less culminated in Tansley's *British Isles and their Vegetation* in 1939 and until the 1955 Poore revival, little sociological work on vegetation classification was completed. Similarly, the Clementsian approach in the Americas gradually became eclipsed by the individualistic hypothesis of Gleason (1926), referred to as the 'individualistic dissent' by Whittaker (1962). The data from Curtis and McIntosh (1951) indicated that in relatively undisturbed localities no two stands of vegetation were alike and that they could all be arranged along a *continuum* according to their degrees of relative similarity. This aspect of vegetation classification *versus* continuum has received much attention and in consequence there is a voluminous literature on the subject. Two papers which are perhaps representative of the more modern views are those of Curtis (1967) and the replies to this paper edited by Dansereau (1968). The major contributors to the aspect of vegetation *ordination* or gradation are seen in the works of the Wisconsin School, Curtis (1959) and the extensive gradient analyses of Whittaker (1951, 1956, 1967). Several of these methods have infiltrated into British ecology in the past decade although much more attention has been paid to multivariate analyses of vegetation and environmental gradients (e.g. Williams and Lambert, 1959, etc.) in their association-analysis techniques, which for the purpose of this chapter would be better referred to as correlation-analysis. The gradient analysis approach to vegetation is dealt with in Chapter 8.

In a short conclusion to this chapter the major vegetation units recognized in the different traditions today are as follows:

(1) The *formation-type* (formation-class) a group of geographically widespread communities of similar physiognomy and life form and related to major climatic and other environmental conditions.

(2) The *formation*—a group of communities in a single region or continent, of similar physiognomy and related climatic and environmental conditions. These two categories are the main ones of the plant geographer whose interests are world vegetation types.

(3) The *association*—various units defined by: (*a*) floristic composition, character and differential species and constancy—the Zurich–Montpellier System which erects a formal hierarchy of Associations, Alliances, Orders and Classes; *the most generally accepted usage of the term* following the 1935 Botanical Congress; (*b*) stratal structure and constancy (= sociation = micro-association); the Northern Tradition including modern British sociology; (*c*) by one or more dominant species where the species dominants are those in the upper vegetation layer, consociation being used for single species dominance, prevalent in British ecology up to 1955, and called dominance-type by Whittaker (1962); (*d*) as a climax regional vegetation unit geared to stability and absence of further development; in the sense of Clements (1928); (*e*) as a stratal unit, the unistratal association used by Lipmaa and other members of the Estonian School.

(4) The *union*, a stratal community consisting of one or more species of related physiognomy and life form.

Other units often encountered include:

(i) The *landscape-type*—an extensive area of land characterized by all major aspects of vegetation and environment.

(ii) The *habitat-type*—a group of communities resembling one another through habitat relationships.

(iii) The *site-type*—a group of communities corresponding to a particular set of site characteristics.

(iv) The *nodum*—an abstract, neutral term applied to a vegetation unit of any rank.

CHAPTER 3

THE PHYSIOGNOMIC, FUNCTIONAL AND STRUCTURAL BASES OF VEGETATION DESCRIPTION

One of the most recent and clearest expositions of the criteria used in vegetation description and classification is that of Fosberg (1967), whose careful definitions of physiognomy, structure and function are particularly enlightening. In the past, these three concepts have been much used and abused in various forms and this has led to a general difficulty in application and interpretation. Many of the earlier plant geographers wavered considerably in their definitions and from the use of their separate and distinct tenets, divergence and convergence of interpretation has been profuse. Warming (1909) states emphatically that 'the leading features upon which the pertinent distinction (between vegetation types) depend are physiognomic'. He proceeds to define seven 'circumstances' that determine the physiognomy of vegetation:

1. Dominant growth forms.
2. Density of vegetation (number of individuals).
3. Height of vegetation.
4. Colour of vegetation.
5. Seasonal relationships.
6. Duration of the life of the species.
7. Number of species present.

In the correct interpretation of the word physiognomy, of these seven characteristics only numbers 4 and 5 are definitely physiognomic. The first three are essentially structural, 6 is functional and 7 compositional. This is perhaps an extreme example, but because of the diverse interpretations of these three basic concepts a detailed explanation is necessary.

Physiognomic characteristics

Physiognomy is the external appearance of the vegetation, and attributes comprise such obvious features as colour, luxuriance, seasonality and overall compositional features which are rapidly determined by an initial, visual assessment. This is basically the definition of Fosberg (loc. cit.), and against it one encounters an interpretation of the type supplied by Cain and Castro (1959) (who do not define structure or function): 'Physiognomy—the form and structure of vegetation; the appearance of vegetation that results from the life forms of predominant plants.' This thesis, which derives its support from the fact that functional and structural characters such as life form and stratification can be directly or indirectly responsible for the physiognomy is followed by many leading ecologists and phytocenologists. For example, Küchler (1967), although providing no strict definitions, implies that each of the 'life-form categories' in his analysis technique is physiognomic in character whilst 'structural categories' are apparently reserved for coverage and height stratification. Later, in the appendix to his book, he heads the whole descriptive technique *Description of Vegetation Structure by Küchler*. On the following page in a heading including *Based on Physiognomic System of A. W. Küchler* a classification of the vegetation of Mount Desert Island, Maine, is outlined using life form and structural categories. There is clearly an argument for the use of the term physiognomy in the sense of Cain and Castro as a broad concept encompassing the synthetic characteristics of function and structure and, since this is the case, all three words and concepts need to be rigidly delimited. Moreover, although the above examples from Küchler may seem rather pedantic, it is the loose usage of the three terms which has led to confusion and has probably been partially responsible for the innate desire of individual workers to devise and develop their own system of vegetation description.

The distinction between physiognomic and functional characteristics

The existence of any plant species is always governed by the adaptation of its form to function efficiently at the integrated requirements of its total environment. Every species is functionally adapted to a greater or lesser extent and some associated morphological features of adaptation are more obvious than others. Where to draw the line to demarcate functional characters which greatly influence the physiognomic appearance of vegetation and those which do not make such a basic contribution is a matter of opinion. The problem, which exists at the two descriptive and classificatory levels, can be solved to some extent by accepting the rule that physiognomic characters which are also functional should only be used in a preliminary vegetation description or an overall vegetation classification, e.g. the *Durisilvae*—broadleaf sclerophyll forest of Rübel (1930). In the actual stand descriptions such as those of Küchler (1949) and Dansereau (1951) these two characters— broadleaf and sclerophyll—can be regarded as functional and applied to individual species to give a semi-quantitative assessment of function.

Fosberg (loc. cit.) states that physiognomy of vegetation partly results from and must not be confused with structure and function, 'which are more exact and objective categories'. He lists 'relative xeromorphy' as a physiognomic character and 'xeromorphy' as a functional character. Presumably, the abandonment of the word relative is the trend toward precision and objectivity, and implies the change from a gross compositional vegetation feature to a precise synthetic character applied to individual species.

The solution to the problem of delimitation of the two categories is far from obvious, but the basis of comprehension probably lies in the distinction between practice and theory-born systems. Many physiognomic characters are derived from extensive fieldwork—indicating that the use of such characters is feasible and practicable. That these characters may also be functional goes without saying.

Many functional characters such as periodicity, life form (*sensu* Raunkiaer) involve fieldwork at different seasons and are therefore impracticable to the majority of expeditionary forces of plant geographers and ecologists. They are also, to a greater or lesser extent, based upon herbarium material.

With such considerations in mind it seems that the major physiognomic and functional characters to be dealt with in detail are best delimited as follows:

Physiognomic
  (a) Life form—i.e. ground form *sensu* Du Rietz.
  (b) Leaf size. (It is true that Raunkiaer based his final leaf-size boundaries on herbarium material but his original observations were based on fieldwork.)

Functional
  (a) Life form—*sensu* Raunkiaer.
  (b) Periodicity.
  (c) Dispersal mechanisms.

Other broad physiognomic characters which are frequently used in vegetation descriptions and classifications and which are more or less self-explanatory, include:

  (a) Forest, woodland, scrub, savanna, grassland, marsh, bog, mire, desert, tundra.
  (b) Trees, shrubs, herbs, grasses (graminoids), erect woody plants, climbing or decumbent woody plants.
  (c) Deciduous, evergreen, broadleaf, needle leaf, sclerophyll, succulent.
  (d) Bryophytes (bryoids), acrocarpous, pleurocarpous, *Sphagnum*, epiphytes and crusts, lichens, fungi, algae.

Physiognomic life-forms

It is clear that the same basic problem is met with in the distinction of life form (i.e. the vegetative form of the plant body) into physiognomic or functional (biological, epharmonic) categories, a rambling subject which is well reviewed by Du Rietz (1931). Several life-form systems are an end in themselves in that they are primarily aimed at a classification of the individual species, and either because of their complexity or for other reasons, have not been generally applied to vegetation classifications. Other more simple, physiognomic systems have been designed with the classification of vegetation in mind and have been used successfully to this end. The classical work of Du Rietz holds the key to an understanding of the development of the many life-form interpretations. In it a chronology of life-form systems is given as a series of historical periods, as follows:

1. The early period of purely physiognomic life-form systems—1884.
2. The period of the early epharmonic life-form systems—1884–1905.
3. The period of Raunkiaer's first life-form works—1905–1913.
4. The period of beginning reaction against the one-sided epharmonic point of view in life-form classification—1913–20.
5. The modern period of re-established purely physiognomic life-form systems—1921–30 (1970).

From the five periods outlined above it is implicit that the purely physiognomic life-form systems were the earliest and latest periods up to 1930, since which date there have been few significant additional or modification proposals. The first of the two periods represents the exploratory attempts of plant geographers to describe the earth's vegetation. The second period represents a reaction to the preoccupation with life-form species classifications which developed to a complex nomenclature between 1884 and 1920 and apart from the work of Raunkiaer (1905, 1916) made little impact on the scene of vegetation classification. For the sake of clarity, the two periods designated as physiognomic need to be expanded in this section whilst the methods of the three intervening periods must be dealt with under the subsequent heading of Function.

The work of Humboldt (1808) was one of the earliest treatises of the first period. He described nineteen *Hauptformen* which were mostly named after some characteristic genus or family, e.g. Palmen-form, Bananen-form, Cactus-form, Gras-form, Heidekräuter, Nadelholzer— or, Palm, Banana, Cactus, Grass, Heath and Needleleaf form. Expansions of this basic system yielded as many as sixty physiognomic types in Grisebach's work of 1875. These sixty types were grouped into seven main categories:

Holzgewächse—woody plants
Succulente Gewächse—succulents
Schlinnggewächse—climbing plants
Epiphyten—epiphytes
Kräuter—herbs
Gräser—grasses
Zellenpflanzen—lower plants (cryptogams)

By way of example of further classification, the woody plants contained thirty *Vegetationsformen* in seven intermediate groups, one of which, *Sträucher* (Shrubs) consisted of such vegetation-forms as *Erica*-form, *Oleander*-form and *Rhamnus*-form. Similarly, the classical sociological text of Hult (1881) contains forty-three vegetation forms of the type listed above, but grouped into ten *grundformer* (ground forms)—coniferous trees, deciduous trees, shrubs, dwarf-shrubs, grasses, herbs, lianas, peat-mosses, other mosses and lichens. The interim years (1884–1921) are marked by the publication of Warming's earliest work on life-forms and that of Du Rietz (1921). During this period controversy arose between epharmonists and physiognomists and Du Rietz proposed to restrict the term life-form to epharmonic characters and to use ground form for a purely physiognomic point of view. The system devised by Du Rietz is well worth repeating because its use has been quite widespread in both Scandinavia and northern Europe (Table 4). The system was in general use in Swedish ecological studies, notably those of Oswald (1923) and Sernander (1925). Other workers have used selected Du Rietz terms in a more complicated system, for example Rübel (1930), whilst individual categories have been used in restricted ecological investigations, e.g. the use of *Aquiherbiden* types by Den Hartog and Segal (1965). In the modern descriptive techniques of Küchler (1949) and Dansereau (1951), their 'Basic Life-forms' and 'Life-form' types are basically those of Du Rietz, and it is also interesting to note that the same author was one of the first ecologists to use short formulae to represent different types.

Leaf size as a physiognomic character

Cain and Castro (1959) point out that there is a general observation that leaves are large in moist tropical vegetation (rain forest), medium sized in temperate woodlands, and small in cold and dry conditions, such as tundra and heaths. This phenomenon can be regarded as a physiological-functional response of the plant and vegetation to mesoclimate in a narrow sense, or as a broad geographical, physiognomic feature reflecting macroclimatic situations.

The *leaf-size spectrum* for the characterization of a vegetation type

TABLE 4

The life-form system of Du Rietz (1921)

A. Higher plants
  I. Ligniden (woody plants)
    (a) Magnoligniden (m)—Trees taller than 2 m
      1. Deciduimagnoligniden (md) deciduous
      2. Aciculimagnoligniden (ma) needleleaf evergreen
      3. Laurimagnoligniden (ml) other evergreens
    (b) Parvoligniden (p)—Shrubs 0·8 m to 2 m tall
      4. Deciduiparvoligniden (pd)
      5. Aciculiparvoligniden (pa)
      6. Lauriparvoligniden (pl)
    (c) Nanoligniden (n)—Under 0·8 m tall
    (d) Lianen (li)—Climbing plants
  II. Herbiden (herbs)
    (a) Terriherbiden—Terrestrial herbs
      9. Euherbiden (h) herbs
      10. Graminiden (g) grasses
    (b) Aquiherbiden—Water plants
      11. Nymphaeiden (ny) rooted with floating leaves (*Nymphaea*)
      12. Elodeiden (e) rooted without floating leaves (*Elodea*)
      13. Isoetiden (i) rooted, bottom rosettes (*Isoetes*)
      14. Lemniden (le) free floating, not rooted (*Lemna*)
B. Moose (Bryophytes)
      15. Eubryiden (b) all mosses and liverworts excluding *Sphagnum*
      16. Sphagniden (s)—*Sphagnum* spp.
C. Flechten
      17. Lichens
D. Algen
      18. Algae
E. Pilze
      19. Fungi

---

was a brain child of Raunkiaer (1916), who believed that leaf size was primarily a measure of climatic effect, or the effect of the aerial environment in general. Raunkiaer's main concern was the structural adaptations of plants to 'the water problem', such as cuticular wax, stomatal protection, etc. Speaking of such adaptations, Raunkiaer says: 'The matter however is so complicated that it is very difficult to reach an exact appraisal of those adaptations in characterizing the individual plant communities biologically. The fact is that in a community which survives dry periods, some species are adapted to their environment in one way, and others in other ways, and we are still unable to determine quantitatively the value of the individual adaptations or the different combinations of different adaptations.' From the adaptations in question, Raunkiaer chooses an obvious morphological character, namely, diminution of the transpiring surface as an index of adaptation to prevailing climatic conditions. But having previously cast doubt on the value of this and other functional characters, it seems that this phenomenon must be accepted as being solely of morphological nature. More recently, Beadle (1953) and Loveless (1961) have shown that edaphic factors, especially soil phosphorus levels are also important in determining leaf size. For these reasons, leaf size spectra classes are taken at their face value, as a physiognomic character rapidly used for assessing the field relationships of stands of vegetation in different geographic regions within the same macroclimatic zone. The six Raunkiaerian leaf size classes are illustrated in Table 5.

TABLE 5

Leaf-size classes

Leaf class	Size range (mm^2) Raunkiaer (1916, 1934)	Webb (1959)
LEPTOPHYLL	up to 25 mm^2	–
NANOPHYLL	25–225	–
MICROPHYLL	225–2,025	225–2,025
(NOTOPHYLL)	–	2,025–4,500
MESOPHYLL	2,025–18,225	4,500–18,225
MACROPHYLL	18,225–164,025	–
MEGAPHYLL	164,025+	–

(The leaf-size class limits are calculated using 25 mm as the upper limit of the leptophyll class, and the subsequent multiplication of this figure by 9, e.g. 9×25 = nanophyll; $9^2 \times 25$ = microphyll, etc., to give classes which correspond with rough estimates based on a continuous size series of herbarium material.)

The only significant addition to Raunkiaer's leaf-size classes is that of Webb (1959) who proposes and defines the name notophyll for leaves of the small mesophyll class. Also, Webb maintains only three main classes, meso-, noto- and microphyll, expressing the view that recognition of macrophyll and nanophyll classes causes unnecessary complications in the production of a field key.

The tedium of Raunkiaer's method of tracing the leaf outline on to centimetre squared paper has been alleviated by a much simpler procedure developed by Cain and Castro (1956). This involves the measurement of the length and maximum breadth of leaf lamina, and then the derivation of the lamina area by the calculation of two-thirds of the rectangular length × breadth area. This is a straightforward procedure in the majority of instances, but three minor complications occur, namely drip tips, leaf maturity and compound leaves.

Richards (1952) has shown that within the tropical rain forest, there is a predominance of entire sclerophyllous leaves of the mesophyll size class. Such leaves are of striking uniformity of size and shape in all forest tree layers, being mainly oblong-lanceolate in shape with a pronounced acumen or 'drip tip'. Drip tips are common in leaves of the trees of all forest layers and vary in length considerably (Table 6).

TABLE 6

Drip tip statistics from wet evergreen forest, Nigeria
(data from Richards, 1952)

Tree layer	A	B	C	Shrubs
Number of species	10	13	14	10
Average acumen length mm	2·9	8·8	12·9	12·4
Number of species without acumen	4	1	–	–

Investigators of leaf-size classes are not always clear as to whether they include the drip tip in the leaf length measurement. Normally the drip tip is excluded, e.g. Grubb *et al.* (1963). Clearly, from the data of Richards the measurement of lamina plus drip tip could produce a much greater figure for leaf area and hence possibly a different leaf class. The problems do not end here. There are marked differences between the leaves of juvenile and mature trees, the former possessing drip tips, the latter often without. Similarly, juvenile leaves are often divided or compound and adult leaves of the same species entire and undivided. Richards cites the classical example of the Malayan species *Artocarpus elastica* which has incised leaves in the sapling stage and entire in the adult tree. There is obviously a case for careful sampling which is based on floristic identification, which in turn questions one of the main advantages of the leaf-size classification—i.e. its use where floristics are incompletely known or inaccessible. The final complication to be dealt with in this section involves the measurement of compound leaves. Cain and Castro (1956) base their leaf-size measurements of compound leaves on the dimensions of the leaflet blades, and for rain forest in Brazil find that compound leaves tend to have smaller leaflets than simple leaves or blades; but this results in only a 10% downward deflection from the main leaf class type.

In spite of these minor complications, Richards's observation 'that

TABLE 7

Tree leaf size spectra of some tropical rain forests

Author	L	NA	M	NO	ME	MA	MG	Rain forest type	Locality
Richards (1952)	–	–	10	84		6	–	Lowland	Shasha, Nigeria
Beard (1946)	–	2	5	86		7	–	Lowland (emergents)	Trinidad
" "	–	–	12	80		8	–	Lowland (canopy)	" "
Brown (1919)	–	–	4	86		10	–	Lowland	Mt Maquiling, Philippine Islands
" "	–	–	6	87		7	–	Sub-Montane	" "
" "	–	–	50	50		–	–	Montane (mossy)	" "
Cain and Castro (1956)	2	2	13	74		8	1	Lowland	Castanhal, Pará, Brazil
Webb (1959)	–	–	16	31	53	–	–	Lowland	Mucambo, Belém, Brazil
" "	–	–	2	39	59	–	–	Lowland	N.E. Australia
" "	–	–	95	5	–	–	–	Montane	" "
Grubb et al. (1963)	–	–	13	26	57	4	–	Sub-Montane	Borja, Ecuador
" "	–	–	9	14	50	27	–	Lowland	Shinguipino, Ecuador

in normal tropical rain forest, at least 80% of the species of trees have leaves of the mesophyll size class' can be seen to be basically correct from the data shown in Table 7. Although there is a lack of standardization in this data (e.g. variation in species number, height of species considered, etc.) the same basic pattern of mesophyll, or in Webb's terms meso-, plus notophyll predominance applies to both the Lowland and sub-Montane rain forest (Grubb *et al.*, 1963), but it is interesting to note the change to microphyll predominance in the Montane rain forest types.

Bark surface and slash

Studies of the physiognomy of the bark of tree boles in the tropical rain forests have proved of limited use in the description of vegetation, but of greater importance in the identification of tree species of the emergent and canopy layers when flowers and foliage are difficult to obtain. A neat demonstration of the use of bark in taxonomy is provided by Whitmore (1962a, b) working on members of the Dipterocarpaceae, a large family of over 100 species in seven genera, many of which are emergents in Malaysian rain forests. Seven bark types are recognized—smooth, dippled, shallow-fissured, deep-fissured, scaly, surface-rotten and laminate—using the two field characters of surface pattern and slash. The first five types can be recognized by surface patterns which involves a visual assessment of the surface configuration, the sloughing pattern, texture and colour of the bark. An oblique tangential section through the inner and outer bark or the slash appearance serves to define the remaining two types.

There seem to have been no attempts to quantify such data and comparisons are difficult. In their work on the rain forests of Ecuador, Grubb *et al.* (1963) record a preponderance (*circa* 80%) of smooth bark species with prolific cryptogamic epiphytes in all sample plots. Similar data from other rain forests enables their conclusion that 'smooth bark is not nearly so common as this in Malayan and Bornean lowland rain forest, nor are bole bryophytes, but it is the commonest surface pattern in Nigeria'.

Vegetation physiognomy and aerial photographs

The appearance of vegetation in the stereoscopic image of an aerial photograph conveys two features to the observer—pattern and texture. Basically speaking, pattern here refers to the recognition of individual objects and their spatial arrangement in a repeated sequence or at random, such as trees in a woodland. This is a structural characteristic. Whereas the individual pattern of trees can be differentiated, the leaves are not individually recognizable and their mass confers upon each tree a particular texture—a physiognomic characteristic. Based

Plate I. Photographic pair for stereoscopic observation on the texture of the vegetation (by courtesy of the Geography Department, University of Hull) Scale = 1 : 5000.

upon the appearance in the stereo-image, five levels of texture can be subjectively delimited, namely very coarse, coarse, average, fine and very fine. The five levels have their main role in the recognition of different vegetation types and in distinguishing the composition and heterogeneity of the latter and in the production of a vegetation map prior to fieldwork.

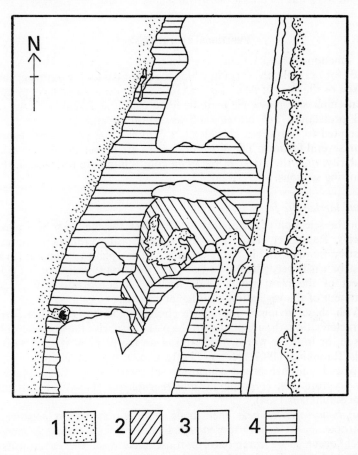

Figure 20. Map drawn from enlargement of Plate I to show areas of different texture. 1, sand; 2, *Festuca rubra* short sward—texture very fine; 3, *Hippophaë* scrub; 4, *Ammophila* dunes—texture coarse, very coarse.

The levels of texture are elegantly demonstrated by the use of a pocket stereoscope in conjunction with Plate I and Figure 20 showing the vegetation of a section of the Spurn Peninsula, E. Yorkshire.

Pattern is seen in the spatial distribution of the shrubs *Sambucus nigra* and *Hippophaë rhamnoides* each possessing their own texture. The levels of texture are shown by the dune grasses:

Very coarse/coarse—*Ammophila arenaria*.
Average—*Agropyron* spp.
Fine/very fine—*Festuca rubra*/Moss short sward.

Functional characteristics

All functional characteristics are associated with periodicity of some kind; for example, the form in which the plant passes the unfavourable season or the time of flowering and fruiting. Some authors have placed great emphasis on a single periodic feature such as Raunkiaer's use of bud protection, but others such as Warming and Du Rietz have attempted to produce a classification of functional plant types based upon several features. The different systems are legion and for the sake of clarity, only the works of the main contributors are reviewed in the following sections.

Raunkiaerian life forms

In the selection of characters to determine the life form of a plant species, Raunkiaer (1907) lists three prerequisites: (*a*) the character must be fundamental in the relationship of the plant to climate; (*b*) it must be easily recognizable in the field; (*c*) it must represent a single aspect of the plant which lends itself to a comparative statistical treatment of the vegetation of different regions.

With these in mind, Raunkiaer chose as the basis for life form characterization, the adaptation of the plant to survive the unfavourable season, or in other words, the position on the plant of the vegetative buds. Raunkiaer (1907) recognized fifteen main types of life form which he placed into five broad categories—phanerophytes, chamaephytes, hemicryptophytes, cryptophytes and therophytes (Figure 21 redrawn from Raunkiaer 1934).

(*a*) *Phanerophytes* bear their buds or shoot apices on negatively geotropic shoots in an exposed aerial position. They are thus plants which occur more commonly in those regions of the earth where prolonged periods of cold, drought and winds are infrequent, e.g. the warm moist tropical regions where they form the bulk of the total species complement. In other regions with progressively longer and warmer dry seasons or severe winters, phanerophytes decrease in proportion to other life forms.

Two subdivisions of type of phanerophyte life form are widely adopted. One is based simply on the height of the buds above the ground and recognizes four categories which unfortunately vary according to authors:

(i) Megaphanerophytes	<30 m	<25 m
(ii) Mesophanerophytes	8–30 m	10–25 m
(iii) Microphanerophytes	2–8 m	2–10 m
(iv) Nanophanerophytes	>2 m	0·5–2 m
	(after Raunkiaer, 1934)	(after Dansereau, 1957)

The second character on which phanerophyte types are classified is the form of covering and protection of the buds, again interpreted

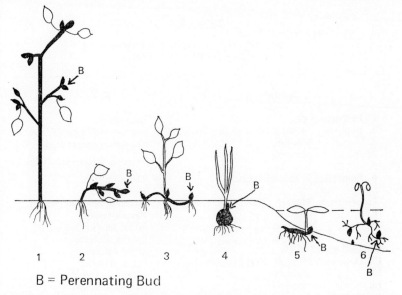

B = Perennating Bud

Figure 21. Examples of the six main Raunkiaerian life forms. 1, Phanerophyte; 2, Chamaephyte; 3, Hemicryptophyte; 4, Geophyte; 5, Helophyte; 6, Hydrophyte.

as a reflection of climatic adaptation. As Raunkiaer (1934) points out, in many instances there is no hard and fast limit to the determination of whether a bud is protected or not. A long series of adaptations to survival of the unfavourable season is readily demonstrable in evergreens, from the unprotected buds of *Eucalyptus orientalis* and *Olea europaea*, through such examples as *Calluna vulgaris* and *Drimys winteri* where the buds are protected by older and underdeveloped leaves respectively, to species such as *Ilex aquifolium* with an elaborate system of bud scales.

A combination of the above two sets of categories and the criterion of whether the plant is deciduous or evergreen gives rise to twelve phanerophytic sub-types (Table 8). In addition, Raunkiaer recognizes three other distinct phanerophyte types:

Herbaceous phanerophytes—resembling large herbs, but differing in that their aerial shoots remain alive for several years without becoming woody. They are found only in the most favourable climates of the constantly humid tropical zone.
Stem succulents—phanerophytes with succulent stems and without proper foliage leaves.
Epiphytic phanerophytes (including hemiparasites).

TABLE 8

Conspectus of phanerophytic sub-types
(modified from Raunkiaer 1934)

Sub-type	Example
1. Herbaceous ph.	Begonia sp.
2. Evergreen Megaph. without bud covering	Sequoia wellingtonia
3. „ Mesoph. „ „ „	Cupressus macrocarpa
4. „ Microph. „ „ „	Veronica (Hebe) salicifolia
5. „ Nanoph. „ „ „	
6. Epiphytic ph. (hemiparasitic)	Viscum album
	Hedera helix
7. Evergreen Megaph. with bud covering	Nothofagus cunninghamii
8. „ Mesoph. „ „ „	Pinus sylvestris
9. „ Microph. „ „ „	Buxus sempervirens
10. „ Nanoph. „ „ „	Ruscus aculeatus
11. Semi-succulent ph.	Many Cactaceae
12. Deciduous Megaph. with bud covering	Fagus sylvatica
13. „ Mesoph. „ „ „	Fraxinus excelsior
14. „ Microph. „ „ „	Corylus avellana
	Salix capraea
15. „ Nanoph. „ „ „	Salix repens
	S. lapponum

(*b*) *Chamaephytes* are woody or herbaceous, low-growing plants with buds produced on aerial branches close to the soil. They are most abundant in the drier or colder climates where they derive protection in the unfavourable season from the withered remains of other plants or a cover of snow. Raunkiaer designates four sub-types to this group:

(1) *Suffruticose chamaephytes*. At the end of the vegetative period the upper parts of the vegetative and flowering shoots die back so that

only the lower portions survive the unfavourable season (Figure 21, type 2, for example).

(2) (3) *Passive and active chamaephytes*. Biologically these two sub-types are the same in that their vegetative shoots are persistent and lie along the surface of the soil. Passive forms produce more or less erect growing apices, but because the shoots produced lack sufficient strengthening they become procumbent, e.g. *Stellaria holostea* and *Cerastium tomentosum*. In active chamaephytes the bud apices and shoots are produced only in a horizontal plane. The most common chamaephytes in the British flora belong to this group, e.g. *Thymus drucei* and *Veronica officinalis*.

Other applications of the life-form classification, (e.g. Clapham, Tutin and Warburg, 1962) recognize woody and herbaceous chamaephyte categories rather than the above three types of Raunkiaer so that *Thymus drucei* is a woody form and *Veronica officinalis* herbaceous.

(4) *Cushion chamaephytes* are really a modification of the passive forms where the shoots are arranged so closely together that they prevent each other from falling over and the solidity of the cushion affords mutual protection, e.g. several arctic-alpine herbs, such as *Silene acaulis* and *Cheleria sedoides*.

(*c*) *Hemicryptophytes* are characterized by the degeneration of the shoots to the level of the ground at the beginning of the unfavourable period, so that only the lower aerial parts of the plant remain alive and bear buds at the level of the soil surface. Most of the world's herbs and grasses and approximately half the flora of central Europe are hemicryptophytes. Raunkiaer erects three sub-types and these are then subdivided into forms with and without stolons:

(1) *Protohemicryptophytes* bear aerial shoots which are elongated from the base with the largest foliage leaves at about the middle of the stem, decreasing in size towards the base where small, scale-like leaves are usually present to protect the bud. Many British species of *Hypericum*, e.g. *H. pulchrum* and *H. perforatum* are examples without stolons, whilst the genus *Epilobium* offers several examples with subterranean stolons, e.g. *E. hirsutum, E. palustre*.

(2) *Partial rosette plants* are commonly biennials with their largest leaves attached to the base of the plant in the form of a radical rosette in the first year of growth, and in the second year form an elongated aerial shoot with foliage leaves and flowers. Many members of the Ranunculaceae fall into this category and two species of *Ranunculus, R. acris* and *R. repens* provide examples of types without and with stolons respectively.

(3) *Rosette plants* is a category used for hemicryptophytes with an elongated aerial portion which is exclusively flower-bearing and where the foliage leaves are attached to the plant in a compact rosette at the

soil surface. Many Compositae are rosette plants, and two common species, *Bellis perennis* and *Pilosella vulgaris*, illustrate the two subdivisions of stolon absence and presence respectively.

(*d*) *Cryptophytes* are plants whose buds or shoot apices survive the unfavourable season at varying depths below ground or at the bottom of water. This group of plants is conveniently subdivided, according to habitat, into geophytes or terrestrial cryptophytes, helophytes or marsh plants and hydrophytes or water plants.

(1) *Geophytes* include land plants with most types of subterranean perennating organs, e.g. rhizomes, bulbs, stem tubers (including corms) and root tubers. Examples are numerous and well known.

(2) *Helophytes* are those cryptophytes which grow exclusively in soil saturated with water or in water itself, and are commonly rhizomatous members of the Gramineae and Cyperaceae with aerial vegetative and flowering shoots, e.g. *Phragmites communis* and *Cladium mariscus*. Other common helophytes from other families are species of *Typha*, *Alisma* and *Sparganium*.

(3) *Hydrophytes* survive the unfavourable season by means of rhizomatous buds or detached vegetative buds (turions) which lie at the bottom of the water. The vegetative shoots are always submerged or floating and only the inflorescences rise above the surface for pollination. Common examples of rooted rhizomatous hydrophytes are to be seen in *Ranunculus* subgen. *Batrachium* and *Nymphaea*, whilst those which survive the winter as turions include species of *Lemna*, *Hydrocharis* and *Utricularia*.

(*e*) *Therophytes* are plants which survive winter as seed, i.e. are annuals where the life cycle usually extends only from spring to autumn. Such plants are characteristic of a hot, dry climate and open, temporary habitats such as cultivated ground. Examples are numerous in the British flora and range from *Senecio vulgaris*, a common garden weed to *Hornungia petraea*, a rare crucifer of calcareous screes and sand dunes.

Applications of the life-form system

I. Raunkiaer. By the development of this system of life-form classification, Raunkiaer succeeded in providing a common basis for the comparison of the floras of diverse regions of the world and for illustrating their relationship to major climatic changes. His initial investigation was a comparison of the floras of the then Danish West Indies islands of St Thomas and St Jan and Denmark (Table 9). Detailed explanation is unnecessary, for it is obvious that the plant climate of Denmark is a hemicryptophyte climate and that of the two West Indian islands a phanerophyte climate. Similarly, the comparison of

TABLE 9

Selected life-form spectra
(from Raunkiaer, 1934)

Region	Number of species	S	E	MM	M	N	Ch	H	G	HH	Th
A.											
Denmark	1,084	–	–	1	3	3	3	50	11	11	18
St Thomas and St Jan	904	1	2	5	23	30	12	9	3	1	14
Seychelle Islands	258	1	3	10	23	24	6	12	3	2	16
Aden	176	1	–	–	7	26	27	19	3		17
B.											
Clova (Scotland)	373	–	–	2	2	5	9	58	7	5	13
Spitzbergen	110	–	–	–	–	1	22	60	13	2	2
Normal spectrum	1,000	2	3	8	18	15	9	26	4	2	13

these two islands with other tropical regions, e.g. the Seychelle Islands reveals the same picture of phanerophyte preponderance. There are, however, certain climatic differences between these latter two regions, the climate of the Seychelles being much wetter. In consequence, in the drier climate of the West Indies there is a marked displacement from left to right in the life-form spectrum when compared to the Seychelles, revealed by a decrease in the percentage of large phanerophytes and an increase in nanophanerophytes and chamaephytes. Further displacement to the right of the spectrum is to be seen in the more arid tropical regions exemplified here by Aden.

In 1916, Raunkiaer added to his statistical investigations, the Normal Spectrum as a yardstick for comparison of actual spectra. This was the result of the sampling of 1,000 species selected objectively from the world flora to provide a random sample. From the data the following distribution pattern was obtained: S 2; E 3; MM 8; M 18; N 15; Ch 9; H 26; G 4; HH 2; Th 13 or Ph 46%, Ch 9%, H 26%, Cr 6%, and Th 13%.

This Normal Spectrum is usually appended to all of Raunkiaer's actual calculations.

Raunkiaer considered the knowledge of biological spectra fundamental to the ability to define plant climatic boundaries or *biochores*. In his work on the plant climate of the cold region of the north temperate zone, he showed that the latter zone is characterized by the high percentage of hemicryptophytes (Table 9, Denmark and Clova), and that of the cold zone is distinguished by the rise in the percentage of chamaephytes towards the north (Spitzbergen). It was this rise in the chamaephyte percentage that Raunkiaer took to be critical in delimiting the different phyto-climatic regions by means of biochores (Table 10).

TABLE 10

Chamaephyte biochores in the north temperate region of Europe
(modified from Raunkiaer, 1934)

Plant climate zone	Chamaephyte %
1. Cold temperate—a hemicryptophyte zone 10% Chamaephyte biochore (10°C June isotherm)	Scilly Isles 3·5% Shetlands 7% Faeroes 10·5%
2. Boreal—hemicryptophyte/chamaephyte zone 20% Chamaephyte biochore (4·44°C June isotherm)	Iceland 13% W. Greenland (64°–71°) 19% Spitzbergen 22%
3. Arctic—chamaephyte zone 30% Chamaephyte biochore	N.W. Greenland (74°+) 29%
4. Arctic nival—chamaephyte zone	Jan Mayen I. 32%

Extracting data from the floras of diverse regions of the north temperate zone from Alaska to Labrador and Novaya Zemlya to the Scilly Isles, Raunkiaer was able to construct three chamaephyte biochores, two of which show close correspondence to climatic lines. The 20% chamaephyte biochore is thus shown to correspond fairly closely with the June isotherm of 4·44°C (40°F) and similarly, the 10% biochore is coincidental with the 10°C (50°F) June isotherm.

In addition to latitudinal changes in plant climate, Raunkiaer demonstrated changes with altitude. With data collected mainly from the Alps he illustrated the change in the percentage of chamaephytes with the increase of altitude, recording high percentages of over 50% above 3,000 m in many central alpine localities.

The decrease of the altitude level of biochores in more northerly latitudes is shown neatly in his tables for Glen Clova, Scotland and the Faeroes, based on data from Willis and Burkhill (1904) and Ostenfeld (1908) respectively (Table 10). In Clova, the highest chamaephyte percentage of 27 is attained only above 1,000 m, whereas in the Faeroes, some 5° to the north this figure is achieved at 700 m. These figures are extremely significant when the overall spectrum figures of the two floras are compared and seen to be more or less the same.

TABLE 11

Changes in chamaephyte percentage with altitude and latitude
(modified from Raunkiaer, 1934)

Region	Percentage distribution of species among life forms								Number of species
	MM	M	N	Ch	H	G	HH	Th	
Clova (Scotland)									
Above 1,000 m	–	–	–	27	64	9	–	–	11
900–1,000	–	–	2	25	52	14	–	7	44
800–900	–	–	3	22	60	11	–	4	72
700–800	–	1	5	11	67	11	1	4	170
600–700	1·5	2	4	15	62	9	1·5	5	206
500–600	3	3	4	13	63	8	2	4	182
Faeroe Isles									
Above 700 m	–	–	3	27	62	8	–	–	37
„ 600 m	–	–	3	24	65	5	–	3	63
„ 500 m	–	–	4	23	64	5	–	4	78
Whole flora (Clova)	2	2	5	9	58	7	4	12	373
„ „ (Faeroes)	–	–	1·5	10·5	56	12	10·5	9·5	254

This section of Raunkiaer's work culminated in the production and characterization of a number of plant climates with four major series:

1. A *Phanerophyte* climate in the tropical zone where the precipitation is not deficient.
2. A *Therophyte* climate in the regions of the sub-tropical zone with winter rain.
3. A *Hemicryptophyte* climate in the greater part of the cold temperate zone
4. A *Chamaephyte* climate in the cold zone.

The main plant climates and their subdivisions are delimited by biological boundaries—biochores—based on exact life-form percentages and analogous with climatological lines such as isotherms.

II. Other applications—Macroclimatic. It is convenient to divide other applications of the life-form system into three categories according to the magnitude of climatic effect, i.e. at the macro-, meso- and microclimatic levels. Raunkiaer's own use of the system was mainly macroclimatic and many authors have followed suit with comparisons of similar climatic vegetation types from widespread geographical regions (Cain, 1950). The example provided in Table 12 is for temperate broadleaf deciduous woodland in North America and Britain and apart from the characteristic phanerophyte-hemicryptophyte canopy-ground layer relationships the variation in cryptophyte percentage towards the southern and drier extremities of the zone is interesting.

TABLE 12

Life-form spectra for temperate broadleaf deciduous woodlands in North America and Britain

Type	Locality	Species number	Ph	Ch	H	Cr	Th
1. Mixed	Tennessee	113	36·3	4·4	30·1	25·8	3·4
2. Mixed	Cincinnati	127	33·6	3·9	34·4	23·4	3·9
3. Aspen	Michigan	310	22·9	3·9	47·1	16·1	10·3
4. Aspen	Alberta	170	25·8	1·8	48·2	17·1	7·0
5. Beech	South England	–	23·0	7	56	15	–
6. Ash	Peak District	120	22·5	7	56	13·5	1
7. Mixed	Teesdale	190	18	14	58	15	1
8. Ash	West Scotland	75	17	5·5	61	15	1·5

1–5. From Cain (1950).
6, 7. Shimwell (unpublished).
8. Derived from data of McVean and Ratcliffe (1962).

Continuing on a macroclimatic scale, several American plant geographers have applied Raunkiaer's life forms to the classification of basic climatic types by Köppen (1923). This latter system creates five major climatic categories as follows:

A. Tropical forest climates,; coolest month above 64·4°F.
B. Dry climates (BS—steppe and semi-arid; BW—desert and arid).
C. Mesothermal forest climates; coldest month above 32°F but below 64·4°F; warmest month above 50°F.
D. Microthermal snow-forest climates; coldest month below 32°F.
E. Polar climate; warmest month below 50°F.
   (ET—tundra, warmest month above 32°F; EF—continuous frost.)

To these major divisions are added various subsidiary categories, e.g. Am—tropical *monsoon* forest climate with a short dry season, or BWh—*hot* and dry desert with all months above 32°F as opposed to BWk—*cold* dry desert with at least one month below 32°F. The relationship of this system to life forms is expounded by Cain and Castro (1959) and Table 13 illustrates the type of picture which emerges using the data of Raunkiaer (1934) and Braun-Blanquet and Marie (1924) for the hot desert climate type (BWh).

TABLE 13

Life-form spectra for the hot desert climate type (BWh) of Koppen (1923)

Locality	Species number	Life-form classes						
		S	MM	N	Ch	H	Cr	Th
1. El Golea, Sahara	169	–	–	9	13	15	7	56
2. Ghardaia, North Africa	300	0·3	–	3	16	20	3	58
3. Libyan Desert	194	–	3	9	21	20	5	42
4. Oudjda Desert	49	–	–	–	4	17	6	73
5. Oudjda semi-desert	32	–	–	–	59	14	–	27
6. Aden	176	–	7	26	27	19	3	17
7. Death Valley, California	294	3	2	21	7	18	7	42

1, 2, 3, 6, 7 data from Raunkiaer (1934).
4, 5 data from Braun-Blanquet and Marie (1924).

It is clear from Table 13 that there is great variation between life-form spectra within the BWh climatic class, even in the same geographical region. The use of the system on such a macroclimatic scale is thus questionable. It is also indicative that even the major vegetation boundaries cannot be defined by climate alone and further recommends the use of the life-form system on a much finer scale. Both points are discussed subsequently.

Mesoclimatic. At a mesoclimatic level the superficial relationships of climate to the life-form spectrum are less precise and more inter-dependent on a number of environmental factors. Geological and pedological factors emerge as features which form a template upon which climatic effects may be superimposed. Thus the main chamae-phyte zone of Britain can be broadly correlated with the upland environment of the older geological formations which possesses its own peculiar climate (Figure 22). At this level it is difficult to cite a single climatic feature as being responsible for changes in life-form percentages, but rather the synthetic climatic characteristics which constitute the *climate type.* Walter and Lieth (1962) have worked out a composite climate diagram which presents available climatic data on a single

D

Figure 22. The life-form zones of the British Isles.

diagram (Figure 23). Correlation of meteorological stations with similar climate types enables the delimitation of climate-type zones, to which the vegetation can in turn be referred. It is of some significance that the driest part of Britain with its sandy soils and therophyte-dominated vegetation falls within a single climate-type zone (VI_1).

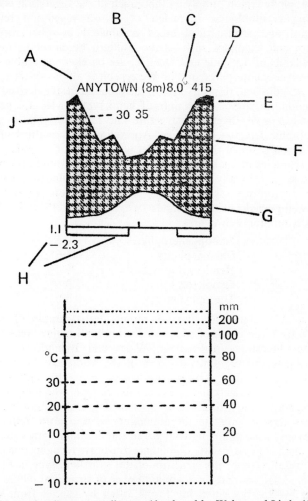

Figure 23. Composite climate-type diagram (developed by Walter and Lieth, 1967). A, location of meteorological station; B, altitude in metres; C, mean annual temperature (°C); D, humidity; E, period with precipitation over 100 millimetres; F, precipitation graph (mean monthly data); G, temperature graph (mean monthly data); H, absolute minimum temperature and duration below 0°C; J, duration of observation period in years.

The application of life-form types at this level enables recognition of three main mesoclimatic vegetation types in Britain—hemicryptophyte, therophyte and chamaephyte—and several subsidiary types some of which are reviewed below. *Hemicryptophyte Vegetation.* In his monograph Tansley (1939) refers to 'the British Isles, a hemicryptophyte region', implying that much of the vegetation is composed of hemicryptophytes. A review of British vegetation reveals that natural and semi-natural areas dominated by phanerophytes are sparse and examples such as the Chiltern beechwoods, or the pine woodlands of Speyside and Deeside, are barely comparable with their extensive counterparts in the Wesergebirge of north-west Germany and Scandinavia, respectively. On the other hand, the derived grassland communities composed mainly of hemicryptophytes are perhaps the most extensive, diverse in structure and luxuriant than in any other geographical region of Europe. Approximately one-third of the total vegetation cover of the British Isles is dominated by hemicryptophytes and examples of such vegetation types are to be found at virtually any altitude and in any region. A fairly typical life-form spectrum of hemicryptophyte vegetation may be derived from an overall figure for well-grazed basic grasslands (*sensu* Tansley, 1939) in the British Isles from Shimwell (1968).

Nanophanerophytes	0·5%
Chamaephytes	11·6%
Hemicryptophytes	79·4%
Geophytes	3·5%
Therophytes	5·0%

The above figures are based on representative samples of basic grasslands from the chalk of southern England, all major areas of carboniferous limestone in England, Wales and Ireland, the magnesian limestones and the Jurassic series of the Midlands. These grasslands constitute the first sub-type of hemicryptophyte vegetation in the British Isles—namely the *grass/rosette* sub-type, which term is more or less synonymous with the term 'permanent grass'. Within this grassland type several changes in the abundance of the more frequent rosette herbs and accompanying changes in the dominant grass are apparent and distinct grass/rosette dominance zones are recognizable (Table 14). An overall view of the structure of this grassland type in the different regions suggests that there are a number of structural niches which are filled by closely related life forms, often by species from within the same family. The essential structural components are as follows:

(1) Gramineous: (*a*) caespitose
          (*b*) subsidiary
(2) Rosette composite

(3) Partial rosette: (a) Compositae
                  (b) Umbelliferae
                  (c) Violaceae
(4) Protohemicryptophyte: (a) Rubiaceae
                            (b) Hypericaceae
(5) Ubiquitous/Constant—at least one member of the following families: Campanulaceae, Rosaceae, Papillionaceae, Labiatae, Dipsacaceae, Plantaginaceae, Gentianaceae and a stoloniferous composite.

TABLE 14

Grass/rosette hemicryptophyte vegetation types on calcareous soils in Britain

Dominant grass	Dominant rosette	Distribution	Climate type	Soil Parent material
Bromus erectus Brachypodium pinnatum	Cirsiumacaulon Leontodon taraxacoides	East Yorkshire East Midlands South-east and South Central England	VI_{1b}, V(IV)	Chalk Jurassic (oolite and lias)
Festuca ovina Helictotrichon pratense	Leontodon hispidus	North Wales South Wales Peak District	$V(VI)_2$, VI(X)	Carboniferous limestone
Sesleria caerulea	Leontodon hispidus	East Durham Craven and Teesdale Pennines	VI_2, VI_{3a}, VI_{3b}	Permian (magnesian) Carboniferous limestone

The regional replacement of one species by another is never absolute and two related species of the same life form can be found occupying different parts of the same structural niche in a homogeneous vegetation type. Thus, chalk grasslands frequently contain both *Leontodon taraxacoides* and *L. hispidus*, and occasionally both *Galium pumilum* and *Asperula cynanchica* occur in the same Cotswold oolitic grassland. Often within regions, related species tend to replace each other in related structural types, for example, the alternation between *Viola hirta* and *V. riviniana* in many regions. Nevertheless, the broad geographical changes remain fairly obvious (Table 14). Apart from the obvious changes in caespitose grass dominant perhaps the most notable change in the rosette niche is the loss of *Cirsium acaulon* in the more northerly regions. This loss is apparently not compensated by the presence of a closely related life type other than *Senecio jacobaea* or *Centaurea nigra*, and increases in the gregariousness of otherwise ubiquitous rosette types such as *Leontodon hispidus*, which expand to fill the niche.

The second sub-type of hemicryptophyte vegetation can be labelled broadly as the *tall herb/tall grass* sub-type. This includes all hay meadow communities with an abundance of tall rosette and partial rosette herbs such as *Heracleum sphonydlium*, *Knautia arvensis*, *Geranium pratense*, and the more palatable tall meadow grasses like *Helictotrichon pubescens*, *Arrhenatherum elatius*, *Cynosurus cristatus*, etc. These communities are poorly studied but there is some evidence of geographical replacement in community niches, so that in Teesdale the more southerly species *Geranium pratense* and *Cirsium palustre* are more or less replaced by their northern counterparts *G. sylvaticum* and *C. heterophyllum*.

Therophyte vegetation. Natural therophyte-dominated vegetation is restricted to the driest regions of Britain in the region of the Breckland of East Anglia where the flora shows a closer relationship to that of the north European plain than to the rest of Britain. The zone delimited by climate-type VI_{1a} (Figure 22) can be called the *Phleum/Silene/Veronica* zone characterized by the rare therophytes *Phleum arenarium* (in one of its few inland localities), *Silene gallica*, *Veronica verna* and *V. triphyllos* and the continental hemicryptophytes *Phleum phleoides* and *Silene otites*. Other more widespread therophytes include *Aira praecox*, *A. caryophyllea*, *Catapodium rigidum*, *Teesdalia nudicaulis*, etc.

The more common therophyte communities are those of arable land which tend to have their maximum development after harvest of the crop. In addition to phanerogam communities in which *Pao annua*, *Veronica* spp., *Kickxia* spp. and *Papaver* spp. play an important part, there are the ephemeral cryptogam communities which develop in November and December and quickly reach maturity before the fields are ploughed. In a survey of some twenty arable fields on diverse soil types in the East Riding of Yorkshire, the following bryotherophytes were the most common: *Bryum bicolor*, *B. erythrocarpum* agg., *Barbula fallax*, *B. unguiculata*, *Phascum cuspidatum*, *Pottia truncata* and *Riccia glauca*.

It is interesting to note that many of these species have a highly specialized reproductive system. Many bryotherophytes produce rhizoid gemmae or tubers, while in others the sporophyte is cleisto-carpous, having no special dehiscence mechanism, thus increasing the chances of the survival of spores in an intact capsule.

Champaephyte vegetation. While hemicryptophyte-dominated vegetation covers one-third of the total land surface of the British Isles, approximately another third is covered by chamaephyte vegetation in the shape of the upland heaths and moorland dominated by *Calluna* and other ericaceous shrubs, *Empetrum*, *Sphagnum* and fruticose lichens. Lowland heaths on the sandy soils of north-west Europe, especially those of Holland and Germany, are dominated by *Calluna vulgaris*, *Genista anglica* and *Empetrum nigrum* and frequently form a vegetation mosaic with therophyte communities. Such areas are generally poor in the British Isles due to burning and over-grazing, but in western upland regions the ericaceous chamaephyte communities of acidic soils become an increasingly important feature of the vegetation. In northern England and Scotland, *Erica tetralix*, *Vaccinium myrtillus*, *V. vitis-idaea* and *Empetrum nigrum* are common chamaephytes, while in the arctic-alpine regions of north and central Scotland, *Empetrum hermaphroditum*, *Arctostaphylos uva-ursi*, *Arctous alpina*, *Betula nana* and *Loiseleuria procumbens* are widespread.

On calcareous soils the change to a chamaephyte region is first noticeable through the increase in the importance of the role of *Dryas octopetala* in community structure with a consequent loss of several hemicryptophytes, e.g. in regions such as the Burren, western Ireland. In other base-rich localities in central Scotland, species of *Salix*, especially *S. reticulata*, *S. herbacea* and *S. myrsinites*, dominate rockledge communities along with cushion chamaephytes like *Silene acaulis*, *Cheleria sedoides* and *Saxifraga oppositifolia*.

Microclimatic. The percentage of hemicryptophytes in the grasslands of any region can vary greatly, due to slight differences in aspect and microclimate. Quantin (1935) recognizes two types of hemicryptophyte vegetation in the Jura mountains depending on exposure and evaporation from the habitat—(*a*) semi-dry or mesophilous, (*b*) xerophilous grading into a third type of chamaephyte/therophyte community in extremely dry conditions (Table 15).

As the percentages of chamaephytes and therophytes increase at the expense of the hemicryptophytes there is an observable increase in average temperature and average daily evaporation which are in turn related to an increase of pH and percentage carbonates and a decrease in the humus content.

In general, small changes in temperature and an increase of adverse habitat factors such as the frequency of frosts affect the percentage of hemicryptophytes and therophyte- or chamaephyte-dominated vegetation develops. As yet only broad generalizations of this type can be applied to the balance between life-form zones and a more clear definition of causal factors must await accumulation of data on plant to microclimate relationships similar to that provided by Newman (1967) in his work on *Aira praecox*.

The validity of the life-form system

Two pieces of research underline the value and validity of the

TABLE 15

Life form, microclimate and habitat relationships
(modified from Quantin, 1935)

Vegetation type	a	b	c	Vegetation type		a	b	c
Life form				*Habitat factors*				
Chamaephytes	10·6	18·6	42·5	pH	(max)	7·3	7·9	7·9
Hemicrypto-								
phytes	76·3	49·3	25·0		(min)	6·9	7·1	7·5
Geophytes	7·8	8·7	2·5	% CO_3	(max)	28·2	67·1	51·5
Therophytes	5·3	8·7	37·5		(min)	4·4	12·0	12·7
Parasites	–	1·5	–	% Humus	(max)	21·4	17·1	4·6
					(min)	11·4	7·1	1·7
Average temperature (°C)				*Evaporation* (cm²) *daily average*				
Spring	7·5	9·7	12·8	Spring		2·6	3·3	5·4
Summer	15·3	20·3	23·4	Summer		4·6	6·5	8·3
Autumn	12·0	16·8	17·6	Autumn		2·4	4·3	6·5
Winter	0·7	5·8	6·1	Winter		1·5	3·2	4·4

Raunkiaerian concept of relationship of biological life form to climate. The first is found in the work of van Leeuwen (1936) and concerns the re-colonization of the island of Krakatau in the East Indies after the eruption in 1883. Van Leeuwen's work between 1886 and 1934 illustrates the development and stabilization of a life-form spectrum which reflects the prevailing climatic conditions, i.e. the development of a stable phanerophyte-dominated flora and vegetation in keeping with similar tropical islands, e.g. Seychelles (Table 16).

TABLE 16

The re-colonization of Krakatau; successive life-form spectra
(data from van Leeuwen, 1936)

Date	Species number	S	E	Ph	Ch	H	Cr	Th
1886	15	–	–	59	7	20	7	7
1897	49	–	–	70	6	14	6	4
1906	73	–	–	78	6	8	3	5
1934	219	–	–	65	6	12	2	15
Seychelle Islands (Raunkiaer, 1934)	258	4	3	57	6	19	5	16

The second piece of research involves experimental study. Under suitable climatic conditions, many chamaephytes fall within the nanophanerophyte category, hence Clapham, Tutin and Warburg (1962) refer *Calluna vulgaris* to this latter category. This distinction is not simply based on height, but more on the position of the apices. In a detailed investigation into ecotype formation in *Calluna*, Grant and Hunter (1962) collected data on growth habit, 'maturity types', flowering period and frost resistance in plants raised from seed collected at varying altitudes. Their results, based on the experimental cultivation of the ecotypes, indicated that there was a continuous variation in growth habit, from erect (a nanophanerophyte life form) to prostrate (a chamaephyte), and that the percentage of the various growth habits changed with altitude. In addition, they showed that 'maturity type'— or the time taken for the plant to flower and fruit—was related to the length of the growing season and the site of origin of the plant, i.e. populations of an earlier maturity type occurred in areas with a shorter growing season. While these data may seem to obscure the limits of some of the life-form classes, by illustrating the sensitive genotypic response of a plant species to climatic and environmental factors it tends to validate Raunkiaer's whole concept of the description of vegetation types based on biological life form.

Direct modifications of the Raunkiaerian system

Because of the completeness of this system there have been few changes of Raunkiaer's original categories. Several workers have attempted to integrate the biological types with other life-form systems and these, being indirect modifications are considered in a subsequent section (Rübel and Ellenberg). Perhaps the only major modification of Raunkiaer's scheme of life-form classification is that of Braun-Blanquet (1932) which introduces a series of Latin binomials for the different life forms and also adds some categories to accommodate cryptogams which are not readily incorporated in the higher plant categories. Many of the binomials have an obvious derivation, e.g. *Phanerophyta herbacea*, *Chamaephyta suffrutescentia*, *Hydrophyta natantia*, but others such as *Hemicryptophyta caespitosa* or tufted hemicryptophytes, represent an elaboration of Raunkiaer's subdivisions.

The main cryptogamic categories introduced are as follows:

Therophytes
    Thallotherophyta   slime moulds
    Bryotherophyta   annual mosses and liverworts (e.g. *Riccia glauca*, *Funaria hygrometrica*, *Pottia truncata*)
    Pteridotherophyta   *Anogramma leptophylla*
Cryptophytes
    Geophyta mycetosa   mainly Basidiomycetes

Hemicryptophytes

Hemicryptophyta thallosa	liverworts—*Marchantia polymorpha, Conocephalum conicum* lichens—*Peltigera* sp.

Chamaephytes

Bryochamaephyta reptantia	most pleurocarpous mosses
Chamaephyta lichenosa	fruticose lichens, e.g *Certraria*
Chamaephyta pulvinata	cushion plants, including perennial acrocarpous mosses
Chamaephyta sphagnoidea	*Sphagnum* sp.

Braun-Blanquet (1932) visualizes the system of life-form categories as an insight into the basic functions of vegetation, mainly as a reflection of annual periodicity. Since vegetation description is usually carried out during the season where there is more or less maximum vegetative and floristic development, and seldom in the unfavourable season, the mere recording of biological life form enables a basic understanding of seasonal changes. Examples of the application of this system are to be found in the work of Braun-Blanquet and Tüxen (1952) for many types of Irish vegetation, and in many other studies by these and other authors.

The contributions of Warming and Iversen

In the chronology of life-form study periods listed previously, the date for the start of the second period, 1884, marks the first contribution of Eugenius Warming to the life-form problem. From this original system, modifications and alternative classifications appeared in 1908, 1909, 1918, 1919, and 1923 along with criticisms of Raunkiaer's work in 1908. In his work of 1884, the plants were classified primarily according to their life span and other characters such as the power of vegetative propagation, the hypogeous or epigeous character of the shoots, the evergreen or deciduous habit, etc. His two major groups were thus hapaxanthic plants (monocarpic or in other words with one flowering period in the life cycle), and perennial or polycarpic plants. By 1909, the tide of practical opinion demanded a more concise and complete life-form system which incorporated all types of plant, both phanerogam and cryptogam, autotrophic and heterotrophic. The hapaxanthic and pollacanthic categories thus formed a slightly less conspicuous part of a more definite life-form classification in Warming's *Oecology of Plants* (1909). A detailed system of the life forms of ligneous plants followed in 1916, a modified form of the herbaceous system in 1918, and a further detailed variant of this latter in 1919. The culmination of Warming's work was the publication of his comprehensive monograph of life forms entitled *Okologiens Grundformer* in which the life-form system is presented in much greater detail and the main sub-divisions are based upon the relationship of the plant to water.

TABLE 17

The life-form system of Warming (1923)

(*a*) Examples of main class divisions

A. *Autotrophs*
  I. Aquatic plants—*Hydatophytes* (7 classes omitted here) 1–7
  II. Aerial plants—*Aerophytes*
    (*a*) Autonomous
      1. Epiphytoids—water absorbed by aerial assimilators not from the soil:
        α whole surface assimilation   8. *Atmophytes*
        β rain water absorbed by specialized structures as parts of the plant   9. *Ombrophytes*
      2. Chtonophytes—terrestrial water absorbed from soil by roots:
        α Water absorption restricted in some way—
          (i) Plants of physically dry hard substrates (rock, etc.)   10. *Chylophytes*
          (ii) Saline soil plants   11. *Halophytes*
        β plants of porous—well-aerated soils—
          (i) Herbs, broad leaves   12. *Agrophytes;*
                graminoid   13. *Poiods*
          (ii) Plants with ligneous stems   14. *Xyloids*
    (*b*) *Not autonomous*—dependent upon others for support, e.g. lianas   15. *Klinophytes*

B. *Allotrophs*
  I. Living on dead organic substances   16. *Saprophytes*
  II. Living on live organisms   17. *Parasites*

(*b*) Example of sub-divisions of a class: Agrophytes

  I. Hapaxanthic herbs (sub-divided into summer annuals, winter annuals, biennials and polyennials)
  II. Pollacanthic, non-crassipedic herbs without travelling runners
  III. Crassipedic types (i.e. swollen stem bases) with xylopodia (woody base)
  IV. Crassipedic types with sarcopodia (fleshy bases)—bulbs, tubers
  V. Pollacanthic herbs with above-ground runners
  VI. Pollacanthic herbs with slender underground runners, internodes long (stolons)
  VII. Pollacanthic herbs with thick fleshy underground runners (rhizomes)

The main features of this and the previous systems of Warming are reviewed in detail in Du Rietz (1930) and only the bare outline of the 1923 system are reproduced above in Table 17, a and b.

The intricate system of Warming has had few adherents in recent decades, but there have been a number of developments of aspects of the system, most of which have involved the incorporation of physiological adaptations. Bakker (1966) has begun a detailed study of Dutch hapaxants with the background idea that vegetation should be considered the most susceptible 'registration apparatus' of the habitat, 'since the response of a plant to habitat conditions finds expression in its life forms'. Based on such morphological and physiological characters as life span, stage of development during winter, absence or presence of seed dormancy period, influence of low temperature on flowering, etc., he describes eleven types of hapaxant, classified in three life-span categories:

A. $1\frac{1}{2}$–4 months life span   Ephemerals
B. 1 year   Annuals
C. 2 years   Biennials

This is perhaps the most recent development of the Warming life-form system and, like its predecessors, although it remains a useful morpho-physiological classification of plant species its use in vegetation description has been and will remain of limited application only.

In the true Danish tradition, yet another life-form system of *Hydrotypes* was introduced by Iversen (1936). Fundamentally, it forms an expansion of Warming's use of the relationship of plants to water. The hydrotypes are based largely upon the physiological-anatomical adaptations of plants to various levels of water availability. Four main anatomical characteristics are used in Iversen's system:

(a) Too much water reduces the oxygen content of the soil and the oxygen for respiration must be carried to the roots by *aerenchyma*.

(b) Drought causes wilting which in turn causes damage of stem and leaf vascular tissues unless they are surrounded and protected by woody tissues or *sclerenchyma*.

(c) Several plants possess water reservoirs in stem and roots with a specialized cell structure confirming *succulency*.

(d) Other plants possess thick cuticular wax or hairs on their leaf surfaces or have inrolled leaves which reduce evaporation and constitute *xeromorphy*.

Along with root length these four features help to characterize the major divisions of the life-form system of Iversen (Table 18).

Zonneveld (1960) suggests a system of sclerotypes, where the three scleromorphic classes—scleromorphic, mesoscleromorphic and a-scleromorphic are not determined by laborious sectioning of material, but by study of their relative collapsibility or rigidity after immersion in boiling water for a given period of time.

From the above discourse it becomes obvious that there is a trend away from the use of functional characters purely as an indication of prevailing climatic conditions, and a trend toward environmental,

TABLE 18

Life-form system of Iversen (1936)—Hydrotypes

A. Terriphytes (land plants)	
Seasonal xerophytes	With no strongly developed system but with special xeromorphy (succulency) or water-conserving structures which enable them to survive dry periods
Euxerophytes	Well developed root system
Hemi-xerophytes	Plants without special structures or strong root system but with strong scleromorphy resisting damage by wilting
Hygrophytes	Plants without any xeromorphic structures or root development, and also without scleromorphy, thus badly damaged by wilting
B. Telmatophytes	Plants with assimilating parts adpated to air but also with aerenchyma available to provide (and store) air for root respiration
C. Amphiphytes	Plants with adaptations to both the atmosphere and the hydrosphere
D. Limnophytes	Plants that are only adapted to the hydrosphere, floating or rooted to the bottom of rivers, lakes, etc.

ecological specialization. This is the basic distinction between the Raunkiaerian and Warming life-form system. Whilst the system of Warming was attempted to be omni-applicable those of Iversen, Zonneveld and other workers not mentioned here are obviously devised with a particular aspect of the functional specialization of vegetation types in mind.

TABLE 19

A classification of tropical rain forest life forms
(from Richards, 1952)

A. *Autotrophic*
  1. Plants not dependent on others for mechanical support (trees, shrubs and herbs)
  2. Plants dependent on others for mechanical support
     (*a*) Climbers and similar forms
          (i) Climbers on 1st and 2nd storey trees
          (ii) Climbers on undergrowth trees
     (*b*) Epiphytes
          (i) Shade epiphytes
          (ii) Sun epiphytes
B. *Heterotrophic* (Saprophytes and parasites)

Almost without exception, the systems described in the above section have proved too cumbersome for use in extensive field surveys, being adopted mainly as a laboratory technique for analysis of data. The need for simplicity in classification is emphasized by Richards (1952) in his classification of the constituents of tropical rain forests 'according to their methods of solving the light problem' (Table 19).

A synthetic trend in life-form studies

Drude and Du Rietz. Until the 1920s there had been few attempts to synthesize the diverse life-form systems and create a universally accepted system. Drude (1890) recognized the essential difference between the use of morphological (indifferent) characters and the biological ones, and though he criticized the physiognomic systems, the main types of his classification involved both functional and physiognomic attributes. Forty years later, in his introduction to *Pflanzengeographische Okologie* (1928) he presented a new edition of his life-form system which was again based on physiognomy and function, and incorporated several modifications of Raunkiaerian terms, such as xylo- and podochamaephytes.

By 1930, Du Rietz was faced with a complex of systems based on either or both physiognomic and functional characteristics, and one of the first necessary tasks of this author was to reject or accept either or both for use in a unifying system. He made 'no attempt to distinguish between epharmonic and indifferent characters' and further, only made use of characters 'having an obvious importance for the characterization of vegetation physiognomy'. Six main types of life form, or, more correctly, types of character suitable for consideration were recognized:

1. *Main life forms* (grundformen)—as used by Du Rietz (1921) (see Table 4).
2. *Growth forms*—a term here restricted to life form based primarily on shoot architecture (Warming).
3. *Periodicity life forms*—based on seasonal, vegetative periodicity (many authors).
4. *Bud height life forms*—synonymous with Raunkiaerian life forms.
5. *Bud-type life forms*—Raunkiaer.
6. *Leaf life forms*—entirely based upon leaf, form, size, structure, etc. (most authors but Raunkiaer, 1916, in particular).

Such an ambitious synthesis involving an enormous volume of characters and an extremely derived terminology was doomed to failure at the outset. Although the main life-form system was designed for use 'not only by professional botanists but also by general geographic travellers' and the terms were made as simple as possible, the result is still rather overwhelming and the classification little more than a subdivision of Raunkiaerian types. For example, the woody plants (holoxyles) are divided into trees, shrubs, dwarf shrubs, woody cushion plants and woody lianas on height and habit characters. Each group is divided again according to height using the same height limits as Raunkiaer. And so the dichotomies and indentations proceed.

As a physiognomic-functional classification of plant species the life-form system is unrivalled and its completeness is paralleled only in standard taxonomic literature. As a contribution to vegetation science it was proved to be invaluable if only in a negative sense, in that it was a life-form system to end all life-form systems and to discourage further complexities of terminology and interpretation. As a field descriptive technique, the use of any of its component systems has been extremely limited.

Rübel and Ellenberg. Rübel (1930) developed a system of life forms which he applied in his classical work *Pflanzengesellschaften der Erde.* This system was based on Raunkiaer's types with the use of several Du Rietz (1921) divisions. The main divisions included Magniligniden (trees), Parviligniden (shrubs), Sukkulenten, Epiphyten, Herbiden, etc. Each subdivision of the first two main divisions involved the addition of a prefix to the division name and also the donation of the appropriate Raunkiaerian term with a suitable prefix, thus:

Pluviimagniligniden—Ombromacrophanerophytes—Evergreen, rain forest
Laurimagniligniden—Daphnomacrophanerophytes—Evergreen, soft leaved
Durimagniligniden—Skleromacrophanerophytes—Evergreen, hard leaved
Aestimagniligniden—Theromacrophanerophytes—Summer green, broad leaved
Hiemimagniligniden—Cheimomacrophanerophytes—Rainy green, broad leaved
Aciculimagniligniden—Belonidomacrophanerophytes—Needle leaved

Ellenberg (1956) incorporated the system of Braun-Blanquet (1932) with the salient features of the above system of Rübel to provide a life-form classification system which has proved to be a complete and valuable reference work. Most central European phytosociologists are fully conversant with the terms of the system and because of the copious amount of literature produced by workers of this origin, Ellenberg's classification is reproduced partially in Table 20 and in full in Appendix I.

Periodicity and phenology

Many broad periodic characters of vegetation such as seasonality have been used in life-form systems, e.g. evergreen, rainy green or summer green. On the other hand, the phenological periodicity of flowering and fruiting is generally studied separately and presented as a specific study for a particular vegetation type. The use of the phenological diagram has been quite widespread in the central European

TABLE 20

The revised life-form system of Ellenberg
(Ellenberg and Müller-Dombois, 1967)

Autotrophic plants
  Kormophytes (vascular plants)
    Self-supporting plants
      Phanerophytes, chamaephytes, hemicryptophytes, geophytes and therophytes
    Plants requiring support
      Lianas, hemi-epiphytes and epiphytes
    Free-floating water plants
      Errant vascular hydrophytes
  Thallophytes (non-vascular cryptogams)
    Plants attached to the ground
      Thallo-chamae-, thallo-hemi- and thallo-cryptophytes
    Plants attached to others
      Thallo-epiphytes
    Free-floating thallophytes
      Errant thallo-hydrophytes, kryophytes, edaphophytes and chemo-edapho-
      phytes
Semi-autotrophic plants
      Vascular semi-parasites and thallo semi-parasites
Heterotrophic plants
  Kormophytes
      Vascular parasites and saprophytes
  Thallophytes
      Thallo-parasites and thallo-saprophytes

---

schools of phytosociology. The method, involving the plotting of the
flowering period of species in a succession of graphs appears to have
been introduced by Schennikow (1932) and further developed by
Ellenberg (1939). The work of Preis (1939) contains a neat exposition
of the basic pattern exhibited by the seasonal aspects of a dry continental
grassland (Figure 24), where Raunkiaerian life forms instead of single
species and actual numbers rather than percentage figures are plotted.
The peak of spring annuals and earlier flowering characterizes this type
of grassland and distinguishes it from the Atlantic type of grassland
exemplified by data from a chalk downland.

After the recommendation of Tüxen (1962) the vertical axis of
phenological diagrams has normally been based upon the *Gruppenwert*
(g.v.)* calculation of Tüxen and Ellenberg (1937). In its present
application this involves the assessment of the percentage flower cover
for each species in a number of quadrats, the calculation of an average

* Group value.

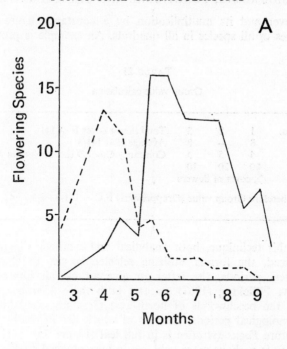

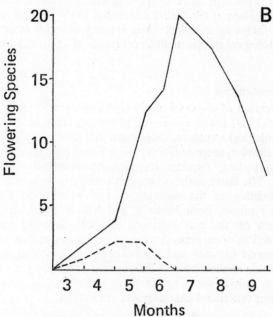

Figure 24. Life-form periodicity in Continental steppe and Atlantic grasslands,
A, data of Preis, 1939; B, data of Shimwell, unpublished, from East Yorkshire.
– – –, Therophyte; ——, Hemicryptophyte

flower cover and its multiplication by a constancy figure for the occurrences of all species in all quadrats. An example is provided in Table 21.

TABLE 21

Group value calculation

Quadrat No.	1	2	3	Total flower cover F = 141
Species A	8	–	8	Average $\bar{F} = 141/3$
B	5	5	5	Constancy C = 8/9 (i.e. no species A in Q.2)
C	80	20	10	
	% cover of flowers			

Therefore Group value (Gruppenwert) $\bar{F}.C = 141/3 \times 8/9 = 42$

Using this technique, both analytical and synthetic diagrams can be produced, the former utilizing selected species to characterize phenological periods, the latter to give an overall impression of phenology. Füllekrug (1967) features both these approaches in his paper on the beechwoods of north-west Germany, delimiting three main phenological periods, the first of which ties in neatly with the period before *Fagus sylvatica* is in full leaf (Figure 25). Such phenological data is perhaps more valuable in autecological studies such as the *Biological Flora of the British Isles* series in the *Journal of Ecology*, but it also forms an excellent basis for experimental ecological and microclimatological studies on different stands of a particular vegetation type.

Dispersal mechanisms

A classification of dispersal mechanisms was produced by Molinier and Müller (1939) based upon seed (diaspore) types, dispersal agents and morphological structure. Dansereau and Lems (1957) substituted a so-called simpler, more uniform system based only on seed type, but which, nevertheless, requires reference to dispersal agents in its application. The latter authors recognize ten types to which they give names (consisting of the neutral suffix *chore* plus an appropriate prefix) and symbols. Both Molinier and Müller and Dansereau and Lems remark on the predominance of wind dispersed (ptero- and pogonochores) in open grass-dominated vegetation and an increase in animal dispersal (desmo- and sarcochores) in woodland and forest. Apart from this application and the possible correlation of small-scale pattern with ballochore range (expulsive mechanism), prospects for the use of this functional characteristic are limited.

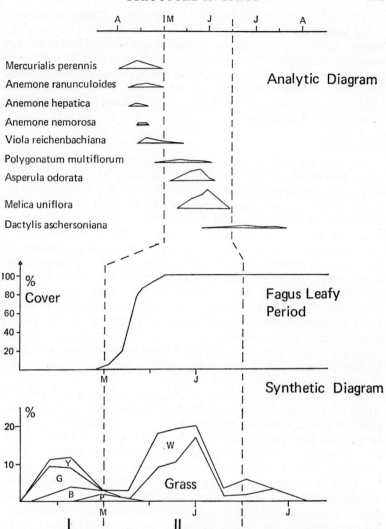

Figure 25. Phenological periods in a German beechwood. Y, yellow; G, green;
B, blue; P, purple; W, white (modified from Füllekrug, 1967)

Structure in space

The structural features of vegetation are basically those characteristics
related to the spatial distribution of the biomass. Structure may be
defined by three components: (1) vertical structure—the vertical

arrangement of the species into layers or strata; (2) horizontal structure —the spatial distribution of individuals of the species which confer a *pattern* on each species and the vegetation as a whole; (3) *abundance* of each species derived from calculations such as counts of individuals per unit area (*density*) or a subjective *cover* assessment and standing crop measurements based on the dry weight of the above-ground vegetation (*biomass*), and its relation to the volume of space occupied at ground level (*basal area*).

Stratification

The layering of vegetation to form distinct strata is a common phenomenon of all but the simplest communities such as floating mats of *Lemna* spp. A cursory examination of any woodland of the north temperate zone reveals a stratification which is composed of five layers—the canopy of trees, an understory of shrubs and saplings, a herb or 'field' layer, a moss or 'ground' layer and the subterranean root systems and soil microflora layer. This structure is constant for most types of forest and woodland with the exception of tropical rain forest which frequently has a layer of emergents above the main canopy. These layers are illustrated in the profile diagram from Richards (1936) of mixed forest on Mt Dulit, Sarawak (Figure 26), in which are recognized three tree layers: (1) emergents with a marked gap between their crown bases and the crowntops of the second story layer; (2) canopy, a more or less continuous cover of trees of uniform height; (3) undergrowth trees, characterized by deep narrow crowns of conical shape in a discontinuous layer.

The method of representation was originated by Davis and Richards (1933) who constructed scale profile diagrams based on narrow sample strips of forest. These are commonly of convenient dimensions 60 m long by 8 m wide and are known as clear felling plots. Vegetation lower than an arbitrary height is cleared and the following features of the remaining trees documented—location, trunk diameter, total height, height to first main branch, lower limit of crown and crown width. Felling enables accuracy, and extrapolation of data to surrounding areas gives a meaningful picture of forest structure. The profile diagram has been applied extensively to British woodlands (Tansley, 1939), but the accuracy which comes with felling is usually sacrificed in the interests of conservation for posterity.

With respect to the height classes of the forest profile, the emergents are nearly all above 30 m and correspond to Raunkiaer's megaphanerophyte category. Canopy layers of rain forests vary between 15 and 30 m and in consequence, analytical surveys of individual species heights have involved the use of different height limits. Raunkiaer (1934), Küchler (1949, 1967) and Dansereau (1951) all vary slightly in their

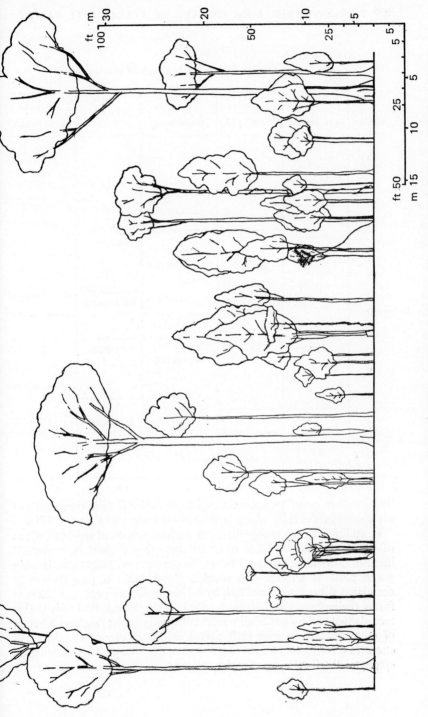

Figure 26. Profile diagram of mixed rain forest on Mt Dulit, Sarawak (from Richards, 1936, courtesy of Blackwell Scientific Publications).

TABLE 22

A comparison of tree and shrub height classes of various authors

Metres	RAUNKIAER	KÜCHLER (A)	DANSEREAU	CAIN and CASTRO	KÜCHLER (B)
35	Mega-	t Tall	Tt Trees tall	Tall tree stratum	8
30					7
25	Meso-	m Medium	Tm Trees medium		
20					
15				Intermediate tree stratum	6
10 8	Micro-	l Low	Tl Trees low		
5			Ft Shrubs tall	Low tree— High shrub	5
3					4
2 1	Nano-	S	Fm	Low shrub	3, 2, 1
0		Z	Fl		

limits and in order to facilitate rapid comparison, their demarcations are shown in Table 22, along with those of Cain and Castro (1959).

In addition to the profile diagram, various graphical representations of the data collected from clear fellings are prevalent in ecological literature, the most common being the diameter to height relationship graph plots of selected tree species. Height data is also frequently correlated with climber and epiphyte distribution, of both sun tolerant forms (heliophytes) and shade tolerants (skiophytes). Richards (1939) records two main types of climbers: tall woody lianas reaching a height of more than 15 m with leafy apices at the canopy level, and small climbers growing to a height of 8 to 9 m in deep shade. Commenting on epiphyte distribution at Shasha, Nigeria, Richards remarks on the paucity of the Shasha flora compared to that of Moraballi Creek, British Guiana (a maximum of thirty-five species as opposed to 115), with only 16% of the trees bearing epiphytes and a maximum number of eleven species per tree. Further comparative data is given by Grubb *et al.* (1963). Data on epiphyte location is normally presented with reference to height characters such as height to the first branch and height to the crown base. Table 23, based on data of Richards (1939), illustrates the method of recording for three individuals of three tree species of different heights. This data is in no way a representative sample of Richard's total data and merely serves as an example of presentation.

In many rain forests plank-like projections known as *buttresses* arise from the principal lateral roots and extend for some way up the trunks of several forest trees, mainly of the families Bombacaceae, Leguminosae and Moraceae. Cain and Castro (1959) figure the plank buttresses of *Ceiba pentandra* of Brazil with planks extending up the

TABLE 23

Epiphytes on three tree species from clear felling plots of mixed rain forest at Shasha, Nigeria

(data from Richards, 1939)

Tree species	Total Height (m)	Height to first branch	Height to crown base	Epiphytes
Lophira procera	45·4	27·8	36·0	Trunk: *Asplenium africanum* at 4 m, *Angraecum subulatum* at 20 m Large branches: *A. subulatum, A. distichum, Bulbophyllum oreonastes, Platycerium stemaria, Polypodium phymatodes* Small branches: *Bulbophyllum oreonastes* and 3/4 spp. orchids
Berlinia auriculata	32·3	13·1	20·1	Orchid at 5·8 m Branches at 17–20 m. *Angraecum subulatum*, three other spp. of orchid and young fern Upper branches: *Polystachya odorata, A. subulatum* and four other spp. of orchid; *Platycerium stemaria*
Diospyros insculpta	17·7	8·5	11·9	Young *Asplenium africanum* and *Polypodium* irioides on trunk just below first branch

trunk to 10 m and radiating up to 20 m from the base of the trunk proper. This phenomenon is regarded by some authors as a response to mechanical stresses by shallow-rooted tree species. This theory, however, has been questioned on the grounds that development of buttresses is not correlated with height or crown size, or prevailing wind direction, and the occurrence of the best developed buttressed trees in sheltered floodplain situations has led to alternative suggestions as to their function. Further discussion of this topic, however, is not desirable in this context and until positive evidence of function is acquired, buttresses are best regarded purely as structural features. Comparative measurements of buttress dimensions do not appear to be readily available and there is little quantitative data on height, basal volume and associated environmental effects.

Stratification of root systems has received considerable attention in studies on grasslands (Weaver, 1920), the mechanical erosion effects of bird-sown trees such as *Sorbus* and *Taxus* on cliff faces (Jackson and Sheldon, 1949), the water budget of desert plants (Cannon, 1911) and with respect to the calcicole-calcifuge problem. A common representation of root stratification is shown in Figure 27, an example of species distribution along a microtopographical gradient with root systems related to drainage. The example is of a flush complex at Mid-Garraries, Kirkcudbright, where *Carex-Campylium* runnels alternate with *Sphagnum-Polytrichum* hummocks on a 10–15° slope.

Pattern

The phenomenon of pattern has been discussed in the introductory chapter at an intensive study level with respect to the plant community, the intricate nature of vegetation and the levels of homogeneity. There are, however, certain gross features of horizontal structure which need to be discussed in this section. Perhaps the largest scale of pattern which depends upon the overall morphology of a vegetation type or community is whether the individual plants in the community are so spaced as to form a continuous lateral contact. This situation is known as *closed* vegetation, and in stratified vegetation at least one or all layers are touching or overlapping. Where there is space between individuals which can be colonized and where the space is not more than twice the diameters of the predominant individuals, the term *open* vegetation is applied. *Sparse* vegetation refers to any situation where there is a greater amount of ground space than in the previous case, so that substrate not vegetation dominates the landscape. For these three types, Fosberg (1967) proposes the term *primary structural group* and proceeds to use them as the basic categories in his key to the major vegetation types of the world.

Quadrat maps are one of the simplest forms of illustrating the

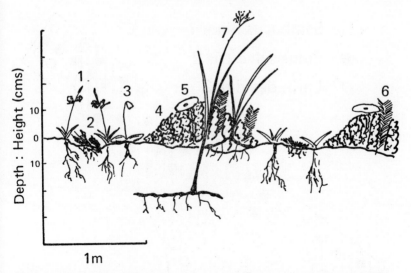

Figure 27. Root stratification along a microtopographical gradient in a flush complex. 1, *Carex demissa;* 2, *Campylium stellatum;* 3, *Pinguicula vulgaris;* 4, *Sphagnum palustre;* 5, *Hydrocotyle vulgaris;* 6, *Polytrichum commune;* 7, *Juncus acutiflorus.*

spatial pattern of vegetation and have been widely used as alternatives to the clear felling plots of tropical studies. The construction of such a map requires the measurement of inter-species distances in a defined area and plotting them on a grid. In this way, simple environmental patterns are often revealed to which experimental investigations can be applied to produce ecological answers. Figure 28 shows the distribution of *Sambucus nigra* in a mixed coniferous plantation of *Larix decidua* and *Pinus sylvestris,* where a marked association of *Sambucus* with *Pinus* is apparent. The answer is simple. The evergreen *Pinus* provides autumn roosting shelter for passerine birds whose diet is mainly fleshy fruits and the excreta provide the seed source. The experimental ecologist analyses the soils to find a higher phosphate and nitrate level under the *Pinus* than the *Larix* and is content.

This is an example of obvious pattern with an elementary answer and as Kershaw (1964) points out, such patterns are readily visible and explained without recourse to quantitative methods. Kershaw cites a less obvious example of environmentally determined pattern from his work on upland grasslands where there is a close relationship between the pattern of *Agrostis tenuis* and soil depth. Associated species such as *Dactylis* and *Lolium* occur more frequently on areas of deeper soil, while *Agrostis* shows a pronounced correlation with the shallower soils. This is partially explained by the difference in the mean rooting

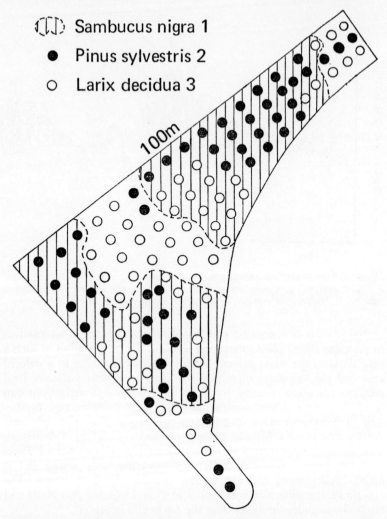

Figure 28. Eco-sociological pattern in a mixed conifer plantation. 1, *Sambucus nigra;* 2, *Pinus sylvestris;* 3, *Larix decidua.*

depths of the species concerned, indicating that small variations of soil depth are manifested by a species and vegetation pattern.

The establishment of permanent quadrats and the continual annual recording of plant locations gives an insight into the short-term migratory and stability patterns of species and the successive changes of species complement and spatial structure of vegetation. The classical work on this approach to vegetation change is that of Watt (1962) who monitored the relationships between *Festuca ovina* and *Hieracium pilosella* as phytometers for the study of compositional changes in grassland after the removal of rabbit grazing. The situation to which permanent quadrats are applied needs to be carefully selected to give maximum rewards and in general short-term problems associated with the instability of open vegetation types are usually chosen. Figure 29 shows changes in the patterns of *Minuartia verna, Thlaspi alpestre* and *Agrostis tenuis* on a lead mine spoil heap in Derbyshire after fluorspar extraction. The most noticeable feature is the closing of the habitat and the gradual restriction of the biennial rosettes of *Thlaspi* with smaller changes in the *Agrostis-Minuartia* relationship.

Cover, abundance and dominance

Abundance and cover relate to the floristic composition of the vegetation and are basically estimates of the numbers of individuals of the component species in a vegetation type, i.e. their densities. The determination of density is, in most cases, extremely tedious and the substitution and use of an appropriate cover-abundance scale forms a more realistic method of vegetation description to most phyto-sociologists. Many investigators have found it convenient to erect five abundance classes, some of which are illustrated in Table 24. Tansley and Chipp (1926) set the scene on a variety of adjectives which vary considerably from worker to worker. Böcher (1933) used six classes and gave each a number, while Hanson (1934) goes as far as to assign density ranges to each abundance class. The categories of Tansley and Chipp are often modified with prefixes such as 'very' and 'locally' and Class 5 is often represented as 'dominant', which in this sense is taken to mean predominant. The use of 'frequent' also donates a certain amount of misunderstanding to the nature of the scales. Apart from such unfortunate words, there is ample scope for over-estimation and inexactitude. Species of a tussock habit are commonly rated as abundant when in fact there are only two or three individual plants and in terms of Hanson's density ranges would only rate as scarce or rare. Moreover, the assessment of species to classes varies from person to person, a problem which is discussed by Hope-Simpson (1940) in his investigation of the causes of errors in the subjective assessment of abundance. His data, involving 'ordinary' and 'specially careful' recording, shows clearly that a subjective assessment, no matter how careful, is subject to large error and will only give an approximate indication of abundance.

In common usage in phytosociology are scales of cover values which when assigned to the component species of a vegetation type in some

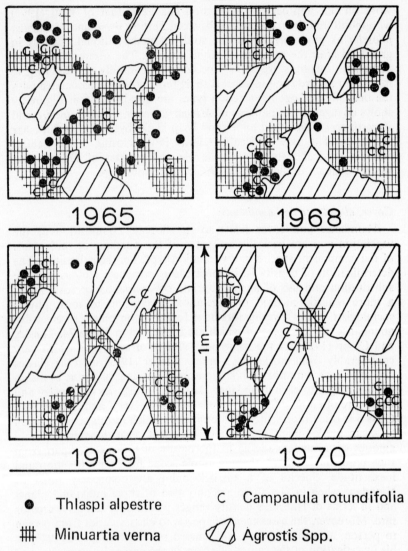

Figure 29. Permanent quadrat studies on an unstable habitat.

ways reflect their importance or dominance. The term cover refers to an estimate of the area of coverage of the foliage of the species in a vertical projection on to the ground. There are several scales which have been widely applied, but all differ slightly in cover class delimitation. The

TABLE 24

Variation in abundance scales

Class	TANSLEY AND CHIPP (1926)	BRAUN-BLANQUET (1932)	HANSON AND LOVE (1930)	BÖCHER (1933)	HANSON (1934)	
						plants/m^2
1	Rare r	Very sparse	Very scarce	1 Rare	Scarce	1–4
2	Occasional o	Sparse	Scarce	2 Uncommon	Infrequent	5–14
3	Frequent f	Not numerous	Infrequent	3 Here and there	Frequent	15–29
4	Abundant a	Numerous	Frequent	4 Common	Abundant	30–99
5	Very abundant va	Very numerous	Abundant	5 Very common	Very abundant	100+

four main systems in use are shown in Table 25 where a rapid comparison of class limits is possible.

Arising from the assignment of cover values is the phenomenon of the dominance of individual species over others in the various strata of the vegetation. Dominants are basically those species which have the greatest total biomass or plant body weight and in complex

TABLE 25

Variation in cover scales

Class	DOMIN	BRAUN-BLANQUET	HULT-SERNANDER	LAGERBERG-RAUNKIAER
+	A single individual	Less than 1%	–	–
1	1–2 individuals	1–5	0–6·25	0–10
2	Less than 1%	6–25	6·5–12·5	11–30
3	1–4%	26–50	13–25	31–50
4	4–10%	51–75	26–50	51–100
5	11–25	76–100	51–100	–
6	26–33	–	–	–
7	34–50	–	–	–
8	51–75	–	–	–
9	76–90	–	–	–
10	91–100	–	–	–

communities different strata commonly have different dominant or co-dominant species contributing the major biomass. The mere predominance of a species over all others in terms of biomass also implies a certain amount of sociological or physiological dominance. If a tree shades a shrub, simply because of its canopy area physiological dominance factors of light and temperature environment could be involved. However, this may not be the case, for the shrub may simply be occupying a certain amount of ecological and sociological niche space existing at its growth optimum under the environmental conditions dictated by the community as a whole.

The demonstration of physiological dominance involves long and complex experimental work on the relative performance of the dominated species and hardly proves to be a feasible project. A more indicative study of dominance effects is to be found in the investigation of the spatial niche occupied by selected species underneath and outside a closed canopy. A good example for this purpose is the panicled sedge, *Carex paniculata* where the actual volume of tussock can be easily calculated from the formula $\pi r^2 h$ where h is the height of the tussock and πr^2 the basal area. Table 26 gives data for three stands of *C. paniculata* under an alder canopy and at the woodland fringe. There is a significant difference in performance of the individual tussocks of the sedge away from the alder canopy, although the total basal area of the plant is less in the woodland due to simple spatial pressure from the alder tree bases.

The situation where one or more species apparently limit the performance of others by their superior biomass and greater competitive ability is often termed ecological or physiological dominance (Cain

TABLE 26

Individual performance in *Carex paniculata* populations

Number of individuals (10 m²)	Mean individual size (cc)	Range
A. *Outside alder canopy*		
10	189,550	108,300–292,500
12	194,300	154,900–236,000
7	185,750	142,300–227,450
B. *Under alder canopy*		
9	55,700	27,900–82,500
11	50,500	29,300–78,250
8	59,450	28,200–85,700

and Castro, 1959). In general ecology, the species with the greatest foliage cover in each stratum is used as an indication of ecological dominance or, in forestry, the foliage cover or crown class are used as measures of community dominance. Foresters classify the component forest trees as dominant, co-dominant, intermediate and suppressed, based on their crown height with reference to the strata of the forest. In this case, dominance in tree species refers to whether their crowns are in the superior forest stratum, irrespective of their contribution to the stratum cover. Intermediate trees are those whose crowns are partially overtopped by the canopy, for example, a coniferous species such as *Picea* with its conical shape in a deciduous canopy, while suppressed trees are those which form a stratum below the canopy.

A final important feature frequently used as a measure of ecological dominance in both forest and herbaceous vegetation is basal area. In woody plants, basal area refers to the cross-sectional area at breast height (approximately 4·5 ft above the ground) and expressed in square feet per acre. The oft-quoted example of the use of basal area as a descriptive tool is that provided by the data of Korstian and Stickel (1927) on the effects of the chestnut blight *Endothea parastica* on the composition of a mixed *Castanea Quercus* woodland in the eastern United States (Table 27). Over a period of fourteen years the total basal area is greatly reduced, due to the failure of the *Castanea* to regenerate, while individual species of *Quercus* either increase their basal area by expanding their growth into the new canopy space or by invading the vacant spatial niches previously occupied by the chestnut tree bases. To illustrate which of these two methods of growth is responsible for the change of basal area, the number of individuals of

TABLE 27

Changes in basal area in a chestnut-oak forest after a blight epidemic
(data from Korstian and Stickel, 1927)

Species	Basal area, 1910–11	1924	% Increase
Castanea dentata	64·18	–	–
Quercus montana	5·25	14·48	176
Q. rubra	9·13	16·45	80
Q. coccinea	2·75	3·35	20
Q. alba	1·97	2·07	5
Other species	0·95	2·63	54
Total	84·23	38·94 ft²/acre	

each species are usually classed into pre-selected diameter categories (Cain, 1935).

In herbaceous vegetation, the basal area of a species is measured at ground level. Such a measurement will give the actual ground space occupied by a species, whereas a simple estimated cover value reflects the spread of the aerial vegetative growth. The difference in basal area and cover value in two different life forms is immediately apparent

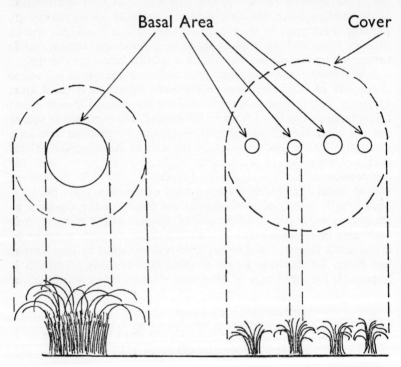

Figure 30. Basal areas in herbaceous species—*Carex appropinquata* and *C. diandra*

from Figure 30 using two related species of *Carex* as examples. *C. appropinquata* is a densely tufted hemicryptophyte with a large basal area and a total coverage which only sightly exceeds the basal area. *C. diandra*, on the other hand, is a rhizomatous helophyte producing small clumps with a small total basal area but a large total cover, greatly exceeding the basal area.

The total basal area of all tree species in mature woodland growing under similar environmental conditions frequently assumes a similar figure in different stands. Further, perennial herbaceous species also tend to reach an optimum basal area related to that of the canopy dominant. A good example is to be found in the *Alnus-Carex paniculata* upland poor fen woodland where the basal areas of the two species appear to be related (Table 28), not necessarily in terms of individual plant basal areas but in total and the ratios of total basal area. This striking pattern is repeated in a further example of poor fen woodland where *Salix cinerea* and *S. capraea* form the canopy and *Carex approprinquata* occupies the tussock sedge niche. How far this relationship is applicable to all poor fen woodlands remains to be seen.

TABLE 28

Basal area relationships in some mature poor fen woodlands
(figures calculated in cm^2 per 10 m^2)

A.	*Carex paniculata—Alnus glutinosa*			
	1.	*Carex* 12044 : *Alnus* 18566	Basal area ratio	0·649
	2.	11625	17956	0·647
	3.	12174	18320	0·665
B.	*Carex appropinquata—Salix* spp.			
	4.	11574	17430	0·664
	5.	11921	17820	0·669

Localities: 1 and 2, Youlgrave, Derbyshire; 3, Waldridge Fell, Co. Durham; 4, Malham Tarn, Yorkshire; 5, Pulfin Bog, Yorkshire.

Besides ecological dominance Cain and Castro (1959) list two other types of dominance, namely physiognomic and family dominance. Both types are basically the same implying dominance in terms of cover either by members of a particular life form or of a certain family, which categories are often one and the same thing. The north-west European heaths, for example, are dominated by ericaceous shrubs of the same xerophytic chamaephyte life-form type which fall mainly within the family Ericaceae. Family dominance, however, is incomplete in that representatives of other families, e.g. *Empetrum* dominate the heaths in certain regions. The term family dominance was introduced by Richards (1952) with reference to the preponderance of species of the Dipterocarpaceae in the upper stories of tropical rain forest—a common feature in primary forests without a single dominant species.

Raunkiaer's area of dominance

Raunkiaer (1928) was perhaps the first person to define dominance

E

in terms of frequency (see Chapter 1). By the term *dominance-area* of a species, Raunkiaer understood the region or regions throughout which the species occurs as a frequency dominant, i.e. with a frequency per cent over 80 determined by $1/10$ m^2 plots. Similarly, co-dominants are those species which occur as frequency dominants within the dominance area of that species, the degree of co-dominance being expressed as a percentage of the observations. There are clearly more points which need to be clarified with respect to the calculation of the dominance-area of a species. It is considered better, however, to restrict their discussion to a later section where the concepts of frequency dominance can be interpreted as an integral part of the descriptive method.

Dominance-diversity relationships of plant communities

The dominance-diversity relationships of land plant communities have been extensively studied by Whittaker (1965, 1970) who has illustrated the variation in dominance concentration and the different types of dominance-diversity curves formed by the ranking of component plant species according to their importance values. The calculation of the importance values for a species is usually based upon the species contribution to the productivity of the community which reflects its ability to utilize a fraction of the community's resources of light, water, nutrients, etc. Productivity—the amount of organic matter produced, the net annual production or simply biomass have been used as measurements of importance. These features enable some appreciation of the species, structural and functional, place in a community in relation to other species. They illustrate the species niche and its dominance relative to the diversity of the community.

One of the simplest measurements based on the quantitative relationships of species is Simpson's Index $C = \Sigma(/yN)^2$ where the dominance concentration of the layers of a community (C) is derived from the sum of the squares of importance values (y) divided by the total importance value (N) of the community. The data of Whittaker (1965) using net production as importance values indicates dominance concentration figures of up to 0.9 in the tree stratum of forests where a single species is strongly dominant (e.g. coastal redwood forests) with figures as low as 0.12 in cove forest/heath vegetation types. Similar patterns emerge from the shrub and herb layers of the forests in question. The example of dominance concentration quoted in Table 29 is taken from Coppins and Shimwell (1971) from nine sample plots which represent the four stages of the life cycle of *Calluna vulgaris*—pioneer, building, mature and degenerate. In this instance, the importance values are based on biomass calculated per 10 m^2 from $10-20 \times 20$ cm sub-samples. Dominance concentration values for the cryptogam layer can be seen

TABLE 29

Dominance-concentration values for the different life-form components in *Calluna* heath of various ages
(from Coppins and Shimwell, 1971)

Life form	Plot number 1	2	3	4	5	6	7	8	9
Algal mat	0·121	0·0025	0·0169	–	–	–	–	–	–
Lichen, crustose	0·103	0·028	0·140	–	–	–	–	–	–
Lichen, squamulose	0·020	–	0·175	0·002	0·001	0·277	0·281	0·116	0·567
Lichen, cladinous	–	–	–	–	+	–	–	0·001	0·00008
Musci, acrocarpi	0·033	0·382	0·119	0·874	0·292	0·104	0·103	0·244	0·048
Hepaticae reptantia	0·107	0·038	0·010	0·00001	0·089	0·007	0·007	0·053	0·0002
Other	–	–	–	–	0·117	0·004	0·001	–	+

Plot 1, 2, 3, pioneer; 4, 5, building; 6, 7, mature; 8, 9, degenerate.

to be lowest in life-cycle stages where environmental stresses are greatest, e.g. in the pioneer stages when niche spaces are abundant and never filled to a maximum, and in the mature/degenerate transition stages when the *Calluna* canopy opens and the stable niches are altered. The highest values are seen in the degenerate stage where there is a concentration of dominance in the squamulose lichen life form, and in the building stage where a low species number and high biomass of *Pohlia nutans* produce an abnormally high figure. As *Calluna* is the only phanerogam in these heaths, its dominance concentration is 1·00 throughout.

Having illustrated the dominance concentration of communities the next step involves a demonstration of how the niche spaces of the community are divided and how the importance values of the species are related quantitatively. Whittaker has shown that when species are arranged in a sequence from most to least important they form a continuous progression from dominants through intermediates to rare species. A number of geometric models to fit such progressions have been postulated and curves expressing the two main hypotheses are shown in Figure 31. These are as follows:

(i) The niche pre-emption hypothesis supposes that niche sizes are initially determined by the success of certain species pre-empting a greater part of the total niche space, thus restricting less successful species to what remains. If a species occupies a fraction k, say 75% of niche space contributing a corresponding amount to community productivity this species is the dominant of the communtiy. If the second species occupies a similar fraction of niche space unoccupied by the first, and the third a similar fraction to that occupied by the second of the remainder, etc. a geometric progression of importance values results to which the geometric series of Motomura (1932) has been applied

$$I^v_x = N(1-k)^{x-1} \quad \text{or} \quad Iv = Ac^{x-1} \qquad k = 1-c$$

where x is the number of the species in the sequence from most to least important, N is the total biomass, c is the ratio of the importance value of a particular species to that of its predecessor in the sequence, A is the importance value of the community dominant and $k = 1-c$. When plotted in the manner of Figure 31 a straight line results.

(ii) The log normal distribution of Preston (1948) which considers that the niche space occupied by a species is determined by a large number of factors interacting to affect the success of the species in competition for niche space. If this is the case, the relative importance of species should give rise to a normal or so-called bell-shaped distribution. This phenomenon is more apparent in communities of high species diversity such as open calcareous grasslands or the spoil-banks of a

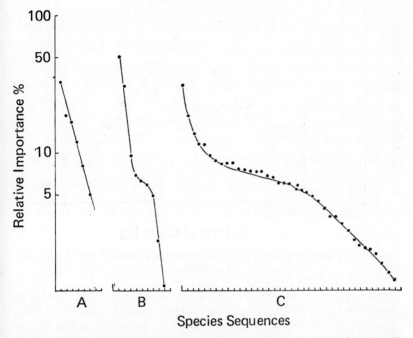

Figure 31. Some examples of dominance-diversity curves. A, geometric series of Motomura from pioneer stage of *Calluna* heath (biomass); B, modified sigmoid curve from stable, mature stage of *Calluna* heath (biomass); C, flattened sigmoid curve from species-rich limestone grassland (cover).

quarry. Figure 32 exemplifies this type of distribution pattern for a species-rich limestone grassland based on coverage derived from 400 point contacts in 4 m². The values are divided into ranges and recorded in the manner described in Chapter 1. The scale is logarithmic with the species grouped by octaves of cover percentages. A modal range of importance values has the largest number of species with decreasing species numbers in the ranges either side of the mode. When the data for such communities are plotted in the manner of Figure 31C a sigmoid distribution of moderate slope throughout results.

Whittaker (1965) has shown that dominance-diversity curves from plant communities are not all alike but represent a series of intergrading types. Curves approximating a geometric series are of common occurrence especially in pioneer and ephemeral communities and those which have rigorous environments and relatively few species. The characteristic shape of the dominance-diversity curve here is a steep, straight line. Most communities, however, have a small group of dominants, a larger number of moderately important species and a

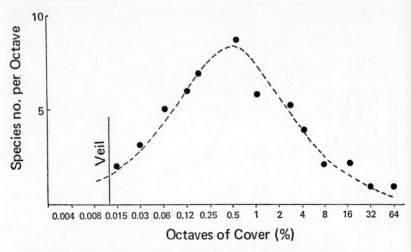

Figure 32. A lognormal plot of Curve C in *Figure 31* according to Preston's lognormal hypothesis.

small number of rare species. These curves are commonly sigmoid in form, but various departures from a simple sigmoid curve result from community variables such as the number of dominants, a low species number and the competitive relationships of species. Samples from communities of high species diversity (commonly anthropogenic base-rich grasslands and scrub) produce the flattened sigmoid curve seen in Figure 31C. The more common type is the intermediate sigmoid curve B from communities of intermediate species diversity, an example taken from the mature stage in the life cycle of *Calluna* heath.

This method of representation of the interrelationships of species in terms of dominance-diversity curves enables a realistic view of the place of each species in the vertical and horizontal niche space. It also demonstrates that a community is a functional system of interacting species each with their own structural and ecological niches, or in other words that a study of these structural and functional niches and importance value progressions mirror the general structure of the vegetation in space and time.

CHAPTER 4

STRUCTURE IN TIME

Structure should not merely be considered and studied in its static, spatial aspect. It needs to be investigated in its progressive, dynamic aspects if some basic concept of vegetation is to be achieved. This was realized at the turn of the twentieth century and for the first thirty years of this century most ecological and phytosociological investigations involved reference to vegetation as a dynamic entity with a progressive trend known as succession (Clements, 1928; Tansley, 1939; Whittaker, 1953). The period between 1930 and 1960 saw the detailed analyses of permanent quadrats and the development of the pattern and process theories for the plant community (Watt, 1947, 1957); theories which via the information theories of Margalef (1958) and Ross Ashby (1958) have culminated in the production of the ecological systems theory of Van Leeuwen (1966), but, first we must start in the nineteenth century.

Clements, Moss and Tansley

The process of succession of vegetation or, in other words, the changes in the structure and composition of vegetation with time are reported and commented upon in some of the earliest botanical literature. Studies on the subject gained impetus in the middle of the nineteenth century from the evidence of ancient forests submerged beneath the peat supplied by such works as those of Vaupell (1857) on the invasion of beech in Denmark, and Blytt (1876) on the climatic changes relative to peat formation in Norway. Hult (1885) and Warming (1891), also Scandinavians, were among the first persons to be credited with the recognition of the fundamental importance of development in vegetation and to make systematic studies of regions and habitats on this basis. Hult maintained that the distribution of vegetation types could only be understood by tracing their development from the first pioneer colonies of land laid bare or produced by ice retreat to a stable climax vegetation. It was largely these basic studies which inspired Clements (1904, 1916) to produce and develop his treatise on *Plant Succession and Indicators* in 1928 and the detailed concepts and causes of succession.

121

The paper of Clements (1916) provides a good starting-point for discussion.

Clements's approach to vegetation description and classification involves the concept of vegetation as a functional organism. 'As an organism the climax formation arises, grows, matures and dies. Its response to the habitat is shown in processes or functions and in structures which are the record as well as the result of these functions' (Clements, 1916, p. 1). Vegetation is thus considered to develop in structure and organisation so that all its components form a complex relationship in which all the components combine to produce a climax vegetation type under the strict environmental conditions dictated by macroclimate. This process consists of several essential functions—initiation, selection, continuation and termination which are in turn under the influence of five successive or interacting processes—denudation of the habitat, migration, ecesis (adjustment to environment) competition and stabilization. The motive force behind the development of vegetation is to be found in the species-populations, life forms and their interaction with the habitat. In the colonization of bare ground or open water the habitat will be progressively better utilized by species which are able to function more efficiently under the condition of the habitat. Each functional invasion modifies the habitat conditions and new species-habitat niches are formed. This change in function is reflected in the vegetation which passes from a pioneer stage through intermediates to the climax stage where the functional relationships of the vegetation to the habitat are maximal.

The movement from pioneer to climax is practically continuous, but there are periods where some stabilization is apparent and well-defined stages or communities are recognizable. For these, Clements adopted the word *sere*, the Latin root of the word *ser* meaning join or connect. A sere is thus a unit succession comprising the development of a vegetation type from denudation to stablization of a habitat. Each step in the sere is known as a seral stage and every complete sere ends in a *climax*. The climax represents the close of the general development of the sere but its recognition is only possible by the careful scrutiny of the whole seral process. It was Clements's opinion (1928, p. 107) that the real climax was determined by climate and that features of climate were the overriding factors in the formation of a monoclimactic vegetation type in a particular region. Clements, however, pointed out that there were various interposing agents which prevent complete development and produce subordinate climaxes or subclimaxes. These latter are formed mainly by restrictions imposed by the edaphic factors of soil or by the activities of man. In addition, Clements (1928, p. 110) recognizes developmental stages known as preclimax and postclimax with which the normal potential climax is in contact as a series of zones.

The *Taxus-Corylus* woodlands of the limestones of Killarney, Ireland (see Chapter 5) are a preclimax to the adjacent *Quercus-Fraxinus* woodlands, while the *Corylus* woodlands of the Burren, Co. Clare, are a temporal and climatic postclimax of this same woodland climax. Similarly, the species-rich ash woodlands of the Peak District limestones are a spontaneous postclimax of a *Quercus-Tilia-Alnus* climatic climax affected by man.

The climax communities of the world represent the types of vegetation which are visualized when the main geographical divisions such as desert, prairie, wet tropics, Mediterranean, etc., are considered. Clements (1928) lists fourteen climaxes for North America, two grassland, three scrub and nine forest climaxes, all of which are illustrated in Table 30. They coincide with broad geographical categories —for example, the *Stipa-Bouteloua* climax or prairie and the *Quercus-Ceanothus* climax or chaparral.

TABLE 30

Climax vegetation types of North America
(after Clements, 1928)

Grassland climaxes
  Stipa-Bouteloua climax—prairie
  Carex-Poa climax—tundra

Scrub climaxes
  Atriplex-Artemisia climax—sagebrush
  Larrea-Franseria climax—desert scrub
  Quercus-Ceanothus climax—chaparral

Forest climaxes
  Pinus-Juniperus climax—woodland
  Pinus-Pseudotsuga climax—montane forest
  Thuja-Tsuga climax—coast forest
  Picea-Abies climax—subalpine forest
  Picea-Larix climax—boreal forest
  Pinus-Tsuga climax—lake forest
  Quercus-Fagus climax—deciduous forest
  Isthmian forest
  Insular forest

In many regions these climax types are restricted in distribution even fragmentary or absent. Many agencies are constantly causing their alteration or complete destruction so that conditions are produced for new successions to be initiated. Natural factors such as tidal deposition, land slides, glacier retreat and volcanic eruption are continually producing new habitats where new seral development can commence, via pioneer communities through a number of seral stages to a climatic climax. This natural succession from pioneer to climax is known as primary sere or *prisere* (Figure 33a) and is taking place on such natural bare habitats as the Island of Surtsey, south of Iceland. In this locality the natural succession will follow a *xeroseral* development or a sere of dry rock exposures commencing with lichen pioneer communities and developing through to the climatic climax of *Betula* scrub-woodland. The flora and vegetation will adjust to the climate, just as that of Krakatau did after its eruption in 1883 (van Leeuwen, 1936, see Table 16).

At the other end of the moisture scale, a *hydroseral* development is initiated in the open waters of lakes and tarns, on their submerged sands and silts. These habitats are first colonized by floating aquatics followed by rooted species. Gradually, the lake is infilled by dead

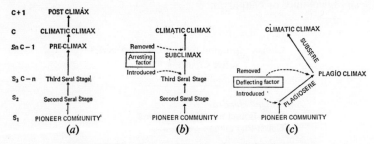

Figure 33. The three main types of seral development.

plant matter and becomes shallow, allowing sedges and reeds to become established. Further accumulation of humus and decaying vegetation increases the process of terrestrialization and eventually raises the soil surface above the water level enabling marsh and fen plants to invade the habitat, decay and form fen peat and continually alter the soil surface. Water-tolerant shrubs and trees such as *Salix* and *Alnus* settle down among the fen plants and gradually form a closed canopy of fen woodland or carr. The rise in soil level is accelerated by the annual deposition of dead leaves and twigs and winter flooding results in the deposition of mineral silts so that a rich humus-mineral soil is formed. Subsequently, the surface layers of the soil become dry enough for the growth of mesophytic trees and the hydrosere, like the xerosere, culminates in the development of climax forest.

Clements (1928, p. 112) has said that 'zonation is the epitome of succession' and it is by study of zonation around lakes that the stages of the hydrosere can be detected or postulated (Figure 34). If this is correlated with the zones of a peat boring taken down through the climax mesophytic forest the same pattern of zonation should appear

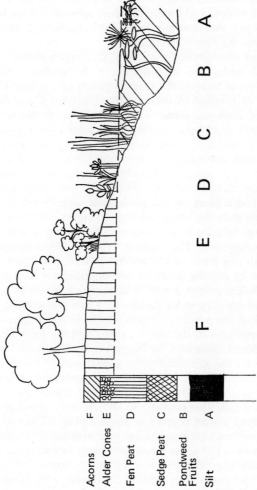

F   Acorns
E   Alder Cones
D   Fen Peat
C   Sedge Peat
B   Pondweed
     Fruits
A   Silt

Figure 34. Hydroseral succession and the zonation around a lake. A, zone of floating aquatics; B, rooted aquatics; C, reedswamp; D, sedgemarsh; E, fen-carr; F, mixed mesophytic woodland.

with (*a*) a layer of lake silt at the bottom, succeeded by (*b*) a layer of aquatic remains recognizable by *Potamogeton* fruits; (*c*) a band of *Phragmites-Carex* peat in which the *Phragmites* rhizomes and *Carex* nutlets are visible; (*d*) fen peat composed of sedge and moss remains (species such as *Paludella squarrosa* or *Camptothecium nitens*); (*e*) fen peat with *Alnus* cones; and finally the drier mineral/humus soils with *Quercus* leaf litter and acorns. Thus, both horizontal and vertical zonation indicate the nature of succession.

But the climax mesophytic forests only develop on relatively favourable soils and under the influence of a favourable mesoclimatic situation. One could imagine the redevelopment of a climax beech and oak woodland in England in two or three centuries if man was completely removed from the scene. There are, however, many areas on which the climax woodland would not develop; on exposed rocky dalesides or in marshy tracts alongside lakes and rivers where exposure and high temperature ranges and the continual flushing of river water and silt, respectively, are physiographic and edaphic factors which arrest the development to climax. To such vegetation types which are kept at a more or less permanent stage below the climax, Clements (1916) gave the name *subclimax* (Figure 33b), because they represent a stage inferior to the climatic climax in the xeroseral and hydroseral development. The various types of subclimaxes seldom correspond with any stage of the normal prisere. The constant intervention of man by grazing flocks or burning does not stop succession but tends to deflect it into a new course, which proceeds subject to these biotic and anthropogenic factors to a new stable state different from any stage of the prisere (Godwin, 1929). For example, the chalk and limestone pastures of the South Downs and the Peak District are derived from forest clearance by Neolithic herdsmen and stabilized through scrub to the grassland *plagioclimax* along a *plagioseral* or deflected succession (Figure 33c). When the biotic and anthropogenic deflecting and arresting factors are removed the vegetation will resume its progress towards the climax. A shorter sere developing from partially destroyed climax vegetation, from a plagioclimax or from an earlier seral phase in which progress has been arrested by anthropogenic and biotic activity is known as a secondary sere or *subsere*.

These then are the basic concepts of the dynamism of vegetation expounded by Clements (1916, 1928) and adhered to by Tansley (1939) and many other British ecologists. The work of one other man deserves mention in this section, however, namely that of C. E. Moss, the father of British ecology. In his attempts to classify British vegetation, Moss (1910, 1913) took the view that the climate of the British Isles was so uniform that there was a justification in classification of the vegetation within the country on the bases of soil characteristics. Moss's formations were thus based on primary edaphic divisions similar to the edaphic formations of Schimper (1898) and were roughly equivalent to the subclimaxes of Clements. But unlike Schimper, Moss stressed the successional aspect. Within each formation, which corresponded to a distinct habitat (edaphic plus mesoclimatic factors) Moss recognized a 'progressive succession' leading to an edaphic climax ('chief associa- tion') which was permanent and stable; and 'retrogressive succession' from this edaphic climax. Thus, the 'chief association' of the limestones of the Peak District was ash woodland while the scrub and grassland were 'subordinate associations' in the progression to and retrogression from the climax. The causes of retrogression are both natural and man induced. The retrogressive moors described by Moss (1913) result from continuous erosion by the elements and drainage; the retrogressive hazel scrub is affected largely by the anthropogenic and biotic factors of coppicing and overgrazing.

Criticisms of Moss's ideas of retrogressive development came mainly from Clements (1916, 1928, p. 147). Here he stated emphatically that 'succession is inherently and inevitably progressive' and that 'regression, an actual development backwards, is just as impossible for a sere as it is for a plant'. He then proceeds to dissect the work of Moss on peat moor retrogression (1913) commenting on the 'convincing picture of the normal destructive action of erosion in producing new areas for succession', so that the apparent retrogression of the moor resolves itself into a progressive movement on new soil. Moss, placing more weight on the edaphic environment, has thus described retro- gression of the habitat, without taking due notice of Clements's organismal vegetation concept. Similarly, the retrogression of scrub was merely the continuous deflection of succession accompanied by habitat degeneration due to excessive grazing and erosion. The legacies of the work of Moss were basically twofold. On the one hand, his ideas on the nature of vegetation and its classification formed the basis of the collaborative work *Types of British Vegetation* (Tansley, 1911) and also influenced Tansley's later thoughts although he realized and was critical of their limitations (Tansley, 1939). On the other hand, Moss set the seeds of dissent concerning the nature of climax, intimating that the climatic monoclimax theory was too great a generalization and that polyclimax, related to edaphic variables, was more realistic. From these courses, the British polyclimax theories developed mainly under Tansley. The basic causes of the rift between the monoclimactic and polyclimactic schools of thought are expressed neatly by Selleck (1960) who states: 'The rift occurs either in the assumption that, given sufficient time, climate is the overall controlling factor of vegetation, or in the length of time considered adequate for stabilization to occur.' Thus, given a sufficiently long period of time, edaphic factors become reduced to a minimal level by the action of climate or that within the four score years and ten of man's lifespan, climatic factors are largely overridden by edaphic factors and the continuous flux of species populations with a cyclic longevity less than man. One of the basic dichotomies of thought thus rests on man's comprehension of temporal stability; on a philosophical somewhat conjectural tack the monoclimax theory is quite acceptable; with a realistic approach by the investigator seeking to classify the vegetation units, the polyclimax theory is a much more feasible proposition. The monoclimax theory has been supported by several somewhat erroneous interpretations of pollen diagrams and by observation of environmental changes such as the decrease of carbonates from the surface layers of dune soil over a period of 600 years (Olson, 1958). The shorter temporal nature of the polyclimax theory falls well within man's immediate comprehension and use of permanent quadrats enables a deep understanding of a particular climax in terms of structure, composition and population flux (Watt, 1947).

The nature of the climax

As an opening to the problem of the nature of the climax, two approaches to climax definition may be distinguished. Most of the logic behind the monoclimax theory has been based on the physiognomy and structure of vegetation in terms of growth forms and life forms and their relationship to climate (see Chapter 3). The other approach considers the climax as a population of plants and animals of different species and their mutual interrelationships in terms of properties other than, or in addition to, physiognomy and structure, e.g. floristics, population dynamics and productivity. Interpretation of the climax in terms of populations and productivity has thus been sought by Whittaker (1953) in his review of his own work on natural communities and previous contributions to the climax theory. But before his theories receive any further consideration some evidence which has a direct bearing on the nature of succession and climax problems must be summarized. The evidence falls neatly into six major points.

1. *Succession.* The succession of populations on recently disturbed or deposited sites has five main characteristics. First, the change of populations through and between the stages of succession is frequently continuous and population stabilization is often minimal. Secondly, as Faegri (1933) has pointed out, because of local climatic and edaphic differences the rates of change between different stages in the same succession or in related successions, toward the same climax type are extremely variable. Thirdly, successions vary in composition of the stages due to local environmental variables which cause different

species populations to appear in similar successions in the same area. Fourthly, population changes may be irregular, again due to environmental fluctuation and finally, depending on the availability of certain species populations at certain stages, certain successions may be telescoped so that whole stages may be omitted.

2. *Climax convergence.* Convergence of successions to a similar physiognomic climax was fundamental in the theories of Clements (1916) and formed the basis of the recognition of the major vegetation types of the world. In like manner, the convergence of seres within a region to the same physiognomy is also a familiar feature.

3. *Climax irregularity.* But, in spite of seral and climax convergence on similar sites and in similar regions, no two stands are identical. Intraclimax and intraseral irregularity on a smaller scale is apparent in most examples, where various conditions of microclimate, microrelief and nutrient status are manifest as a local patchiness.

4. *Climax patterning.* On a larger scale than the previous category, because different combinations of species populations occur in different environments of any region, the vegetation forms a complex pattern of populations. Attempts have been made to link all the vegetation types of a particular region into a single successional scheme, but the parallel variation in vegetation with edaphic factors has been the basis of many objections to the monoclimax theory.

5. *Climax continuity.* It is frequently observed or assumed that vegetation communities are separated by boundaries called ecotones, but also commonplace is the observation that the transition between two types may be gradual so that it becomes impossible to recognize actual discontinuity. Raunkiaer (1908) probably originated the idea of continuity in his application of life forms to distinct zones or biochores. But it was perhaps the individualistic hypothesis of Gleason (1926) from which the idea that vegetation forms 'a complex continuum of populations' (Whittaker, 1953) was wholly derived. It is suggested by Whittaker that climax populations change continuously along continuous gradients and that observed discontinuities are produced by local environmental discontinuities or incompatibility between species.

6. *Climax instability.* The spatial irregularity (expressed in category 3) of populations may be an expression of temporal instability or related to the causes of observed discontinuities along the continuous gradient (category 5). Cyclical fluctuations have been widely observed in animal populations and more recently cyclical phenomena in species populations and community complement have been reported and reviewed (Kershaw, 1964).

Whittaker (1953) concludes that an 'adequate conception of climax and succession must be consistent with these lines of evidence' and that the evidence supports the ideas that the climax is 'a population balance determined by the conditions of its site'. In the light of these six points, the monoclimax theory clearly becomes untenable. The first major assumption of this theory is convergence to identity—as the vegetation develops it modifies the environment and results in a climax of equal mesophytism on different sites (i.e. the climax of both xerosere and hydrosere is mixed mesophytic forest). In this respect the mono-climax theory may be rejected as being contrary to the bulk of existing evidence, which although it falls in favour of the theory in terms of dominant growth and life forms is far outweighed by data of edaphic, microclimatic and other special climaxes.

Secondly, the monoclimax theory is based upon difference within identity and as Whittaker says, this 'may be rejected as a semantic device which begs the question and obscures the problem'. The necessity to subdivide the monoclimax as Clements suggested is simply a reflection of the difference within identity. The designation of the verbose nomenclatural types previously mentioned, such as preclimax, postclimax, subclimax, etc., and association, consociation, fasciation, etc. (see Chapter 2), surely recognizes the non-convergence of climaxes. The two assumptions are thus seen to be in conflict, since they require that the climaxes of two different sites are the same, but different. But this conflict may be somewhat reconciled by the acknowledgement that convergence is only partial. But if the convergence is only partial, dissimilar climaxes will be found on different sites and a mosaic will be apparent; but this is the basis of the polyclimax theory! Therefore, differences between monoclimax and polyclimax are basically differences in semantics (Cain, 1939) since ecologists may describe stands of vegetation by different terms. The assumption that the polyclimax is a 'mosaic of plant communities whose distribution is determined by a corresponding mosaic of habitats' (Tansley, 1939, p. 216) is infinitely simpler than the monoclimax theory which involves suppositions, climatic regulatory processes and vegetation convergence in spite of environmental differences. The monoclimax theory involves a subjective choice of a supposed climatic climax and the relation of other vegetation types to it by rather speculative methods whereas the polyclimax permits greater realism in that communities of a particular edaphic situation can often be observed as an actual zonation, e.g. chalk woodland, scrub and grassland are essentially different from siliceous woodland scrub and grassland (Moss, 1913).

It may have been noticed that this and the last section ended with reference to two human traits—man's recognition of his own temporal instability and his preoccupation with semantics. But Whittaker (1953) concludes that the difference between climax conceptions is far from being a mere semantic quibble and that the choice of one of the two climax theories 'is one of the most important decisions to be made by synecologists and one for which adequate bases of choice exist'. To the monoclimax and polyclimax he adds a third theory of climax.

The climax pattern hypothesis of Whittaker

In the place of monoclimax and polyclimax theories Whittaker (1953) makes three suggestions about the nature and structure of climaxes and their relativity: (1) that the climax is a stable stage of community populations, structure and productivity and that the dynamic balance of its populations, structure and productivity are determined by the factors of its site; (2) that the climax is a shifting pattern of populations which correspond to a similar pattern of environmental gradients; (3) that all factors of the mature ecosystem which affect populations also affect climax composition; that there is no absolute climax for any site or region and that the climax composition is only meaningful if its relative position along environmental gradients is considered.

It should have become obvious from these three suggestions that Whittaker's hypothesis involves the rejection of the units of vegetation which were an integral part of all monoclimax and polyclimax theories, e.g. association, consociation, preclimax, subclimax, etc., and that the main approach to definition of climax is via gradient analysis, i.e. the relationships of gradients in climax populations to environmental gradients. In this present section, gradient analysis is rather out of context, and the subject is taken up in detail in Chapter 7. But to tidy up the ideas surrounding climax and to fully understand the derivation of Whittaker's concept of gradient analysis, mention must be made here of the factors involved in his three corollary propositions associated with climax, namely climax determination, relativity and recognition. These are presented in a truncated form below.

(a) Climax determination—climax composition is determined 'by all factors which are intrinsic to, or act upon, the population on a sustained or repeated basis and do not act with such severity as to destroy the climax populations'. Such factors include:

1. The genetic characteristics of the populations involved and their powers of genecological response.
2. Meso- and microclimatic factors.
3. Site topography.
4. Physical and chemical edaphic factors.
5. Biotic factors associated with food chains and community productivity.
6. Fire—manifest in life-form adaptations.
7. Wind—manifest in physiognomy changes.

8. Other factors—mainly unnatural factors such as suplhur dioxide pollution, or periodic factors such as infrequent inundation with salt water.
9. Flora and fauna—variations in the indigenous flora or migratory habits of fauna.
10. Chance—mainly chances of dipersal.

(*b*) *Climax relativity*—'the climax population has meaning only relative to the environmental conditions of the site' and there are three aspects of climax relativism:

1. Climax and succession—distinctions between climax and seral stages are only those associated with relative instability such as new populations invading the site.
2. Climax and seral species—some species are both seral and climax and variation in the importance of the species is seen from site to site and region to region.
3. Climax and seral types—similarly types of seral stage, association, etc., may be seral in one circumstance and climax in another.

(*c*) *Climax recognition.* Several criteria have been used in the traditional recognition of monoclimaxes. Whittaker examines these and divides them into two series, one which he considers to be untenable, the other applicable in the light of his climax pattern hypothesis.

(*i*) *Untenable criteria*

(1) Unity of growth form. Clementsian approaches require that all climax dominants belong to the same major life form and the attempts of Weaver and Clements (1938) to exclude *Tsuga canadensis* from the list of dominants in eastern deciduous forests of the USA for having the wrong growth form negate this criterion.

(2) Area of climax. The monoclimax is thought to be a climax of a definite geographical region so that climax could be recognized by similarity over a large area. Any field worker will know that this is something of an over-simplification of the real picture (cf. Braun, 1950).

(3) Convergence of different successions to identity is seldom observed and partial convergence involves an assumption of what the true monoclimax really is (cf. Braun, 1950).

(4) Convergence is achieved through both biological and physiographic processes, but the latter takes too long to have any real bearing on climax as a biological entity (Domin, 1923).

(5) Maturity of soil profile is not in itself a criterion of the climax state as has often been assumed (cf. Braun, 1950).

(6) The most mesophytic vegetation type of an area should not be

regarded as the climax type without due regard to hydrology, micro-climate and soil moisture (Oosting, 1948).

(7) The increase in tolerance or ecological amplitude of the component species of a sere is no indication of trend towards climax. Climax dominants are not always the most tolerant species (cf. Graham, 1941).

(8) Vegetation stature usually increases with succession to climax (Weaver and Clements, 1938) but there is no reason why vegetation of lower stature should replace forest as climax.

(ii) Applicable criteria

(1) Regularity within the stand—climax stands often exhibit a marked uniformity of structure and composition on sites of uniform conditions, whereas seral stands lack uniformity.

(2) Regularity between stands—stand-to-stand similarity on similar sites can often be recognized and forms an important part of the concept of the climax state as similar population balances in similar environments. Such stand similarity may be evidence of their having reached the climax state.

(3) Reproduction—the relationships between age and numbers of individuals in forests frequently forms a J-shaped curve similar to that illustrated by Raunkiaer (1928) with many young and few old trees. Climax forest types may thus be identified by the J-type of relationship between canopy trees and their progeny. But these types of age distribution do not apply to seral stages or unstable climax stands (Whittaker).

(4) Trends—if, on a number of similar sites, rocky grassland, scrub on brown rendzinas and woodland on brown calcareous soils are studied as Moss (1913) did, it is reasonable to fit them together into a series on the basis of increasing community productivity, stature and soil maturity. The trends serve to suggest relations between vegetation types and the probable climax.

These then are the basic premises upon which Whittaker (1953) founds his climax pattern hypothesis, a theory which treats vegetation in more fundamental terms than even the polyclimax concept; and in his own words (1967) 'it is but a step from the polyclimax to the climax pattern interpretation'. The major approach to the theory is via gradient analysis, an approach which is dealt with as a method of vegetation description in its own right in Chapter 7.

Pattern and process in vegetation

Not all investigations into the nature of structural changes with time have been inclined towards succession and the determination of the nature of the climax. Since 1940, there has been an increasing amount of research on the fluctuation or short-term changes (say twenty years) in stable and unstable vegetation types and several theories have been proposed around the results of analyses of permanent quadrats. The results have shown that at no time is a stable vegetation type a static entity. There is a continuous flux of populations as each generation reproduces and ages, and these features, intrinsic to the vegetation itself, occur in a directional and cyclical process which is manifest by changes in population composition, structure and productivity which are in turn manifest in the pattern of the vegetation. The changes induced by the processes of the vegetation are repeated many times throughout whole vegetation types and becomes apparent either via the study of permanent quadrats or by reference to a series of population phases at several points at the same time. Both approaches have been used and will be exemplified in the subsequent text.

On the experimental side, a third approach aimed at an understanding of the dynamic interrelationships of species populations is to be found in the work of Harper and his collaborators (Sagar and Harper, 1961; Harper; 1964; and Putwain and Harper, 1970, are perhaps the most relevant papers). The background theory to their work is that in Nature, each species population has a distribution in space and time which is dependent on the physiological properties of its individuals and in particular with reference to interference between species populations. By experiment, Harper has shown that in the presence of associated species, interference frequently restricts a species population to a smaller unit volume of community space than that which the physiological tolerances of the population allow. This is illustrated by examples from data on species populations of *Plantago* spp. and *Rumex* spp. freed from the presence and interference of neighbouring grasses by the use of selective weed killers. Sagar and Harper (1961) showed that such a treatment produced responses in *Plantago* which enabled increased vegetative reproduction, increased production of flower heads and of seeds per inflorescence. With *Rumex*, removal of all sward species except *Rumex* resulted in marked increases in size of the populations of *Rumex* when compared with controls. Also, the removal of non-gramineous sward species did not produce a response in the *Rumex* population, and in one situation neither the removal of grasses, nor the non-gramineous herbs effected an increase in mature individuals of *Rumex acetosella*. Thus responses and performances vary, but they do give pointers towards the formal definition of the niche relationships of species populations in the dynamic flux of plant communities. When interfering species are removed the fundamental (potential) niche can be appreciated and compared with the realized (actual) niche where interference remains.

As yet this type of study is in its infancy—Putwain and Harper (1970) regard it 'as a rather crude start to the analytic study of niche structure in a plant community'—but it is only work of this type which will enable a better understanding of the copious amount of data from permanent quadrats which has been amassed over the last twenty years.

To turn to the results of the analyses of permanent quadrats it will be found that their use was advocated as long ago as 1905 by Clements, but that they really only came into fashion in British ecology in the mid 1930s when Watt began his observations on various plant communities in the Breckland of East Anglia (Watt, 1947, 1955, 1962). For the most part, observations have been related to overall changes in the quadrats relative to all species populations, but occasionally investigations have centred upon a single species population. The work of Wells (1967) on the unpredictable orchid *Spiranthes spiralis* is exemplary, but further consideration of this aspect must be left with this reference. However, the type of quadrat map derived from permanent recording is shown in Figure 29 where the temporal relationships of *Thlaspi alpestre* and *Agrostis tenuis* are under study.

Based on data from permanent quadrats and reference to series of population phases at several points at the same time, Watt (1947) was able to cite a number of now classical examples of cyclic fluctuations which reflect pattern and process in vegetation. The reception of Watt's theory was varied but as a theory it has produced a deeper insight into those cyclical processes which characterize vegetation. Examples of pattern and process are broadly divisible into two groups: (a) intrinsic vegetation processes unaffected by external physiographic or biotic factors; (b) intrinsic vegetation processes partially controlled or accelerated by these same factors.

Perhaps the best-known example of unaffected cyclical changes are those found in *Calluna* heath, which vegetation has received a good deal of attention in Britain (Watt, 1947, 1955; Barclay-Estrup and Gimingham, 1969; Barclay-Estrup, 1970; Coppins and Shimwell, 1971). The life history of *Calluna* has been divided into four phases referred to as pioneer, building, mature and degenerate (Watt, 1947) and each life-cycle stage corresponds with a particular age range within a total range which is usually between 0 to 36 years (Table 31). Watt (1955) showed that the accompanying species are limited in their distribution to certain phases of the community and Barclay-Estrup and Gimingham (1969) indicate changes in species pattern with age over a period of three years in a Scottish heath which has not been burnt for about 100 years or grazed for at least thirty. Coppins and Shimwell (1971) approach the problem by selecting representative stands of the four phases of ungrazed *Calluna* heath after burning—an approach which enables a clear view of the directional process to community maturity

TABLE 31

Age relationships of the Calluna life cycle

	Pioneer	Building	Mature	Degenerate
Age range	0–6 (10)	7–15	(14) 15–25	(23)–30 (33)
Habit, etc.	Low bushes, open community	Dense vigorous growth; period of maximum cover	Canopy becoming discontinuous due to branch spreading	Death of central branches of bush giving open patches surrounded by decumbent branches

from an actual starting-point and also indicates the cyclical change to degenerate and back to a heterogeneous pioneer phase. Their results show that few species are exclusive to certain phases of the life cycle, but that in terms of biomass most species reach a maximum expression in a particular life-cycle phase. Figure 35 depicts the distribution of the major species through nine representative stands and the phase characteristics may be summarized by the grouping of species into three indicator groups and one sundry group as follows:

(*a*) Species reaching their optimum development in the pioneer phase and early years of the building phase; further divisible into:

(i) those that colonize the bare humus after burning and also reappear sporadically throughout or only in the degenerate phase, e.g. *Cephalozia bicuspidata, Lecidea uliginosa, L. granulosa, Cladonia floerkeana, C. coccifera.*

(ii) those which are restricted to the pioneer phases and do not reappear in the degenerate phase—the algal mats of *Zygogonium ericetorum* and *Coccomyxa/Gleocystis* spp.

(*b*) Species of bryophytes attaining their maximum development in the building and early mature phases, namely *Pohlia nutans, Calypogeia fissa* and *Cephalozia connivens* with a dearth of other species of substantial biomass values.

(*c*) A group of species of heterogenous life forms achieving maximum biomass in the mature and degenerate phases, defined by four components:

(i) a major increase in the squamulose lichen biomass mainly in the form of *Cladonia crispata*, but also via *C. chlorophaea* and *C. fimbriata*;

(ii) the replacement of *Pohlia* by other acrocarpous mosses— *Orthodontium lineare* and *Campylopus flexuosus*;

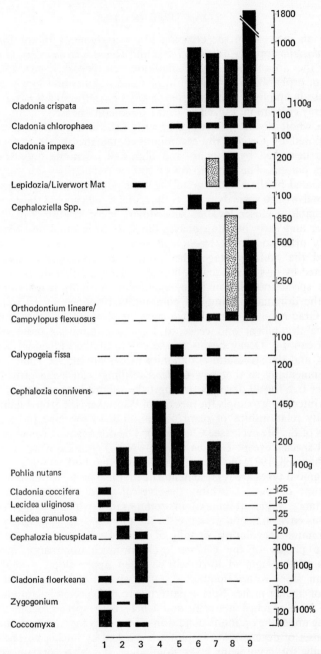

Figure 35. Biomass of cryptogams under different aged *Calluna* canopies at Skipwith Common, Yorkshire (from Coppins and Shimwell, 1971). 1, pioneer; 2, pioneer; 3, pioneer/building; 4, building; 5, building; 6, mature; 7, mature; 8, mature/degenerate; 9, degenerate.

(iii) an increase in the creeping liverwort biomass in the form of *Cephaloziella* spp., *Lepidozia setacea* and *Odontoschisma denudatum*;

(iv) the appearance of the cladinous (fruticose) lichen *Cladonia impexa*, probably as an indication of stand stability and longevity. An increase in epiphytic *Lecanora conizaeoides* gives the stands a characteristic grey coloration, especially in the degenerate phase.

The whole diverse pattern of species distribution and performance is imposed by *Calluna* the community dominant and in the case of heaths described by Coppins and Shimwell often the only vascular species present. The progression of species populations, although not represented by the actual fate of the same populations over a period of twenty-five or so years does reflect the social significance of the dominant and enable a closer look at subordinate niche relationships. In the pioneer and early building phases the *Calluna* is small and the community open so that a free-for-all develops. As the canopy closes toward the building stage, the biomass of cryptogamic species is restricted by loss of incident light and at *Calluna* maturity each subordinate species-population is in equilibrium with its neighbours and with the dominant, using the community resources at an optimum rate. Gradually, as the *Calluna* ages, it degenerates, the branches become decumbent and cease to form a continuous more or less closed canopy. Decay begins at the centre of the individual *Calluna* plants, and the structural community niches are altered spatially and environmentally so that new *Calluna* seedlings can begin growth and another free-for-all develops amongst the subordinate species.

It is interesting to apply the theories of Whittaker (1965) on dominance-diversity relationships of plant communities to the nine plots of the phases of the life cycle (see Chapter 3). The dominance-diversity curves for all species except *Calluna* are shown in Figure 36 (from Coppins and Shimwell, 1971) in which it may be seen that both simple geometric and sigmoid types are represented. The first five plots, with the possible exception of plot 3 exhibit the typical geometric progression of importance values (biomass) in communities which have relatively rigorous environmental pressures in the pioneer and stabilizing phases of community evolution. It falls of some significance that the three mature phases of the *Calluna* cycle produce dominance diversity curves of the sigmoid form with a steep upper slope, a shallower 'plateau' slope with a number of similar spatial niches and a terminal slope of smaller niches. Such a pattern is to be expected in communities which have reached maturity and where each species population is existing under the optimum conditions dictated by the canopy dominant. The onset of degeneration in plot 8 is marked by intense competition, not only for cryptogam layer dominance but also for subordinate niches where related life forms compete for a particular structural niche.

However, the basic sigmoid curve is still maintained, although further degeneration of *Calluna* causes a free-for-all, and a return to the general geometric progression is apparent in plot 9.

In a developmental sense this is a directional process; in terms of cryptogam biomass it is also directional. The dominance-diversity relationship are partially cyclical, a cycle relative to the longevity of the *Calluna*; and in terms of species complement the process is also

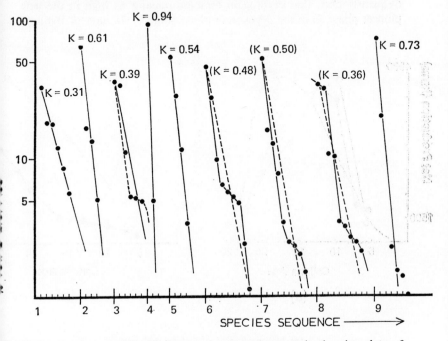

Figure 36. Dominance-diversity curves for the cryptogams in the nine plots of *Figure 35* (from Coppins and Shimwell, 1971).

partially cyclical in that species of the pioneer phase crop up in the degenerate phase. Watt (1947) also suggested that the productivity of each phase appears to follow a generalized cycle of increasing productivity to the mature phase, followed by reduced productivity towards the degenerate phase. This general pattern is shown in Figure 37*a* where the net production of young shoots per annum is plotted against age. In both the data of Barclay-Estrup (1970) and that from East Yorkshire (Shimwell, unpublished) the maximum production is reached in the building phase not the mature phase as Watt suggests. Barclay-Estrup also reports that net production of *Calluna* in relation to total biomass is greatest in the pioneer and decreases sharply throughout the life cycle (pioneer 52%, building 29%, mature 19% and degenerate 14%). Total biomass, however, was shown to be at a maximum in the mature phase by Barclay-Estrup, and this was perhaps Watt's implication in his generalized productivity curve.

Having reached the mature and degenerate phases of the life cycle, British heathlands are either burned to encourage new *Calluna* growth, or are left to enter the second cycle. If the latter situation occurs, one frequently finds that cryptogam biomass remains as high in the new pioneer phase as in the degenerate phase (Figure 37*b* data of Barclay-

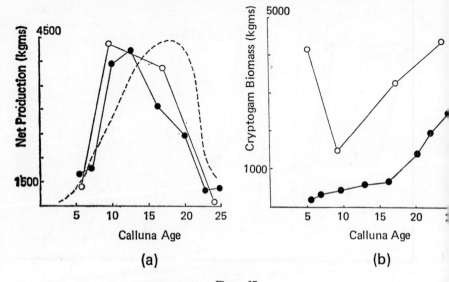

Figure 37.

(*a*) Net production of *Calluna* with age.
●,   data of Shimwell (unpublished) from E. Yorkshire (kgm/hectare/year);
○,   data of Barclay-Estrup (1970);
- - -,   productivity curve of Watt (1947).

(*b*) Cryptogam biomass relative to *Calluna* age.
○,   data of Barclay-Estrup from unburnt heaths (cyclical);
●,   data of Coppins and Shimwell from burnt heaths (directional).

Estrup, 1970), but that it will decrease at building and increase again at maturity and degenerate to produce the familiar cyclical pattern.

As one might expect it is in the pioneer and degenerate stages that the heath is most susceptible to the invasion of other species. Watt (1955) examined this relationship in detail between *Calluna* and *Pteridium aquilinum* on uniform podsolic soils at Lakenheath Warren, in Breckland. His main approach was to grid two areas of advancing bracken and to record the coincidence of fronds with various phases of the *Calluna* life cycle. In both cases there was a marked correlation between the density of fronds and the phases of the *Calluna* cycle (Table 32), i.e. the pioneer and degenerate phases were more liable to invasion from the *Pteridium*. If this is expressed in terms of a dynamic and competitive relationship, at border invasion zones more fronds are found where the defences of the heather are weakest and where the heather is strongest (at building and maturity) the competitive ability of *Pteridium* is suppressed.

The relationship of the *Pteridium* occupying the pioneer and degenerate phases of the *Calluna* cycle is not merely one where an opportunist is taking advantage of the conditions when the opponent is weakest. *Calluna* and *Pteridium* show what has been termed 'phasic

TABLE 32

Frond distribution of *Pteridium* in two areas of *Calluna* heath of different ages
(data from Watt, 1955)

Stage of Calluna *cycle*	P	B	M	D	*Total fronds*
Frond density (10/ft²)	1·80	1·65	1·33	3·27	88
„ „ „	5·00	1·53	0·99	2·27	218
Mean L/p ratio	3·20	2·70	2·12	2·57	–

interdigitation' (Watt, 1955). Watt (1945) had previously shown that a 'Pteridetum' also has a life cycle of the same major phases as *Calluna* and that the phases could be recognized by a number of characters, amongst them the lamina to petiole (L/p) ratio. This was high in the pioneer phase, falling to a minimum in the mature and rising slightly in the degenerate. Watt found that the sequence of L/p ratio was the same as in the corresponding phase of the *Calluna* cycle, i.e. high L/p ratios characteristic of pioneer *Pteridium* occurred in pioneer *Calluna*, etc. (Table 32). This is phasic interdigitation. However, the relationship does not end here. Watt showed that a *Pteridium* bearing a frond of the pioneer type after appearing in the pioneer *Calluna* phase, then proceeded to produce a sequence of building, mature, degenerate frond types when the corresponding phases of the *Calluna* cycle were reached. It thus seems that there is a parallel development of *Calluna* bush type and *Pteridium* frond type which are correlated with stem and rhizome age.

Watt (1947) also demonstrated that the number of fronds rises from pioneer to mature. In its advance upon *Calluna*, bracken tends to show the marginal vigour common to most rhizomatous plants, with younger parts of the rhizome massed parallel to one another and invading along a continuous front. The hinterland behind the wave of advance has lost its rhizome alignment, the plants are smaller and the density less in this the building phase. In the open areas of this phase the heather becomes established, reaches the building phase when its marginal

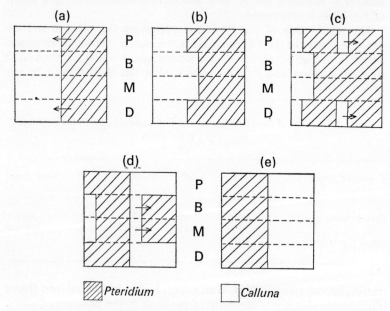

Figure 38. Diagrammatical representation of the pattern and process relationships of *Calluna* and *Pteridium*.

vigour is greatest and suppresses the *Pteridium*. There is thus a continuous flux back and forth across an area, not in a set geometric pattern but in a here-and-there mosaic pattern which represents the continual intrinsic relationships of two species populations. The system is schematized using a theoretical start and end point in Figure 38 where (*a*) the *Pteridium* invades pioneer and degenerate *Calluna*; (*b*) and (*c*) the *Pteridium* follows the invasion with the *Calluna* in its building phase beginning to invade the hinterland behind the invading bracken in its building phase. It is this stage which is always observed at any one point in time, and the whole process is only fully understood by recording permanent quadrats which illustrate the continuous flux and never show real start and end points.

The second group examples of pattern and process, those of intrinsic vegetation processes partially controlled or accelerated by physiographic factors is well known, but two types are particularly well-documented: (i) the hummock-hollow cycle of actively growing raised bog; (ii) the erosion, frost-hummock and polygon cycles of high latitudes and altitudes.

The hummock-hollow cycle of actively growing raised bog manifests itself as a regeneration complex of bog pools, hollows, lawns (or flats), small and large hummocks, which have in the past been linked together as a series of temporally related dynamic community phases. These communities are often visible in sequence if a peat boring is taken down through the bog, which activity led Osvald (1923) and Godwin and Conway (1939) to relate the phases of the cycle by examination of their stratigraphical succession. The initial phase is represented by *Sphagnum cuspidatum* which colonizes small pools of water forming a carpet. This carpet is in turn invaded with either *S. pulchrum* or *S. subsecundum* forming a lawn. Gradually, via invasion of *S. magellancium* and/or *S. papillosum* low hummocks are built up and colonized by *Calluna vulgaris, Erica tetralix, Eriophorum angustifolium* and perhaps more naturally by *Andromeda polifolia* and *Oxycoccus palustris*. The hummock dries and this phase is usually recognizable by the presence of *Hypnum cupressiforme* and *Cladonia arbuscula*. As the *Calluna* ages and dies the hummock becomes eroded and an open pool of water appears via precipitation erosion and the activity of adjacent process. All these stages are represented in sequence as stratified peat deposits. The initial stages of the cycle are dependent on the intrinsic properties of the community and of growth of the species, which in most *Sphagnum* species is greatest in the pools, less on lawns and least on hummocks (Clymo, 1970). However, the latter degeneration phases of the cycle are dependent on extrinsic, environmental factors governing erosion.

Recently, more attention has been devoted to the type of pool-hummock complex which is reputedly caused by these cyclical processes, in bogs in western Scotland where the complex is best developed (Pearsall, 1956; Ratcliffe and Walker, 1958; Boatman and Armstrong, 1968; Goode, 1970). The bogs are characterized by a wide range of pool, lawn and hummock types from open water and *Menyanthes* types through to tall *Rhacomitrium* hummocks which cannot possibly be conceived as a cyclical series. Several theories have been put forward to explain the nature of the environmental influence which accentuates the phases of the regeneration cycle. In the past, Scandinavian workers have favoured differential freeze-thaw processes as an explanation of the phenomenon of hollows and hummock patterns, although Sjors (1965) has emphasized that pronounced patterns also occur in areas where frost action is insignificant, e.g. the temperate Atlantic climate of north-west Britain. Pearsall (1956) suggests that the pattern complex is characteristic of bogs in an advanced stage of development and that they are in reality a mass of semi-fluid peat which on even slight slopes will slide slowly downhill. The vegetation skin on the lower parts of the slope will wrinkle, whereas the upper part, under tension, will be torn. On Pearsall's Strathy bog the shape is half-domed with peat movement away in three directions so that the 'wrinkles' on the lower slope are long and narrow pools and ridges and the tears on the upper surface, isodiametric pools. Whilst much the same pattern exists on the Sliver Flowe bogs, Kirkudbright, Ratcliffe and Walker (1958) reject Pearsall's explanation. They find that stratigraphical evidence points to a marked increase in *Sphagnum* activity above humified *Sphagnum-Molinia* peat and pool mud. They therefore suggest that the bog surface has recently been flooded, inducing increased *Sphagnum* growth, this feature pointing to the fact that the pool systems are associated with renewed growth rather than with degeneration as Pearsall suggests. The deeper pools with steep vertical sides are considered to be secondary and associated with the lowering of the water table and wave action.

Boatman and Armstrong (1968) showed that the direction of elongation of larger pools over part of Rhiconich Bog, Sutherland, is related strictly to the slope of the peat surface, irrespective of variations in slope of underlying mineral ground. They consider that this evidence should preclude the possibility of pool origination as wrinkles or tears due to gravitational effects. In fact, if the linear patterns are a response to such forces, the greater slope should be manifest as a greater exaggeration of pattern. Goode (1970) showed that this was not the case on the Silver Flowe Bogs and concluded that pattern initiation was due to the flooding of a relatively dry peat surface. Stratigraphical evidence also indicates that the patterned surfaces occur only where the peat is more than 2 m deep; that the large-scale pattern is developed only within the upper peat layers and that it replaces a small-scale hummock-hollow relief, superimposed on a relatively level surface of humified peat; and that the differences in ecological conditions between the hummocks and hollows are due mainly to the water table fluctuations in the two phases, i.e. a greater fluctuation in the hummocks (Goode, 1970). So what seemed to be a neat explanation of cyclical phenomena has become a complex of interpretations. Three points are, however, quite clear: (a) that in the early phases of the cycle the patterns are due mainly to intrinsic vegetation processes mainly those which surround the growth of *Sphagnum*; (b) that the later phases are dependent upon fluctuations in water table and progressive dessication; and (c) the complex pattern of extensive pool-hummock bog surfaces is related to surface flooding which exaggerates the end points of the hummock-hollow cycle.

The environmental factors responsible for cyclical vegetation processes at high latitudes and altitudes are more apparent than in bog vegetation. Factors such as frost and wind erosion, frost heave and solifluction are all readily observed and related to the vegetation mosaics which develop. The most simple examples of cycles are to be seen on exposed sites where wind and water erosion cause the movement of isolated patches of vegetation along the wind or stream direction as the vegetation itself is eroded on the windward side and regenerates to leeward. An excellent example of this phenomenon is seen in the work of Barrow *et al.* (1968) on an *Epacris petrophila*

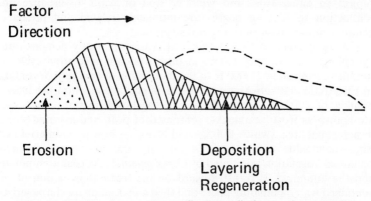

Figure 39. Wind erosion as a factor affecting cyclical vegetation processes.

community on alpine lithosols at 6,500 ft in the fjaeldmark regions of Mount Kosciusko, Australia. The vegetation, dominated by the dwarf chamaephyte *Epacris*, occurs as a series of isolated clumps interspersed on an uneven rocky terrain. As the *Epacris* ages, it is eroded on the windward side, some of the eroded fragments being deposited on the leeward side of the clump, where the partially buried *Epacris* stems undergo layering. If this anchoring by layering occurs before the base of the plant is completely eroded, the *Epacris* clump and associated flora continue to develop. A similar phenomenon has been reported by Burges (1951) in a *Rhacomitrium-Empetrum* heath at higher altitudes in the Cairngorms, Scotland. Whereas Barrow *et al.* (1968) calculate the mean rate of movement at 1·1 cm per annum, Burges records a figure of approximately 1·7 cm per annum (Figure 39).

Good examples of wind erosion are to be found in the communities overlying the unstable metamorphosed sugar-limestone in Upper Teesdale. Here also in the flushes which are common on Cronkley and Widdybank Fells there are to be found examples of cyclical processes in moss-sedge communities affected by water erosion and solifluction (Pigott, 1956). Polsters of the moss *Gymnostomum recurvirostrum* are commonly invaded by sedge species such as *Carex lepidocarpa*. Erosion of the polster causes a migration of the sedges towards the 'leeward' where they eventually form a barrier to erosion and enable the moss to build up a further polster.

Frequently, a whole complex of factors are known to cause cyclical changes in communities. Anderson (1967) describes a phasic development in *Dryas octopetala* communities near Melgraseyri in Iceland as the communities are invaded first by *Kobresia* (sedges), later by dwarf shrubs (*Betula nana*) and finally by *Betula pubescens*. Each phase is subject to either frost and wind erosion or wind disturbance and solifluction so that all phases are intimately related and present a complex mosaic pattern to the investigator.

The significance of frost action and erosion is neatly demonstrated by Billings and Mooney (1959) in their work on vegetation cycles in tundra vegetation at 11,000 ft in the mountains of southern Wyoming. In the Arctic and sub-Arctic zones the intense frost action produces a number of marked topographical and vegetation features. Two such are known as frost hummocks, composed of peat, and assorted stones in polygonal areas, which Billings and Mooney show to be interrelated. The water table level fluctuates annually and in the wet hollows, hummock formation is initiated by *Carex aquatilis*. As peat accumulates and the hummock is thrust upward by the freeze-thaw action of the contained water a building phase and then a mature phase characterized by distinct species are formed. When the hummock rises to such a level that it protrudes above the snow in the winter months, species on the crest are killed and erosion by the action of wind and wind-borne particles begins. Continued erosion and frost action from below eventually heaves the stones up through the peat so that the hummock degenerates to a ring of peat surrounding a central stone polygon.

The relation theoretical approach to pattern and process in vegetation

As a result of the analysis of permanent quadrats in various vegetation types in the Netherlands over a period of some twenty years, van Leeuwen (1965, 1966) has been able to work out the mutual relationships between spatial structure and dynamic behaviour in vegetation, i.e. between pattern and process. This theory has proved to be of use in research on ecological gradients and border areas and its extensive use by Dutch ecologists (Beeftink, 1966; van der Maarel, 1966, etc.) demands explanation of its fundamental concepts.

The basis of van Leeuwen's Relation Theory is taken from the cybernetical approach to research which recognizes that all problems may be reduced to a consideration of equality or inequality in space and time. Complicated relationships in vegetation, ecosystems and life in general can be reduced to four simple primary elements: (a) equality in space; (b) inequality in space (variety, difference); (c) equality in time (stability); (d) inequality in time (instability). When considered in this aspect, the study of pattern and process in vegetation has a simple aim of tracing the connections between variety-in-space (pattern) and variety-in-time (process). To enable full understanding of the Relation Theory, it is imperative that the explanation follows the logical procedure as outlined by van Leeuwen (1966). This involves first a consideration of just what constitutes spatial and temporal patterns.

Generally speaking, spatial patterns only occur where there is 'inequality', a 'choice' or a negative correlation 'no' are found. For example, within the two relationships 'yes-yes' and 'no-no' there is 'equality' and therefore no pattern. In the relationship 'yes-no' there is inequality, a choice, a negative correlation and therefore 'no'.

All spatial patterns exhibit three related properties: (a) form, (b) size and (c) border type, e.g. (a) geographical distribution of a community, (b) its extensiveness, (c) the form of its external limits, which may be sharp or gradual. It is possible to express the relationships of these three using the couplet 'yes-no' (Figure 40). If the species has a pattern in the form of a narrow ribbon ('yes' in Figure 40, a), in the longitudinal axis of its pattern a large amount of 'yes' is available and therefore a large amount of 'equality'. On the other hand, there is a relatively large amount of 'no' or 'inequality' in the transverse direction of the ribbon. If a score of 1 is registered for a 'yes-no' relationship, a score of 10 is apparent transversely and 2 longitudinally. If a community occurs as a circular patch (Figure 40, b), then the inequality although the same is not concentrated in a single direction being more or less evenly distributed in both directions (in this case 6 transversely and 6 longitudinally). It follows that a distribution pattern resembling a ribbon will always contain more ecological significance than a circular patch because the inequality is concentrated in one direction. As a parallel situation to this a small or fine-grained pattern will be of more ecological significance than a large or coarse-grained pattern.

Elements of equality and inequality are also recognizable in temporal patterns (process). Equality implies permanence or stability; inequality suggests change or instability. Close attention is paid to the relationships between spatial and temporal patterns in what has been called the Basic Relations theory, which recognizes that every relation shows aspects of *connection* and aspects of *separation*. Further, a relation is

F

(a)

No	No	No	No	No
No	No	Yes	No	No
No	No	Yes	No	No
No	No	Yes	No	No
No	No	Yes	No	No
No	No	Yes	No	No
No	No	No	No	No

(b)

No	No	No	No	No
No	No	No	No	No
No	No	Yes	No	No
No	Yes	Yes	Yes	No
No	No	Yes	No	No
No	No	No	No	No
No	No	No	No	No

Concentration and dispersion effects

Y	Y	Y	N	N	N
Y	Y	Y	N	N	N
Y	Y	Y	N	N	N
N	N	N	Y	Y	Y
N	N	N	Y	Y	Y
N	N	N	Y	Y	Y

1

Y	Y	N	N	Y	Y
Y	Y	N	N	Y	Y
N	N	Y	Y	N	N
N	N	Y	Y	N	N
Y	Y	N	N	Y	Y
Y	Y	N	N	Y	Y

2

Y	N	Y	N	Y	N
N	Y	N	Y	N	Y
Y	N	Y	N	Y	N
N	Y	N	Y	N	Y
Y	N	Y	N	Y	N
N	Y	N	Y	N	Y

3

Figure 40. Theoretical example of equality and inequality in space.

open where there is connection and *closed* where there is separation; open relations imply *equality* (in both space and time), while closed relations imply *inequality*. So the *first basic relation of equality* can be depicted as two '*relation chains*' thus:

1. Connection ≈ open ≈ equality
2. Separation ≈ closed ≈ inequality

where the words linked indicate progressions, such as the fact that spatial connection and openness will lead to spatial equality and *vice-versa*; that temporal connection will lead to equality-in-time or stability; that spatial separation or closure will result in spatial inequality and that temporal separation will cause instability and change.

To these basic triplets it is possible to add further words expressing the four relationships and from these to compound relation chains of direct connection. For example, spatial connection is represented by the chain, *open ≈ public ≈ permission ≈ everywhere* (IA). When a museum is open the general public have full permission to move almost everywhere thus rendering them all equal in space. On the other hand, the fact that part of the building is closed by a barrier means a certain spatial area is placed in isolation, suggesting that there is a particular item there which is secret; (2A) *closed ≈ barrier ≈ isolation ≈ secret* expresses spatial separation.

Similarly temporal connection and separation may be represented as follows:

(1B) *connection ≈ stability ≈ conservation ≈ continuation ≈ relic*
(2B) *separation ≈ disturbance ≈ precariousness ≈ change ≈ destruction*

At this point three-word couplets need to be introduced to the relation chains; *order* and *disorder* from thermodynamics; *information* and *noise* from the Information Theory and *co-operation* and *competition* from general individual relationships. The basic relation chains are written thus:

1. Connection ≈ equal ≈ disorder ≈ noise ≈ competition ≈ known
2. Separation ≈ unequal ≈ order ≈ information ≈ unknown

It may seem a paradox that there should be a relationship between *information* and *unknown* but this is explained with reference to the Information Theory which holds that a quantity of *information* is identical with a quantity of *missing knowledge*, a *yes-no* chain. *Life* itself is closely tied up with such *yes-no* apparata and combination of spatial connection and separation are to be found in all types of natural and artificial regulators such as membranes, diaphragms, valves, hedgerows and fences. Combinations of temporal separation and connection are to be seen in progressions such as succession, evolution, adaptation and cyclic process.

Van Leeuwen (1966) originally called the theory the Open and Closed Theory, but soon found that many researchers had difficulty in grasping the application of the two terms *open* and *closed* and using them in a temporal sense. In a spatial sense they are directly opposite the uses of open and closed in the structural sense of Fosberg (1967) for his primary structural groups, and extrapolating from this the open conditions of a desert scrub implies temporal precariousness and disturbance. Van Leeuwen therefore abolished such a title and produced four abbreviations to represent the relationships:

$$1A \quad \bar{a}(s) \qquad 1B \; \bar{a}(t)$$
$$2A \quad a(s) \qquad 2B \; a(t)$$

where variety is represented by a and no variety by $\bar{a}$; space by s and time by t.

The *second basic relation* is that of *inequality* and in this the difference between connection and separation is described. To return to Figure 40, if a connection ($\bar{a}$) (*yes*) is introduced into a matrix in a definite direction then a relationship of separation (a) (*no*) develops transversely at that point. Thus, to get from $\bar{a}$ to a and vice versa an angle of 90° has to be described. In the case of Figure 40, the *yes* quantity has to be introduced down into the page matrix from the eye of the reader. Compare this to a lift moving down into the basement of a building. To leave the lift, the operator has to change direction through 90° to get into the foyer (from *yes* to *no*). The basic relation of inequality may thus be defined as the complementary relationship between equal and unequal, represented by the expression $\sqrt{-1}$ (= to turn through 90°); written $\bar{a} \# a$ where $\bar{a}$ the equality is $\#$ 90° different from but allied with a inequality.

If it is assumed that the relationship between $\bar{a}$ and a is the same as that between space and time, it should be possible to relate $\bar{a}(s)$ to $a(t)$ and $a(s)$ to $\bar{a}(t)$, i.e.

$$\bar{a}(s) \leftrightarrow a(t) \quad 3A$$
$$a(s) \leftrightarrow \bar{a}(t) \quad 3B \text{ the Third Basic Relation}$$

The relationships of these two expressions find numerous examples from all life systems, but restricting the context to vegetation examples several interesting examples are apparent. Situations with unstable environmental conditions such as the zones of the shingle foreshore, mobile dunes, forward salt marsh, arctic and alpine erosion and frost heave habitats are commonly poor in species due to temporal instability:

$$\bar{a}(s) \text{ species poor} \leftrightarrow a(t) \text{ precariousness}$$

On the other hand, stable situations where environmental change is not affected by any major feature and which is slow or not apparent, usually support plant communities rich in species:

$$\bar{a}(s) \text{ species rich} \leftrightarrow \bar{a}(t) \text{ stability}$$

It is at this last formula that the whole work of Nature conservation is aimed, i.e. attempting to maintain *variety in space* and *stability*. The third basic relation thus underlines the fact that *equality in space* and *equality in time* must not be treated in the same manner. The same premise applies to *variety in space* and *variety in time*.

However, a basic paradox is readily apparent in the second main relationship. Equality or connection in one direction is only possible where there is inequality or separation in the complementary direction and where observations ($\bar{a}$) can be made on differentiations (a) in the reference matrix. But if there is a connection in the line of vision it is difficult if not impossible to observe. Taking the lift analogy again, if the roof is not taken off the building then the passage from *yes* to *no* is not observable. In ecological relationships, this paradox is most apparent when horizontal and vertical situations are considered relative to one another. For example, in a deciduous oak woodland a high degree of vertical differentiation (a) in the layering of the vegetation strata and soil horizons is correlated with a high amount of horizontal uniformity ($\bar{a}$) of the vegetation which can obscure the differentiation.

From these three basic relations the effects of *concentration* and *dispersion* in space and time relative to vegetation pattern and process or life in general can be inferred. Spatial concentration implies the introduction of equality or connection internally. Conversely, inequality and separation remain the same or increase externally. Consider the three matrixes in Figure 40. Scoring 1 for a *yes-no* relation, matrix 1 produces a dispersion figure of 12 (six horizontally, six vertically), matrix 2 scores 24 and matrix 3, 60. Along the series to matrix 3, inequality or internal dispersion increases, whilst in the reverse direction internal concentration is built up to give a coarse-grained pattern compared with the fine granulation of internal dispersion. If these are faced with a constant external score of 12, matrix 1 will exhibit greater internal concentration and therefore instability than the other two matrices. Moreover, it will exhibit greater tension at the boundary zone because of equality. Internal and boundary stability increases towards matrix 3 because of increase in dispersion and boundary inequality. In other words, if a single species dominates a large area spatial concentration (instability or convergence) and boundary instability (equality) occur. Alternatively, a hundred species in the same area means internal stability (dispersion or divergence) and boundary inequality.

It follows, therefore, that there are two main types of vegetation environments and two types of boundary environment, described by van Leeuwen (1965):

1. Uniform environment—showing concentration homogeneity in the horizontal plain; equal or open in space $\bar{a}(s)$; the principal environment factors fluctuate widely and are often catastrophic; the environment is unstable or closed in time $a(t)$ (Basic Relation 3A).

2. Variegated environment—showing dispersion or heterogeneous in space $a(s)$; stable in time due to the absence of human interference or other drastic effects, therefore $\bar{a}(t)$ (Basic Relation 3B).

3. Convergent limit (*Limes convergens*) is related to spatial concentration, coarse pattern granulation and straight, regular or well-demarcated border lines producing a longitudinal transect of alternating '*all-or-nothing*' where $a(s) \leftrightarrow a(t)$. Boundaries of this type are found where two or more contrasting situations alternate in time and the unstable internal conditions cause a '*now-and-then*' or '*to-and-fro*' existence. To this situation the ecotone concept is applied.

4. Divergent limit (*Limes divergens*)—opposite to the previous boundary type, this environment shows a fine granulation pattern and faint lines of demarcation so that a longitudinal transect produces a '*more-or-less*' result. Internal stability is high and there is a high dispersion or heterogeneity $a(s) \leftrightarrow \bar{a}(t)$. The gradual environmental gradient lacks drastic changes and the boundary is rich in species, many often only represented as single individuals. The term ecocline, borrowed from experimental taxonomy is applied here.

The two types of border environment are illustrated in Figure 41, in a transect across Chee Tor, a limestone promontory in the Wye Valley, Derbyshire. Good examples of *limes convergens* are to be found on the flood plain of the River Wye where the causes of temporal instability are river-level fluctuation and grazing. The *limes convergens* has its own characteristic species, in this case '*all*' contributes *Epilobium hirsutum*, *Eupatorium cannabinum* and *Petasites hybridus* while '*nothing*' supplies *Agrostis stolonifera*, *Potentilla anserina*, *Rumex crispus*, etc. '*All*' is delimited by small water-level changes but not large enough to normally flood '*nothing*', while the cattle keep to the drier ground of '*nothing*' and trampling effects are seen. The whole area is subject to what may be regarded as the catastrophic effect of winter flooding and scouring.

In contrast, the areas on the promontory are practically ungrazed, relatively inaccessible and exist as a series of gradients between two variegated environments, here called ash woodland and limestone grassland. They present a type of scrub interface boundary referable to the *limes divergens* situation with dispersion or inequality in space, viz. an average species number of 48 per square metre, and equality in time. The presence of uncommon species more or less exclusive to this type of boundary zone indicates longevity and stability in time. Species characteristic of this type of boundary in the Peak District include *Geranium sanguineum*, *Silene nutans*, *Epipactis helleborine*, *Gymnadenia conopsea*, *Convallaria majalis*, *Melica uniflora* and *Trifolium medium*.

It now becomes possible to look at the process of succession from a relation-theoretical point of view. Vegetation develops gradually from a convergent, unstable, species-poor pioneer community through to a divergent climax community with relatively stable conditions. Many

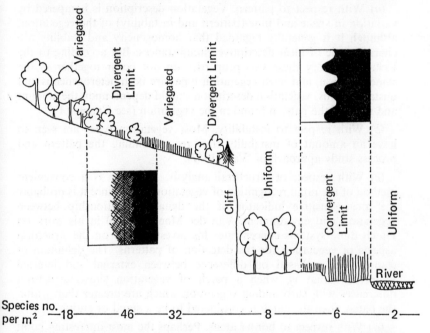

Species no. per m² —18——46——32——————8———6——2—

Figure 41. Limes divergens and Limes convergens habitats at Chee Tor, Derbyshire.

plant species are able to colonize a newly-bared habitat extremely rapidly and by virtue of their ephemeral or short-lived existence they confer a degree of $a(t)$ (precariousness) on the habitat. The characteristic of such species is aggregation $\bar{a}(s)$ with closed monodominant populations being built up by clonal perennation or the rapid reproduction of gregarious inbreeders. The further alteration which constitutes succession is basically a decrease of environmental instability, boosting other local $\bar{a}(t)$ aspects. Margalef (1959) points out that an increase in stabilisation is stimulated by the storage of 'information' as either a function of increasing spatial divergence, growing co-operation to divide the interest of the whole or as the accumulation of peat and humus in the topsoil or simply biomass. This is the directional process of succession $\bar{a}(s) \leftrightarrow a(t) \longrightarrow a(s) \leftrightarrow \bar{a}(t)$ and once the climax community is attained the $\bar{a}(t)$ remains more or less the same and fluctuations in $\bar{a}(s)$ to $a(s)$ produce the familiar cyclical processes.

The Relation Theory of van Leeuwen (1966) when applied to vegetation description raises several interesting points, most of which are still being investigated:

(a) With respect to pattern. Vegetation description is hampered by variation in space and time (pattern and instability) of the vegetation, although it is generally regarded that homogeneity and stability are characteristic of ideal descriptive circumstances. But according to the Relation Theory these two properties do not occur together in the same situation, and most vegetation types are both heterogeneous and unstable. Thus, vegetation description should depend upon the amount and scale of the pattern found in the vegetation (see Chapter 1).

(b) With respect to instability. Most vegetation types are seen to have an amount of instability thus recommending the pattern and process study approach of Watt (1947).

(c) With respect of structural analysis. Structure is a convenient method of the rapid recognition of vegetation types, since it is probably the most sensitive indicator of the dynamic relationship between vegetation and environment. Van der Maarel (1966) in his work on Dutch dune systems concentrates his investigations on the structural aspects of vegetation for the detection of patterns. The definition of stands is then based on difference between external and internal uniformity, that is, when a patch of vegetation shows structural differences with surrounding vegetation which are greater than within the patch, then the patch is considered to be a stand or entity.

(d) With respect to border areas. Perhaps the most interesting point brought out by the Relation Theory is the type and significance of border zones, especially the stable divergent limits. Here the dispersion and stability in time make the ecocline border type extremely important in Nature conservation because it is usually in this habitat that the maximum biological diversity is found. It has already been mentioned that certain species are characteristic of the *limes divergens* (*Silene nutans*, etc.), but here also the rarities and endemic plants are more likely to exist. Beeftink (1965) associated the occurrence of many endemic plants of the salt marshes of western Europe with four areas in each of which a gradual transitional divergent limit between two climatic types occurred. Further, most of the localities in which hybridization and introgression between species populations has occurred are typical *limes divergens* situations (cf. Anderson, 1948). Finally, they sometimes form a key in the understanding of environmental and vegetation structural gradients (van Leeuwen, 1966; Westhoff, 1967) which in turn enables management and conservation proposals to be formulated about the vital $a(s) \leftrightarrow \bar{a}(t)$ relationship.

CHAPTER 5

SOME EXAMPLES OF STRUCTURAL-FUNCTIONAL DESCRIPTIVE AND CLASSIFICATORY METHODS

Most phytosociologists apply structural concepts at one stage or another in their methods of vegetation description, but some show a far greater bias towards structural-functional features than to other features such as floristic composition. This chapter illustrates some of the methods which are considered to be primarily aimed at a structural description. The approaches of Tansley (1939), involving the assumption that vegetation is a quasi-organism, combine both structural and floristic analyses, but because of the primary structural delimitation of stands this approach is reviewed here. Similarly, Raunkiaer (1918) using his life forms and leaf types develops a classification which is entirely structural. On the contrary, Raunkiaer (1910) provides a more thorough floristic analysis and later applies physiognomic and bio-logical structural descriptive methods in the form of life-form analyses. Because of this bias toward floristics Raunkiaer's methods are dealt with later in Chapter 6.

It should be apparent, therefore, that the methods included in this chapter are rather arbitrarily chosen and based more on the content of the foregoing chapters, being a more or less logical extension and application of some of the natural, obvious features of vegetation previously described. No method of description is completely structural, and likewise no method is absolutely floristic. Where to draw the line between the two approaches must thus remain conjectural and depend upon the interpretation of the individual.

To exemplify the various methods a standard set of data is used, namely of woodland types from the carboniferous limestone peninsula at Muckross, Killarney, County Kerry. The data, collected over a period of three years, comprise a series of zones of low woodland on bryophyte-covered limestone pavement beginning with tall *Arbutus unedo* scrub near the shores of Lough Muckross, changing gradually to a mixed *Quercus-Taxus-Ilex* woodland. The woodlands of this region have long been of interest to phytogeographers, being visited by the International Phytogeographical Excursion in 1911. The presence and success of the evergreen shrubs in abundance in the woods, e.g. *Arbutus unedo*, *Ilex aquifolium* and the introduced *Rhododendron ponticum*, provide pointers to the fact that this woodland type is closely related to the hard-leaved sclerophyllous woodland types of the Mediterranean region (Rübel, 1912b).

Within the broad area of structural vegetation description five major traditions may be delimited:

1. The Early European tradition initiated by Warming (1895) and Drude (1905), developed by Brockmann-Jerosch and Rübel (1912) and Rübel (1930) and forming the basis of more detailed phytosociological works in selected regions, e.g. Tüxen and Oberdorfer (1958) and Ellenberg (1963).

2. The Clementsian tradition based on the work of Clements (1916), (1928) and widely applied in the Americas, notably by Braun (1947, 1950).

3. The British tradition, developed from Clements by Moss (1913), Watt (1932) and Tansley (1939) and the basis of most ecological work in the British Isles and Commonwealth.

4. The Tropical tradition, a mixture of approaches with Clementsian and British bases, but notable for the detailed application of Raunkiaer's life-form and leaf-type classes; examples include Loveless and Asprey (1957), Cain *et al.* (1956) and Cain and Castro (1959) from sub-tropical and tropical regions where the floristic analysis of the vegetation presents a formidable task.

5. The Symbolic tradition, developed in North America by Küchler (1949, 1967) and Dansereau (1951, 1957) and involving the graphic representation of vegetation types by letters and symbols.

The Early European Tradition

Most of the earliest descriptions of vegetation by plant geographers were simply physiognomic descriptions using broad categories like scrub, forest, evergreen, deciduous, etc. Warming (1895) considered that the edaphic and climatic effects on the water budget of plants and vegetation was of prime importance and consequently used a system of description and classification related to this angle, e.g. Hydrophytic, Xerophytic, Mesophytic and Halophytic vegetation types. Drude (1905) provides a good example of early physiognomic description and classification. The primary categories are closed and open land formations and aquatic formations. The former is then divided into forests, scrub, grassland and cryptogam vegetation types and these in turn are divided into broad physiognomic groups such as equatorial rain forest or sub-tropical temperate rain forest, prairie or savannah, etc. Description of the vegetation types was always in the form of a short precis taking into account the physiognomy and structure of the vegetation, the dominant species or edaphic relationships.

By 1910, a complex of vegetation nomenclature had already developed and it was at the Third International Botanical Congress that Flahault and Schröter combined their previous opinions to produce the definition of the Association (see Chapter 2). These two workers, from Montpellier and Zurich respectively, were largely responsible for the development of the Swiss–French school of phytosociology. Flahault (1901) had used dominance as the main criterion for describing vegetation units but Brockmann-Jerosch (1907) stressed the value of constant species for characterization. Influenced by the work of Gradmann (1909) which triumphed the need for a floristic approach to description and classification, the 1910 Congress accepted Flahault and Schröter's definition. The work of Brockmann-Jerosch (1907) and subsequent investigations by Brockmann-Jerosch and Rübel (1912) established the procedures of the Zurich–Montpellier school of compiling stand descriptions into community types and of ranking species in the community by constancy. From this early foundation two major dichotomies occurred. On the one hand, Braun-Blanquet (1913) began to develop the aspects of community description and classification by constant and character species (see Chapter 6) while Rübel (1913, 1930) tended to place more emphasis on dominance and functional physiognomy in his treatment of the vegetation of the world in terms of formation types and formations. It is with this latter dichotomy that the present section is concerned.

A project such as *Pflanzengesellschaften der Erde* undertaken by Rübel (1930), because of its magnitude, is bound to have several major discrepancies with respect to the amount of representation received by different vegetation types. Thus, we find that descriptions of various communities in this tome vary between the extremely well documented, floristically well-characterized vegetation of the Swiss Alps to the perfunctory mention of the little-known chaparral-maquis of California and Chile. Ideally, the descriptive method of Rübel is that shown in Table 33, the forerunner of the Braun-Blanquet Association Table. The floristic analysis gives each species a cover abundance value on a ten-unit scale where 1 represents a range of 1–9% cover, 2 = 10–19%, etc. The Associations, recognized largely by an initial physiognomic impression of dominance (unlike Braun-Blanquet, 1932) are then compiled as tables of related stands. For the table, each species is given a constancy value according to their presence in the table. Species with over 50% constancy value are regarded as constants (K); those between 25% and 50% constancy are known as accessory species (A), while species less than 25% are recorded as others. In the data in Table 33 from the *Arbutus* 'zone' around Lough Muckross,

TABLE 33

The floristic table method of Rubel

Life form	Stand number	1	2	3	4	5	6	7	8	9	10	11	12	Presence	Exclusiveness
13 Constant (K)															
MP	Arbutus unedo	10	10	10	10	9	10	9	9	10	10	8	8	12	4
MP	Ilex aquifolium	3	2	5	2	2	2	1	2	–	–	6	7	10	3
H	Brachypodium sylvaticum	–	3	2	1	1	1	2	3	–	2	3	4	10	3
H	Rubia peregrina	1	1	1	1	–	1	–	–	1	1	1	1	9	4
H	Teucrium scorodonia	–	1	–	2	1	1	1	–	1	1	1	2	9	2
M	Isothecium myosuroides	4	–	3	4	4	2	–	5	4	5	–	2	9	2
M	Neckera complanata	5	8	3	–	–	6	9	–	–	2	8	6	9	3
M	Ctenidium molluscum	2	1	–	6	6	–	–	4	6	2	1	1	8	3
Ch	Cotoneaster microphylla	1	–	–	2	2	4	2	–	3	3	4	–	8	1*
Ch	Rubus fruticosus agg.	1	1	–	2	1	2	2	–	1	2	–	–	8	2
MP	Taxus baccata	–	–	–	–	1	2	–	2	1	2	4	3	7	3
H	Solidago virgaurea	1	–	1	–	–	–	1	–	1	1	1	–	6	1†
M	Porella platyphylla	–	2	–	1	1	–	1	–	2	–	–	1	6	2
15 Accessory (A)															
MP	Quercus petraea	2	–	1	–	1	–	–	2	–	2	–	–	5	2
MP	Corylus avellana	–	1	2	–	–	–	–	4	–	2	1	–	5	2
Ch	Calluna vulgaris	–	–	1	2	2	1	–	–	–	–	–	–	4	1‡
Ch	Erica cinerea	–	–	1	1	1	1	–	–	–	–	–	–	4	1‡
E	Clematis vitalba	–	–	–	1	–	–	2	5	2	–	–	–	4	1*
E	Hedera helix	–	1	–	–	1	1	–	–	1	–	–	–	4	2
MP	Betula pubescens	–	–	–	–	–	–	–	1	–	–	1	1	3	2
H	Fragaria vesca	–	–	–	–	–	–	1	1	1	–	–	–	3	2
H	Carex flacca	–	–	–	–	–	–	1	1	1	–	–	–	3	1†
M	Camptothecium sericeum	–	–	1	1	–	–	–	–	–	1	–	–	3	2
G	Pteridium aquilinum	–	–	1	1	1	–	–	–	–	–	–	–	3	2
M	Fissidens cristatus	–	–	–	–	–	–	1	1	–	1	–	–	3	2
Ch	Thymus drucei	–	–	–	–	–	–	1	1	1	–	–	–	3	1†
H	Viola riviniana	–	–	–	1	–	–	–	–	–	1	1	–	3	2
H	Asplenium rutamuraria	–	–	–	–	–	–	–	–	1	–	1	1	3	1†
	21 Others (sp. no.)	6	5	8	8	4	2	7	10	6	5	5	5	–	–
	Total species number	16	16	21	24	19	15	21	24	22	17	18	17	–	–
	K+A species	10	11	13	16	15	13	14	19	16	12	14	12	–	–
	Character species number (4+3)	5	6	5	5	5	6	4	5	4	6	8	7	–	–
	Strange species number (1)	3	–	3	5	3	3	7	7	6	2	4	2	–	–

* Introduced species.   † Species of open limestone.   ‡ Wet heath species.

Killarney, there are thus thirteen constant species and fifteen accessory species. The table is completed by (*a*) the Raunkiaerian life-form type at the left; (*b*) the presence value and exclusive value to the right. This latter is assessed according to the following scale:

5—species exclusive to the community in a particular geographical region (German—*treu*) *exclusive*;

4—species found mainly in one particular community and seldom in others (German—*fest*) *selective*;

3—species preferring one community though occurring in several others (German—*hold*) *preferential*;

2 species indifferent to a particular community (German—*vag*);

1—species uncommon or rarely occurring in a particular community (German—*fremd*) *strange*.*

For example, *Arbutus unedo* is selective for this community in southern Ireland while *Cotoneaster microphylla* is introduced and a stranger. At the base of the table the number of selective and preferential character species, and the number of strange species are calculated for each stand and the numbers inserted in the appropriate columns. Comparison of this Association with other *Arbutus* stands in the Killarney Lakes region, but on non-calcareous soils and strata is illustrated in Table 34 where following Rübel, each species is given its appropriate presence value for each community unit with the appropriate K, A or O appended according to its 'constant', 'accessory' or 'other' status. The table indicates the change in status of species when the two major edaphic types of *Arbutus* scrub are considered, rather than taking the whole regional approach to description. For example, *Rubia peregrina*, *Neckera complanata*, *Ctenidium molluscum* and *Cotoneaster microphylla* are good constant species for the limestone *Arbutus* stands but on a regional basis their status becomes reduced to accessory level (column C). Rübel would tend to include both *Arbutus* types as sub-associations of a single *Arbutus*-Association. This would then be classified into an Alliance, in this case the Alliance Arbution (Rübel, 1930, p. 91) and this into a formation which includes other scrubby Mediterranean maquis vegetation types dominated by species such as *Erica arborea*, *Quercus ilex* and *Myrtus communis*.

A detailed analysis of the type presented above is exceptional in Rübel's volume on the vegetation of the earth. More characteristic is a basic description of the vegetation type, its climatic and edaphic factors, a brief list of physiognomic dominants and/or constants, the geographical distribution of the vegetation type with reference to related types in other geographical regions and finally the relationships of

---

* This system was also adopted by Braun-Blanquet (1932) but slightly modified in its usage (see Chapter 6).

TABLE 34

Constancy values in the *Arbutus* association at Killarney:
Rubel's method

Number of stands	A 12	B 8	C 20
Arbutus unedo	12K	8K	20K
Ilex aquifolium	10K	7K	17K
Brachypodium sylvaticum	10K	2A	12K
Rubia peregrina	9K	–	9A
Teucrium scorodonia	9K	1 O	10K
Isothecium myosuroides	9K	3A	12K
Neckera complanata	9K	–	9A
Ctenidium molluscum	8K	–	8A
Cotoneaster microphylla	8K	–	8A
Rubus fruticosus agg.	8K	3A	11K
Taxus baccata	7K	–	7A
Solidago virgaurea	6K	6K	12K
Porella platyphylla	6K	–	6A
Calluna vulgaris	4A	8K	12K
Erica cinerea	4A	6K	10K
Hypericum pulchrum	2 O	5K	7A
Pteridium aquilinum	3A	5K	8A
Blechnum spicant	–	4K	4 O
Hymenophyllum wilsonii	–	4K	4 O
Eurynchium praelongum	–	4K	4 O
Hylocomium brevirostre	–	4K	4 O

A. Lough Muckross shores (limestone).
B. Upper Lake region.
C. Values for A and B combined.

transitional vegetation types. His ideal descriptive method, proposed at the 1930 Botanical Congress (Rübel, 1931) is illustrated in Appendix II.

Rübel's Alliances are grouped into formations and formation groups. The *Arbutus*-Association fits into the Durifruticeta, a formation group comprising sclerophyllous evergreen scrub types such as maquis, garique, chaparral and mallee and characteristic of the Mediterranean regions of the world. A type of mixed broadleaved, ericaceous scrub dominated by *Buxus sempervirens* and *Erica arborea* represents a transitional (Ger. *Übergang*) community to the next formation, Ericifruticeta, a formation of dwarf heath types. The formation groups are united into formation classes based on physiognomy and function,

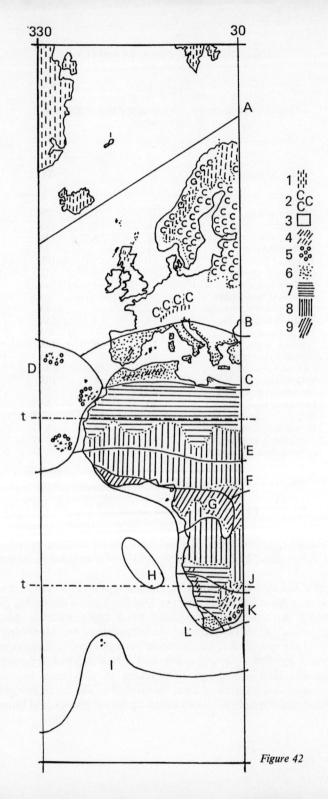

Figure 42

TABLE 35

The classificatory position of the Irish *Arbutus* stands according to Rübel (1930)

LIGNOSA

Formation-class DURILIGNOSA—sclerophyllous woodland and scrub

  Formation-group DURIFRUTICETA—sclerophyllous scrub

    Alliance ARBUTION—Formation—Mediterranean maquis

      Associations: 1. Irish ARBUTETUM unedo with *Ilex aquifolium* and *Calluna vulgaris*
2. Corsican ARBUTETUM with *Erica arborea Phillyrea media* and *Cistus* spp.
3. Dalmatian ARBUTETUM with *Juniperus* spp. and *Phillyrea media*
4. East Mediterranean ARBUTETUM of mixed *A. unedo* and *A. andrachne*

---

rather than structure. Thus the Durifruticeta is joined with the Durisil-vae into the formation class Durilignosa (Table 35), whilst all woody vegetation types are united into a basic structural group, Lignosa. In all Rübel (1930) recognizes twelve formation classes which are defined by structural, physiognomic, edaphic and climatic factors. Nine of these are related directly to climate and are subsequently mapped by him on a world scale. The nine climatic formation classes are as follows:

1. *Pluviilignosa*—tropical and sub-tropical evergreen rain forests.
2. *Laurilignosa*—sub-tropical and warm temperate soft-leaved evergreen forests.
3. *Durilignosa*—Mediterranean, sclerophyll woodlands and scrub.
4. *Hiemilignosa*—deciduous, raingreen tropical and sub-tropical forests.
5. *Duriherbosa*—permanent tall savannah, prairie and pampas grass-lands.
6. *Sicciderseta*—dry sand deserts.
7. *Aestilignosa*—summergreen, deciduous temperate forests.
8. *Aciculilignosa*—evergreen and deciduous needle-leaf forests.
9. Treeless polar and alpine vegetation.

Figure 42. Global transect showing correlations between Good's floristic regions and Rübel's formation-classes.

A, Arctic and Subarctic; B, Euro-Siberian; C, Mediterranean; D, Macronesia; E, African–Indian desert; F, Sudanese park steppe; G, West African forest; H, Ascension and St Helena; I, South temperate oceanic isles; J, East African steppe; K, South African steppe; L, Cape.

1, Polar and Alpine vegetation; 2, Aciculilignosa; 3, Aestilignosa; 4, Duriherbosa; 5, Laurilignosa; 6, Durilignosa; 7, Siccideserta; 8, Hiemilignosa; 9, Pluviilignosa.

These major groups are mapped in the longitudinal transect of the earth between 0° and 330° longitude and 75°N–45°S shown in Figure 42. As an understanding of plant geography was one of the basic aims of this early European school, the floristic divisions of Good (1947) are superimposed on the formation class distributions of Rübel to illustrate the relationships of vegetation and floristics to climate. Details of vegetation types representative of the main formations and formation classes for each of the floristic regions are to be found in Appendix II where the Rübel-type of 'geographical' vegetation description is followed.

From the foregoing work a formation class thus seems to be a geographical vegetation unit which shows a characteristic response to a type of climate or climatic trend. This has been the trend in the production of maps of the world's vegetation although the numbers of formation classes recognized varies between nine and thirty-nine. Apart from the work of Rübel (1930) the most widely accepted classification is that of Schimper and von Faber (1935) who recognize fifteen climatic formations types (= formation class). Whereas Rübel tends to map the natural climatic climax forest types or the potential natural vegetation, Schimper and von Faber differ in presenting finer divisions which are defined more strictly in terms of climate and also by structural aspects of existing vegetation. Thus the area of Africa mapped by Rübel as *Hiemilignosa* is divided by Schimper and von Faber (as re-drawn by Dansereau, 1952) into three formation classes—savannah, steppe and half desert. It may be suggested that this is a progressive increase in the knowledge of vegetation distribution and climatic change, but to map the whole of upland Britain north of the Tees-Exe line as needle-leaf forest and Ireland south of the Sligo-Wicklow line as heath cannot be put forward to uphold this suggestion.

These differences outline the whole problem of plant geographical vegetation description and classification. How much emphasis do we place on climatic factors? How much on edaphic and structural factors? At one end of the scale we find Rübel, at the other Fosberg (1961, 1967) with his 31 formation classes, 62 formation groups and 193 formations; and this is a classification for general purposes! Somewhere in between these two extremes the latest classification of world vegetation is provided by Ellenberg and Müller-Dombois (1966) whose physiognomic-ecological classification is based on the works of both Rübel, and Schimper and von Faber.

The Clementsian Tradition

Of all people, Clements (1928) was perhaps the most certain of what the definition of a formation should be and how it was related to climate. He states emphatically (1928, p. 128) that the formation is an organic entity which arises, grows, matures and dies, existing in a definite region marked by a particular climatic climax. Hence, the vegetation description methods of Clements always follow a pattern aimed at recognizing organismal entities within this climatic climax. Referring back to Table 30 and applying this North American system to Europe, the climax woodland types of Killarney will fit into the Deciduous Forest—*Quercus-Fagus* formation. Further classification of the Killarney woodland types is shown in Table 36. The climax association is in mixed *Quercus* woodland characteristic of acidic soils in western Europe, *Quercium robori-petraeae*. The monodominant *Quercus petraea* woodlands of the Upper Lake and Derrycunihy sites represent a consociation Quercetum, whilst mixed *Quercus-Ilex* stands form a mictium. Various societies are also present. The limestone *Quercus-Taxus* woodlands of the Muckross peninsula would be regarded by Clements as a pre-climax and like the seral *Arbutus* stands would be classified according to the associes-consocies-socies nomenclature. Examples are given in Table 36.

TABLE 36

A Clementsian classification of the woodland types at Killarney, Co. Kerry

---

DECIDUOUS FOREST: QUERCUS-FAGUS FORMATION—temperate deciduous woodlands.

1. Climax ASSOCIATION   Quercium robori-petraeae—west European oak woodlands on acidic soils

      CONSOCIATION   Quercetum petraeae—woodland dominated by *Quercus petraea*—Derrycunihy and Upper Lakes
          Querco-Ilicetum-mictium (co-dominance of *Quercus* and *Ilex aquifolium*)

      SOCIETIES: SHRUB   Vacciniile and Rhododendrile societies of *Vaccinium myrtillus* and *Rhododendron ponticum*
          CRYPTOGAM   Hymenophyllile-Hypnile societies of *Hymenophyllum* spp. and hypnaceous mosses

2. Pre-Climax ASSOCIES   *Quercus-Taxus* woodland on the Muckross Peninsula
      CONSOCIES   *Taxus*-dominated stands (Taxies)
      SOCIETIES: SHRUB   *Corylus* society (Corylule)
          HERB   *Brachypodium sylvaticum* society (Brachypodiule)
          CRYPTOGAM   *Isothecium myosuroides* society (Isotheciule)

3. Seral ASSOCIES   *Arbutus-Ilex* scrub on Muckross Peninsula
      CONSOCIES   *Arbutus*-pure stands (Arbuties)
      SOCIETIES: SHRUB   *Cotoneaster* society (Cotoneastiule)
          Clematis society (Clematiule)
          MOSS   *Neckera* society (Neckeriule)

Because the developmental approach is stressed by Clements, there is no formal floristic approach to vegetation description, other than the floristic analysis of what species are dominants in the consociations, societies, associes, etc. In general, the formation is regarded as a physiognomic group independent of the flora. Applications of Clementsian principles have been mainly in the Americas, and all applications seldom adhere to all of Clements's principles of developmental vegetation. Good examples of applications are to be found in the work of Braun (1947) in her work on the development of the deciduous forests of eastern North America. She recognizes nine climax-types in eastern United States, giving analyses of the history and development of the forests and indicating their relationship to climate.

As a detailed application of the Clementsian system the work of Phillips (1930) on the vegetation of central Tanganyika is exemplary. He recognizes two formations which he then proceeds to analyse in terms of the seral stages of their xeroseral and hydroseral development. One most important feature of his work is a detailed synonymy of the communities he describes, where all previous descriptions of the vegetation types are listed. Following this, a section on floristic and community features accounts for features of physiognomy, structure and floristic lists of the major components of the most prominent consociations and associations. This is perhaps the best example of the application of the Clementsian system *sensu stricto* outside the Americas.

The British Tradition

The British Tradition is characterized by two main periods of investigation; the early work of the members of the British Vegetation Committee of which Moss was the most influential, and which produced *Types of British Vegetation* in 1911; the 1920–40 period where Clementsian categories and developmental approaches were applied with the early work of Moss as a framework, producing a culmination in *The British Isles and their Vegetation*. As it has been explained previously Moss (1910, 1913) based his work on edaphic formations. Thus he described the 'plant formation of calcareous soils' (Calcarion) the 'plant formation of siliceous soils' (Silicion), etc., in his monograph on the vegetation of the Peak District. Within each formation he recognized the progressive succession of different plant communities or 'subordinate associations' towards a 'chief association'. Similarly, the chief association could degenerate to form a 'retrogressive succession' of subordinate associations back to the pioneer stages. Moss's ideas are illustrated in Figure 43, a composite series for the Calcarion of Ireland using the author's own data from Killarney, Galway, and the

Burren, Co. Clare. The Killarney calcareous *Quercus-Taxus* woodlands are postulated as edaphic climax, while the *Arbutus* and *Corylus* scrub represent progressive subordinate associations. The *Corylus* scrub-woodland of the Burren probably represents a retrogressive step from the *Quercus-Taxus* climax.

The second stage of the British Tradition is marked by the excellent work of Watt (1924, 1925, 1934a), on the structure and development of British beechwoods. He follows the Clementsian system, closely

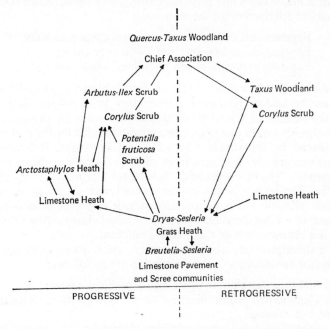

Figure 43. Associations of the Calcarion in Ireland: Moss's method.

classifying the types of woody vegetation of the South Downs. Cots-wolds and Chilterns according to their successional relationships and regarding the pre-climax stages as transitory, culminating in a *Fagus* climax consociation, thus:

GRASSLAND—→SCRUB ASSOCIES—→ASH-OAK
    ASSOCIES—→BEECH ASSOCIES (Pre-climax)—→BEECH
                                    CONSOCIATION

Detailed examinations of the seres revealed minor variations between units and Watt (1924) could thus erect a series of parallel seres, which had fairly distinctive soil types. For example, the pH varied between

5·5 and 7·8 for the five parallel seres, there was a marked variation in loam and humus depths and the mechanical analyses also proved distinctive. The relationships of the habitat as distinguishing features of the parallel seres were further highlighted in Watt's work on the Chiltern beechwoods (1934a, b). Throughout the work, Watt's main consideration was the description of the developmental aspects of the climax consociation and as such he paid little or no attention to classification beyond this level. It was left to Tansley (1939) to classify the three major types and give them association names which reflected the habitat relationship of the woods, e.g.

> Fagetum calcicolum—beechwood on calcareous soils;
> Fagetum rubosum—on deep loams over chalk;
> Fagetum arenicolum—on podsolized sands and loams.

Although Tansley (1939, p. 232) sets out the hierarchy which Clements follows, he seldom reaches a classificatory level beyond association and the suffixes of Clementsian nomenclature are for the most part noticeably absent. In fact, the Clementsian-*ium* suffix for the association is replaced by the standard -*etum* of earlier works. References to formations are also few, and when they do occur, e.g. 'Moss or Bog Formation', 'Heath Formation', 'Salt-marsh Formation', they clearly indicate Tansley's rejection of the monoclimax theory in favour of edaphic, climatic and biotic polyclimaxes. It may thus be evident that the lack of a preoccupation with a hierarchical classification of vegetation is a characteristic of the British Tradition.

The descriptive aspects are, however, much more thorough. The five major components of the description are as follows:

(*a*) Soil type description, often with a profile diagram;

(*b*) Community profile diagram to show vegetation stratification, height and zonation;

(*c*) Structural data such as numbers of trees per acre, maximum and mean height and girths at breast height of the leading dominants;

(*d*) Complete floristic list of all species with each species given an assessment of abundance (see Table 24);

(*e*) Details of observed or postulated succession within the vegetation types described.

A certain amount of information on the Killarney oakwoods is already available through the work of Turner and Watt (1939). Their work is restricted to the acidic woodlands on old red sandstone in Derrycunihy Wood and on Arbutus Island in the Upper Lake, so that the present data from the limestone Muckross Peninsula present an interesting comparison. The data comprise a transect from the water's edge illustrated in Figure 44 as a profile diagram and the table, Table 37

TABLE 37

Woodland transect at Muckross, Killarney: British method

	1	2	3	4	5
Arbutus unedo	d	d	–	–	–
Betula pubescens	r	–	–	f	–
Corylus avellana	o	r	o	va	a
Fraxinus excelsior	r	–	f	f	a
Ilex aquifolium	r	f	f	a	a
Quercus petraea	r	r	–	co–d	co–d
Taxus baccata	o	f	d	co–d	co–d
Sorbus aucuparia	–	–	r	f	o
Clematis vitalba	a	–	–	–	–
Cotoneaster microphylla	f	a	–	–	–
Crataegus monogyna	–	–	–	r	–
Hedera helix	–	o	–	o	o
Lonicera periclymenum	–	–	o	o	o
Prunus spinosa	–	–	r	o	o
Rhamnus catharticus	–	–	–	r	–
Rosa canina	r	–	–	–	–
Rubus fruticosus agg.	o	o	o	o	o
Sorbus hibernica	o	r	–	–	–
Asplenium ruta-muraria	r	–	–	r	–
A. trichomanes	–	r	–	–	–
Brachypodium sylvaticum	a	f	f	a	a
Calluna vulgaris	–	–	f	–	–
Carex flacca	–	r	–	–	–
C. sylvatica	–	–	–	o	f
Erica cinerea	–	r	o	r	–
Euphorbia hyberna	–	–	–	f	f
Euphrasia offinalis agg.	r	–	–	–	–
Festuca rubra	–	–	f	o	–
Fragaria vesca	–	f	f	a	a
Geranium robertianum	r	–	–	–	–
Glechoma hederacea	r	–	–	–	–
Hymenophyllum wilsonii	–	–	–	r	r
Hypericum pulchrum	r	–	o	o	–
Oxalis acetosella	–	–	–	–	f
Phyllitis scolopendrium	–	–	o	o	–
Polystichum setiferum	–	r	–	–	–
Prunella vulgaris	–	–	–	o	–
Pteridium aquilinum	–	o	–	–	–
Rubia peregrina	f	f	–	o	–
Sieglingia decumbens	f	–	–	–	–
Solidago virgaurea	f	r	o	f	–
Teucrium scorodonia	f	f	f	a	–
Thymus drucei	r	–	–	–	–
Vaccinium myrtillus	–	–	–	f	–
Viola riviniana	o	–	f	a	f
Bryophytes	o	a	va	va	va
Epiphytes	r	r	f	f	a

which gives analyses of 10 m² quadrats along the transect. Three scrub-woodland communities are recognizable:

(i) *Arbutus-Ilex* scrub associes—abutting on to a *Calluna-Ulex gallii* heath near the water's edge and characterized by the co-dominance of the two evergreen species. The shrub-stratum is approximately 5 m tall and cover varies between 75 and 90%. The shrubs are rooted in humus-rich skeletal soils in pockets of the limestone pavement which is often bare, with a ground cover of 50%. Alternatively, bryophytes form a 70–100% ground cover with little exposed rock. Average girth at breast height of *Arbutus* 26 cm; *Ilex* 30 cm.

(ii) *Taxus* consocies—*Taxus* dominant, height to 6–8 m, with some subsidiary *Ilex* and *Fraxinus* and *Corylus* to 2 m in damper areas. Herb cover around 20% with some pronounced *Brachypodium* societies; moss cover 80%. Average girth at breast height of *Taxus* 58 cm. *Corylus* and *Taxus* regeneration.

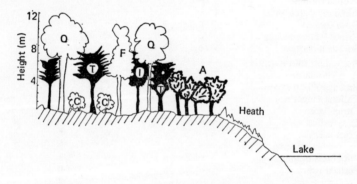

Figure 44. Profile diagram of the Killarney woodlands on limestone. Q, *Quercus*; T, *Taxus*; F, *Fraxinus*; I, *Ilex*; C, *Corylus*; A, *Arbutus*.

(iii) *Quercus-Taxus* association—these two species co-dominant at 10–12 m with *Fraxinus*, and 7–9 m with *Ilex* respectively; intermittent scrub layer of *Ilex* and *Corylus* to 3 m; herb cover 10%, moss cover 100%. Mean *Quercus* girth 112 cm; *Taxus* 64 cm. Some *Taxus* regeneration.

The most pronounced layer societies in all five stands are bryophyte societies several of which are listed below:

(*a*) *Rhytidiadelphus loreus—Thuidium tamariscinum—Thamnium alopecurum* associule of closed woodland on larger boulders and seen in stands 4 and 5;

(*b*) *Isothecium myosuroides—Plagiochila* spp. associule of tree bases (mainly *Quercus* and *Fraxinus* and more or less absent from *Taxus*);

(c) *Ulota—Frullania—Metzgeria* associule of branches and twigs in all zones, though often lacking *Ulota* or *Taxus* and *Arbutus*;

(d) *Isothecium myosuroides—Ctenidium molluscum* associule of limestone pavements;

(e) *Neckera complanata* associule in the more open areas of stands 2 and 3;

(f) *Breutelia—Tortella tortuosa* open limestone associule in open areas of stands 3 and 4.

The observed and postulated succession follows the zonation and is shown in Figure 43 where the succession of Turner and Watt (1939) is

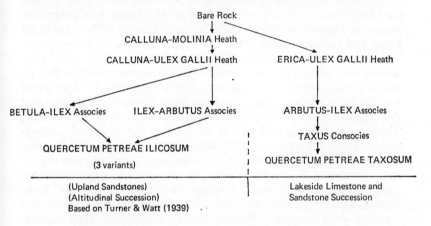

Figure 45. Woodland succession in the Killarney Lakes region.

also reproduced. The two woodland types can be seen to have the same basic succession, proceeding from *Calluna-Ulex* heath through an *Ilex-Arbutus* associes to the climax type. The two successions differ mainly in the presence of a *Taxus* consocies in the limestone sere and a *Betula-Ilex* associes in the more heathy areas of the sandstone sere. The British Tradition would here recognize two edaphic climax types *Quercetum sessiliflorae ilicosum* on acidic soils and *Quercetum sessiliflorae taxosum* on calcareous soils. But again this may not represent the true picture since Turner and Watt (1939) figure a zonation on sandstone on Arbutus Island in the Upper Lake which proceeds as far as a low *Quercus-Taxus* wood, and this points to the fact that like the *Arbutus* the *Taxus* is restricted to a narrow zone around the lakes by microclimatic features. The problem thus presents itself at all levels from macro- to microclimate—edaphic or climatic climax, monoclimax or polyclimax?

The Tropical Tradition

Practically all possible methods of vegetation description and classification have been applied to the vegetation of the tropics. On the whole, a floristic approach has tended to be regarded as of secondary importance in these regions because of the time available, the magnitude of the problem and the numerous features of climate and vegetation antagonistic to detailed recording. Consequently, most approaches involve the use of basic structural, functional and physiognomic features. In the early years, Bews (1917) followed Warming in the definition of formations in sub-tropical Natal and later (1920) applied the Clementsian system. Similarly, Phillips (1930) used the Clementsian system in Tanganyika but on the other hand Chipp (1927) in the Gold Coast and Gillman (1936) in East Africa used a polyclimax interpretation. It was probably Chipp (Tansley and Chipp, 1926) more than anyone who influenced later workers with his recommendation that an intital physiognomic approach to tropical vegetation in description was the most realistic and rewarding. Burtt-Davey (1938) elaborated this aspect in a treatise on tropical woody vegetation types and Richards, Tansley and Watt (1940) further emphasized the structural-physiognomic-functional approach, submitting a list of seventeen main features of tropical forest which need to be investigated if an adequate vegetation description is desired (Table 38). In addition, they added five site-location, seven climatic features, five edaphic and six biotic features along with historical and developmental features which they consider useful secondary characteristics for description. Their essential consideration is that vegetation should be characterized by its own inherent features and not primarily by secondary environmental features such as climatic, edaphic and biotic factors. The field sample is regarded as a representative stand of an association or consociation which is characterized by its structure, physiognomy and total floristic composition, although some species will tend to have a greater indicator value for characterization. Several associations or consociations may be shown to agree in the physiognomy and life-forms of their component species, especially their dominants. Richards *et al.* consider that this vegetation similarity indicates similarity of what they term 'essential habitat'—especially regional climate. Such associations may be united into a formation. Often, within a region of uniform climate, edaphic and biotic factors may prevent the development of typical climatic vegetation by immediately affecting the essential habitat. In this case, associations agreeing in life form and physiognomy are termed edaphic or biotic formations. Finally, formations which exhibit a characteristic similarity of life-form obviously determined by similar climates, irrespective of the geographical location may be united into a formation type. Thus, the tropical rain forests of West Africa, South America and Indonesia may be united into a single formation type characterized by total habitat.

TABLE 38

Features of tropical rain forest useful for forest descriptions
(adapted from Richards, Tansley and Watt, 1940)

Structure
1. Canopy type—open or closed
2. Tree spacing—uniform or irregular with inter-tree distances
3. Stratification—layers enumerated
4. Layer description with height of foliage noted
5. Notes on layer societies

Physiognomy and life form
6. Height and distribution of lianas and epiphytes
7. Trunk characters including data on plank butresses, stilt roots cauliflory bark-type and stem succulence
8. Notes on special life forms in various layers, e.g. tree ferns, cycads
9. Evergreen or deciduous trees; if mixed, then percentage of each
10. If deciduous, time of leaf and leaf-fall
11. Leaves simple or compound
12. Leaf size—use of Raunkiaer's size classes
13. Life forms of field and ground layers
14. Periodicity of field layers, e.g. seasonality
15. Seed production and dispersal type of major trees
16. Vegetative propagation of trees and shrubs
17. Floristic composition of each layer listed separately

Basic site details
1. Community name (from dominants or characteristics)
2. Vernacular name if any
3. Location of stand
4. Sample size and type—not less than an acre square or a belt transect
5. Altitude, exposure, aspect and slope

This approach to classification is adopted by Beard (1944, 1955) in his work on tropical American vegetation types. As a method of description of the representative vegetation of a certain formation he presents 'type specimens' as profile diagrams of the type used by Richards *et al.* (1940). This is accompanied by a type description, adapted from Fanshawe (1952) and analogous to the holotype description of a Linnaean species.

As a representative of the Mixed Deciduous Forest-Formation of Europe, the Muckross woodlands might thus be described by a three-storied woodland with the canopy more or less closed between 10–12m composed of deciduous *Quercus* and *Fraxinus*, a discontinuous understory between 7–9 m, composed mainly of evergreen *Taxus* and *Ilex*; and an intermittent scrub layer approximately 2 m tall composed mainly of *Ilex* and *Corylus*. Illumination at ground level varies with cover of evergreens and the atmosphere is humid. Herb species are rare and constitute about 10% cover; cryptogams are abundant and cover 90–100% of the uneven limestone pavement surface. The

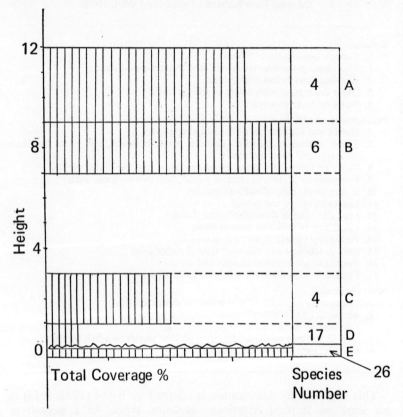

Figure 46. Cover-stratification diagram of the Killarney woodlands.

Quercus trees are somewhat stunted and gnarled and the *Taxus* infected with a black fungus on the trunks which is apparently affecting its vitality.

Beard (1955) follows Richards *et al.* in recognizing his associations as floristic groupings, the formation as a physiognomic grouping and the formation series (type) as a total habitat grouping. The association must bear a floristic name, such as '*Quercus-Taxus* association', the formation a physiognomic name such as 'Mixed Deciduous Forest' and the formation series a total habitat name such as 'Lowland Atlantic Formation'. But, however, this rule can and is broken when well-established names are included into the classification scheme, e.g. Rain Forest.

When it comes to the actual description of tropical forest stands, most modern authors chose all or some of the features enumerated by Richards *et al.* (1940) for their basic approaches. Examples are numerous, but two 'schools' are outstanding in illustrating a sound structural-

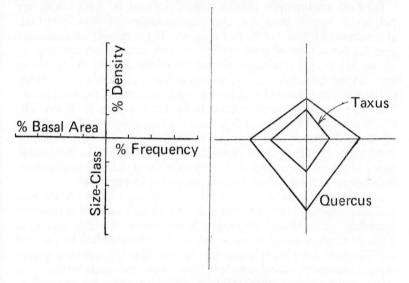

Figure 47. Lutz phytographs.

physiognomic approach. The work of Richards (1952) provides an excellent example of what may be called the traditional approach, whilst the presentation of Loveless and Asprey (1957) gives a good example of the approach to the description of tropical forests other than rain forest. The five main components of their description are: (i) the construction of an accurate profile diagram; (ii) notes on the periodicity of layers with coverage data; (iii) life-form analysis-data as both species and frequency spectra; (iv) leaf size and texture classes; (v) Floristic composition with frequency data based on the analysis of twenty random 5 × 5 m quadrats.

The second school, called the American Tropical Tradition for convenience, follows a similar pattern of description. As a type example there is none better than the works of Cain *et al.* (1956) and Cain and Castro (1959). The basic analysis methods of these phytosociologists are outlined in Figures 46, 47 and Table 39 based on data collected in the *Quercus-Taxus* association at Muckross in twenty 5 × 5 m quadrats.

(i) Cover-stratification diagrams (Figure 46) are a simple graphic representation of the horizontal layers of vegetation and may be used as an alternative to the profile diagram. Coverage may be expressed either as a total percentage figure or according to standard cover classes relative to a vertical scale of layer heights. The diagram is completed by adding the number of species in each stratum on the right-hand side.

(ii) Lutz phytographs (Figure 47), introduced by Lutz (1930) are polygonal figures used for the representation of four structural characteristics for each of the major species. The standard four characters are: (*a*) percentage of total density of trees over 25 cm diameter at breast height; (*b*) frequency percentage of the species of the above size; (*c*) the occurrence of the species in five size classes (see below); (*d*) dominance expressed as a percentage basal area of the total figure. The following size classes were used by Lutz: (1) up to 30 cm tall; (2) 30 cm–4 m; (3) saplings 2–8 cm d.b.h; (4) poles 10–25 cm d.b.h.; (5) 25 cm d.b.h and over. This arm of the polygon is often modified by expressing the distribution through the size classes as a percentage figure of the maximum size of the species in the stand. The phytographs for *Quercus petraea*, *Taxus baccata* are shown in Figure 47.

(iii) Floristic frequency, life form and leaf size (Table 39) are conveniently expressed in table form dealing with each stratum in turn and recording species twice if they occur in two strata. The table enables at a glance a conclusion that the woodlands are characterized by *Quercus* and *Taxus* as the most frequent tree species; that the trees are mainly mesophanerophytes and mesophyllous with the predominance of microphyllous hemicryptophytes in the herb layer and a dense bryophyte carpet. The life-form spectrum, expressed for the species composition and the frequency percentage is as follows:

MP	NP	Ch	H	T	
7	3	6	10	0	Species numbers
45	9	14	32	0	Frequency percentage

The latter figures give a much more realistic representation.

The Symbolic Tradition

Within this tradition, two trends may be recognized: (i) the British–Australian trend beginning with the work of Stamp (1931, 1934) and

TABLE 39

Life form, leaf size and floristics: Tropical American Tradition

	F%	LF	LS
A. *Tall tree stratum*			
Quercus petraea	90	MP	Me
Taxus baccata	80	MP	N
Fraxinus excelsior	15	MP	Me
Betula pubescens	10	MP	Me
B. *Low tree stratum*			
Taxus baccata	40	MP	N
Ilex aquifolium	20	MP	Me
Quercus petraea	10	MP	Me
Sorbus aucuparia	5	MP	Me
Betula pubescens	5	MP	Me
Fraxinus excelsior	5	MP	Me
C. *Shrub stratum*			
Corylus avellana	45	MP	Me
Ilex aquifolium	20	MP	Me
Prunus spinosa	15	N	Mi
Rubus fruticosus	15	N	Me
D. *Herb stratum*			
Brachypodium sylvaticum	55	Hs	Mi
Carex sylvatica	50	Hs	Mi
Oxalis acetosella	50	Chh	Mi
Lonicera periclymenum	45	N	Me
Hedera helix	30	Chw	Me
Fragaria vesca	30	Hr	Mi
Euphorbia hyberna	30	Hp	Mi
Teucrium scorodonia	30	Hp	Mi
Viola riviniana	30	Hs	Mi
Erica cinerea	10	Chw	L
Solidago virgaurea	10	Hp	Mi

(+ 6 other spp. at frequency 5%; 3Ch, 3H)

E. *Moss stratum*

Isothecium myosuroides	90	
Eurynchium praelongum	85	*Life-form categories*
Rhytidiadelphus loreus	70	MP —Micro- and mesophanerophytes
Plagiochila asplenioides	70	N —Nanophanerophytes
Neckera complanata	65	Chh—Herbaceous chamaephytes
Thuidium tamariscinum	65	Chw—Woody chamaephytes
Ctenidium molluscum	40	Hr —Rosette hemicryptophytes
Hylocomium brevirostre	30	Hp —Protohemicryptophytes
Dicranum scoparium	30	Hs —Semi-rosette hemicryptophytes
Rhytidiadelphus triquetrus	15	
Thamnium alopecurum	15	*Leaf-size categories*
Porella platyphylla	15	Me —Mesophyll
Polytrichum formosum	10	Mi —Microphyll
Breutelia chrysocoma	10	L —Nanophyll

(+ 12 other spp. at frequency 5%)   N —Leptophyll

continuing mainly in C.S.I.R.O. Australia in the work of Christian and Perry (1953) and Heyligers (1965); (b) the American trend involving the detailed works of Küchler (1949, 1967) and Dansereau (1951, 1957, 1961).

(i) Stamp was perhaps the first person to suggest the use of shorthand formulae for the rapid precis description of vegetation. He suggested that the types of plants to be enumerated should be given single letters thus: trees A (from the Latin *arbor*); shrubs F (Latin *frutex*); herbaceous plants H (from *herba*); grass G (Latin *gramen*); cryptogams C. A basic vegetation formula would thus be: $xA + yF + zH + uG + vC$ where x, y, z, u and v are the number of individuals per hectare, with the approximate height of each type appended in brackets. A simple formula such as $250A(10) + 300F(2) + \alpha G$ refers to an open woodland type with approximately 250 trees per hectare average height 10 m and 300 shrubs, height 2 m. Elaborations and refinements of this rough formula are readily introduced: (a) for trees and shrubs in more than one storey record $A + A^1$ and $F + F^1$; (b) the character of the trees or shrubs may be indicated by suffixes such as s—sempervirens or evergreen, d—deciduous, c—coniferous, etc., in the square position, e.g. A^s; (c) the letters a, b, c, d, etc., may be used to indicate dominants and x for numerous species. The formula $250\ A^{dc}\ ab(10) + 200\ A^{1c}$ $b(8) + 300\ F^{ds}\ cd(2) + Hx + Gx + Cx$ where a = *Quercus*, b = *Taxus*, c = *Corylus* and d = *Ilex* represents the *Quercus-Taxus* woodlands at Muckross!

After minor modifications in 1934, the formulae of Stamp do not seem to have come into general usage, and the next type of formulae to be described seems to be those of Christian and Perry (1953) for use in a series of vegetation surveys in northern Australia. In this formula, letters and figures are assigned to the tree, shrub and herb strata in an attempt to combine and convey a picture of stratification abundance and spatial distribution. Tall trees are scored A_3, medium-sized A_2 and small trees A_1, and similarly B_3, B_2, B_1, C_3, C_2, C_1, for the three sizes of shrubs and herbs. Heights are recorded for each layer as average values, e.g. A_2^{12}—a medium-sized tree, 12 m tall. Density is also recorded by adding x, y or z for dense, average or sparse with double values for very dense, and very sparse. Muckross would thus receive the formula

$$A_2^{12}\ y,\ A_1^8\ x\ B_1^2\ y,\ C_1^{0\cdot5}\ z$$

One of the most recent developments of structural vegetation formulae has been used by Heyligers (1965). This system represents an alternative to that of Christian and Perry (1953) and is partly a reaction against the more complex systems of Küchler and Dansereau (q.v.). It is essentially a realistic system for vegetation analysis in the field and in aerial photograph interpretation. The three main properties

TABLE 40

Structure symbols of Heyligers (1965)

Main life forms
  H—Herbaceous plants
  W—Woody plants subdivided into
      T—trees
      S—shrubs

  L—lianes
  G—graminoids

Height	T	S	H
v —very tall	> 30 m	> 5 m	> 2 m
t —tall	20–30 m	2–5 m	1·3–2 m
m—medium	10–20 m	1·3–2 m	0·6–1·3 m
l —low	< 10 m	< 1·3 m	< 0·6 m
p —very low, pygmy	< 5 m	< 0·6 m	< 0·3 m

Cover. Six groups to each of which the symbol 'g' may be added to signify grouping or clumping

c—continuous, very dense	76–100% cover
d—dense	51–75%
i —interrupted, discontinuous	26–50%
S—scattered	5–25%
b—nearly barren, very scattered	< 5%
o—odd individuals	–

analysed are the main life forms, height and cover, each being subdivided and given a symbol letter (Table 40). In construction the formulae are virtually the same as in previous examples except that the stratum of highest cover is underlined and downward arrows indicate the main species components of each stratum, thus:

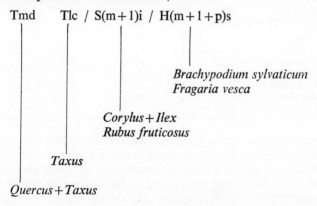

Tmd   Tlc / S(m+1)i / H(m+1+p)s

Brachypodium sylvaticum
Fragaria vesca

Corylus+Ilex
Rubus fruticosus

Taxus

Quercus+Taxus

(ii) The American trend in the symbolic description and representation of vegetation, like the latter trend, is mainly aimed at the graphic

G

TABLE 41

Vegetation structure symbols of Küchler (1949, 1966)

Life-form categories					
Basic life forms	*1949*	*1966*	*Special life forms*	*1949*	*1966*
Broadleaf evergreen	B	B	Epiphytes	e	X
Broadleaf deciduous	D	D	Lianas	j	C
Needleleaf evergreen	E	E	Stem succulents	k	K
Needleleaf deciduous	N	N	Tuft plants	y	T
Aphyllous	O	O	Bamboos	v	V
Semi-deciduous (B+D)	–	S	Cushion plants	q	–
Mixed (D+E)	–	M	Palms	u	–
			Aquatic vegetation	w	–
Herbaceous plants					
Graminoids	G	G	*Leaf characteristics*		
Forbs	H	H	hard (sclerophyll)	–	h
Lichens, mosses	L	L	soft	–	w
			succulent	–	k
			large (400 cm²)	–	l
			small (4 cm²)	–	s

Structural categories			
Height (stratification)	*1966*	*1949*	
>35 m	8	t { tall: minimum tree height 25 m	
20–35	7	minimum herbaceous height 2 m	
10–20	6	m { medium: trees 10–25 m	
5–10	5	herbs 0·5–2 m	
2–5	4	l { low: trees to 10 m	
0·5–2	3	herbs to 0·5 m	
0·1–0·5	2	s —shrubs, minimum height 1 m	
<0·1	1	z —dwarf shrubs, maximum height 1 m	

Coverage (1967) *Density* (1949)	*1949*	*1966*
continuous (>75%)	c	c
interrupted (50–75%)	i	i
parklike, patchy (25–50%)	p	p
rare (6–25%)	r	r
barely present (1–5%)	b	b
almost absent (<1%)	–	a

representation of vegetation types on maps. Küchler (1949) was probably the first American to introduce symbols for this purpose and in 1967 he was able to conclude that the application of these methods to mapping at all scales had been useful in indicating trends and needs within the subject. Unfortunately, like many other investigators before him, Küchler (1966) falls foul of the revision trend so that his early scheme (1949) becomes considerably altered in his 1966 paper.

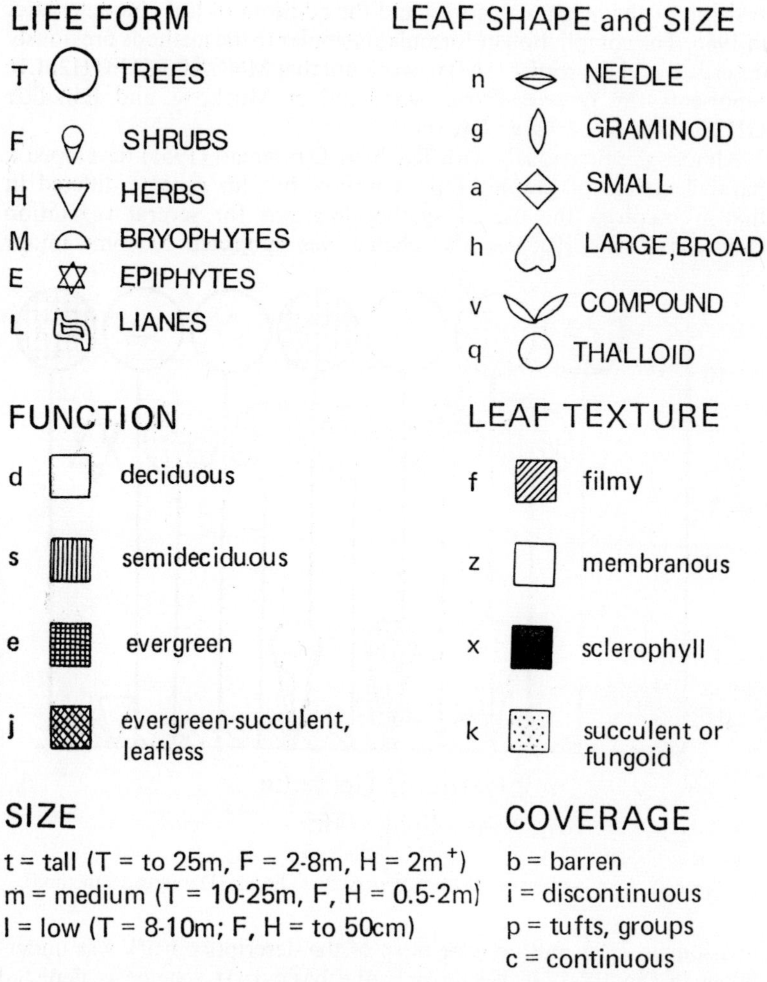

LIFE FORM

T	◯	TREES
F	♀	SHRUBS
H	▽	HERBS
M	◠	BRYOPHYTES
E	✡	EPIPHYTES
L	⬡	LIANES

LEAF SHAPE and SIZE

n	⬭	NEEDLE
g	()	GRAMINOID
a	◇	SMALL
h	♡	LARGE, BROAD
v	∨∨	COMPOUND
q	◯	THALLOID

FUNCTION

d	☐	deciduous
s	▥	semideciduous
e	▦	evergreen
j	▧	evergreen-succulent, leafless

LEAF TEXTURE

f	▨	filmy
z	☐	membranous
x	■	sclerophyll
k	▒	succulent or fungoid

SIZE

t = tall (T = to 25m, F = 2-8m, H = 2m$^+$)
m = medium (T = 10-25m, F, H = 0.5-2m)
l = low (T = 8-10m; F, H = to 50cm)

COVERAGE

b = barren
i = discontinuous
p = tufts, groups
c = continuous

Figure 48. Vegetation function-structure symbols of Dansereau (1951).

Table 41 attempts to clarify the picture by comparing the two schemes. The characteristics for analysis and representation are divided into the broad groups of life-form and structural categories and these are subdivided into basic life forms, special life forms and leaf characteristics, and height and cover respectively. The individual characteristics will either be self-explanatory from the figure or failing this, reference to them will be found in the preceding sections of Chapter 3. The main differences between the 1949 and 1966 schemes can be seen to be the revision of the height categories and the addition of leaf characteristics in 1966. The compilation of formulae is similar to the methods previously described and the reader is left to work out that M6iCXE5cD3iGH2pL1c represents the *Quercus-Taxus* woodland at Muckross and E4hcD2r GH2rL1c(b) the *Arbutus-Ilex* scrub.

Almost simultaneously with Küchler, Dansereau (1951) developed a parallel system of graphic representation but his scheme differed in that it involved the use of symbol-diagrams for several vegetation characters. Like Küchler his scheme was subjected to some minor

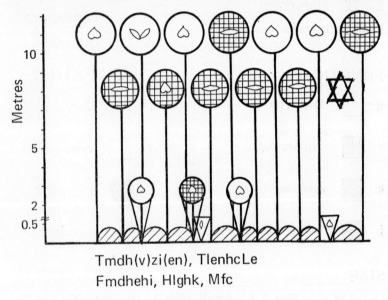

Tmdh(v)zi(en), TlenhcLe
Fmdhehi, Hlghk, Mfc

Figure 49. Profile diagram of the Killarney woodlands (Dansereau's method).

revisions in 1961, but because most of the descriptive work was undertaken in the 1950–60 decade only the basic 1951 scheme is depicted in Figure 48. Again, all the characteristics appear self-explanatory, with the possible exception of some of the leaf-texture characters. For example, *filmy* is a character reserved mainly for the filmy-ferns (*Hymenophyllum*) and most leafy mosses; *membranous* contains the bulk of the flowering plants; *sclerophyll*—hard-leaved evergreens such as *Ilex* and *Arbutus*; and *succulent* or *fungoid* combines the succulent leaf texture of *Euphorbia*, *Sedum*, etc., with thallose liverworts and lichens, e.g. *Marchantia* and *Lobaria*.

The graphic symbols are united into a composite picture with a height (stratification) scale along the vertical axis and an arbitrary horizontal axis where the community is described in letter symbols. Figure 49 illustrates this method of description for the same *Quercus-Taxus* woodlands at Muckross.

Because most of the symbolic methods of vegetation description are mainly aimed at use in vegetation mapping there is often no associated hierarchical classification other than that engendered by the categories themselves, e.g. evergreen broadleaf forest. Dansereau (1957) however, regards the association as his basic unit for graphic representation and classifies these into formations and formation classes after Schimper and von Faber (1935). He further groups the formation classes into four *biochores* or the main subdivisions of world environments, characterized by climatic and general environmental responses of vegetation structure, e.g. forest, savannah, grassland and desert. The choice of words here is unfortunate, since in the sense of Raunkiaer (1934) a biochore is a term used for plant climatic boundaries and is usually used in the context of changing life-form percentages.

CHAPTER 6

FLORISTIC SYSTEMS OF VEGETATION DESCRIPTION

While some authors have adopted a structural and physiognomic approach to vegetation analysis in which only the main or predominant species are noted, others have placed great emphasis upon the floristics of vegetation. This has been the case throughout much of Europe where the number of ecologists is comparatively high and the diverse regions not too inaccessible when compared to the similar situations in the Tropics or North America. Many difficulties are encountered in interpreting the European phytosociological methods especially with difficulties of language and translation and the difficulty expressed by many Americans of a lack of knowledge of the vegetation conditions in the Continent. Further misunderstandings of methods have come with the 'hybridization' of the two main systems, i.e. ecologists choosing parts of one system of vegetation description and using them in conjunction with select parts of the other. There are three main schools of phytosociology in Europe, two of which have strong followings, plus a fourth heterogeneous group which derive their methods from one of the three main schools:

1. *The Zurich–Montpellier School.* The modern bases of this school of vegetation description and classification were set down by Professor Braun-Blanquet in his treatise *Pflanzensoziologie* in 1928 which included a classification of the vegetation of the central alpine and French Mediterranean regions—based on work in institutes at Zurich and Montpellier. The work of Professor Reinhold Tüxen (Hanover) in the 1930–40 period and the subsequent publication of *Ubersicht der höheren Vegetationseinheiten Mitteleuropas* (Braun-Blanquet and Tüxen, 1943) was further developed by Braun-Blanquet in a series of papers under the general title *Die Pflanzengesellschaften Rätiens* in the periodical *Vegetatio* (1948 *et seq.*). Since this time, the school has developed rapidly in north-west Europe, mainly through the industry of Professor Tüxen and his pupils, and the annual Symposia of the International Society for Plant Geography and Ecology at Stolzenau and Rinteln, West Germany, have provided a platform for the development of a generally accepted classificatory system.

184

2. *The Uppsala School*. The fundamental concepts of this school can be traced back to von Post (1862) but major developments came in the 1920s under Professor du Rietz with further modifications since 1930. The culmination of phytosociological investigations in Scandinavia was the recent publication of *The Plant Cover of Sweden* dedicated to Professor du Rietz by his pupils.

3. *Raunkiaerian School*. Raunkiaer (1909), following up his work on life forms proposed a method of vegetation description which would express numerically the frequency of the species in the community independently of personal judgement. His aim was a method which could be used for comparative investigations of vegetation types within the confines of the same geographical flora which at the same time provided a means of incorporating his life-form system. Adherents to this system are uncommon and come mainly from Denmark and Scandinavia in general.

4. *'Hybrid' schools*. Several workers have approached the description and classification of vegetation by a 'middle of the road method' which involved the use of parts of both the Zurich–Montpellier and Uppsala Schools. The three main authors responsible for the different methods were Nordhagen (1936), Dahl (1956) in their surveys of the vegetation of Sylene and Rondane, two montane regions of Norway. In like manner Poore (1955) adopted a similar hybrid method in the description of the vegetation of Breadalbane, Scotland, which was later extended by McVean and Ratcliffe (1962) to the whole of the Scottish Highlands.

The descriptive technique of the Zurich–Montpellier (Z–M) School

A number of accounts of the basic methods of this school of phytosociology are to be found in the literature—Poore (1955), Ellenberg (1956), Küchler (1967), Becking (1957) with further refinements from Moore (1962). Not all these authors have provided a complete breakdown of the techniques, in fact, some of them are considered to be a task to which one must become apprenticed (Webb, 1954) or which must be learned from the master exponents themselves.

There are basically five steps to the description and classification of vegetation according to the Z–M School:

1. Field description.
2. Aggregation of field data into tables to represent the local variations in vegetation.
3. Checking the ecological reality of the units extracted in the field, either by simple field observations or measurement of environmental features.
4. Investigation of similar patterns in other localities thereby

obtaining an overall pattern of variation within a particular vegetation type.

5. The erection, differentiation and characterization of associations.

Field descriptive technique

The field descriptive technique cannot in the first instance be objective. The phytosociologist is frequently governed by some pre-determined choice of topic, whether geographically based, e.g. the vegetation of the Scottish Highlands, whether ecologically based—the vegetation of calcareous soils—or further, autecologically—the plant communities associated with the life cycle of *Calluna*. Each project must of necessity be delimited. Within this area of study the major step in the field technique is the choice of a uniform area for description. As Poore (1955, p. 235) states:

'This is the essential prerequisite of all phytosociological systems, and it is unfortunate that one cannot define with precision what is meant by "uniformity".'

Traditionally, uniformity and homogeneity are related to the physiognomy and ecology of the vegetation. Braun-Blanquet (1951, p. 53) commenting on the choice of uniform stands has stated that they should be 'uniform', especially in relation to floristic make up which in turn will decide the physiognomy of the community'. The stand should also be uniform with regard to relief and soil conditions in so far as it is possible to judge in the field. For example, a grassland quadrat in which there is a small rock outcrop with a number of chasmophyte species should ignore these species and list them as a separate field record. Similarly, the bryophyte-covered bases of woodland trees, or the heavily trampled areas in field gateways, are omitted from the general woodland or pasture quadrat. The uniformity of the vegetation is subjectively assessed by using all the properties which can be directly observed—the traditional *Pflanzensoziologieblick*. The 'picture' or *Aufnahme* (German) *relevé* (French) of the stand of vegetation is then recorded.

The size of the stand chosen for description will depend on the particular type of community under consideration. It must be at least the size of the minimum area of the community which for woodland is generally regarded as above 200 m² and for grassland 10 m². It is customary to carry out several minimal area calculations (see Chapter 1) before each vegetation type is described. In practice, one often finds that due to anthropogenic activities several communities are represented only as fragments, with an area less than the minimal area. If fragmentary and well-developed stands occur in the same locality, only those which attain the required minimal area are chosen for description, but if, in a second locality only fragmentary examples occur, it is legitimate to include these as relevés since they may form an important phytogeographical or ecological link with other types. The fragmentary or heterogeneous nature of such relevés will be revealed when the field records are aggregated in table form. If heterogeneity is represented by a large total species number and a fragment by a low species number when compared to the normal situation, then the relevé will be regarded as atypical and discarded.

TABLE 42

Sample relevé

Relevé No. 1.   *Locality:* Spurn Point, East Yorkshire.   *Date:* 19.9.70.
Grid reference: TA 397107   *Altitude:* 10 ft.

Vegetation	Mature sea buckthorn scrub on exposed dune crest; 2·5 m tall; some evidence of rabbit grazing on smaller side shoots
Cover	Scrub layer, 100%; Herb, 60%; Moss, 20%; Litter, 60%; Bare sand, 20%
Soil type	Surface leached sand with redeposited humus layer at 10 cm
Species	5.1 Hippophae rhamnoides
	1.1 Senecio jacobaea
	2.1 Solanum dulcamara
	+ Rubus fruticosus s.l.
	3.3 Urtica dioica
	+ Rumex crispus
Mosses	1.3 Eurynchium praelongum
	+ Hypnum cupressiforme

TABLE 43

Table of sociability classes

Sociability 1—growing once in a place, singly
    „   2—grouped or tufted
    „   3—in troops, small patches or cushions
    „   4—in small colonies, in extensive patches or forming carpets
    „   5—in great crowds or pure populations

Having delimited a homogeneous stand the next step is the compilation of the data which comprise the relevé. A separate page of the field notebook is given to each relevé. At the top of the page is written the relevé number, locality, date, grid reference, altitude; angle of slope and exposition of the stand, area of the stand, percentage ground cover occupied by the different vegetation layers and other data on soil type, horizons and root layering derived from a soil pit. An example of a relevé is given in Table 42. There then follows a complete list of all species present in the stand. If a species occurs as a mature plant in one layer and as a seedling in another it is recorded twice. When a full species list is to hand, two numbers are donated to each species, one as an estimate of cover-abundance (see Table 5) and the other an index of sociability (Table 43). The inclusion of a sociability figure is considered important as it gives a more complete picture of the community structure and of the communal organization of vegetation. Finally, the total area of the particular vegetation type being described is noted, while additional species outside the quadrat area are appended as a plus in brackets. Contrary to the statement of Küchler (1967, p. 233) the Latin scientific name of the phytocenose is *not* appended to the list. This statement is not only incorrect but sponsors the suggestion that the associations are predetermined. In later stages of field recording, when some idea of the variation in vegetation is understood, it is often common practice to tentatively assign the record to a particular association.

The above recording method is the most rapid for the completion of a relevé. Printed note-pads with the basic quadrat data of environmental factors, etc., cut down the time of recording, but printed lists of species names do not provide a short-cut method. The recording method of the Botanical Society of the British Isles where Latin names of species are represented on master sheets as the first four letters of each word (e.g. *Carex pulicaris* as 'Care puli') have been tried using special cards comprising selected species in alphabetical order. But the twofold mental process of recognizing a species and then locating it in an alphabetical progression proves to require approximately 30% more time overall.

Tabulation of Data

When a representative sample of relevés has been collected—a number which varies with the problem but is usually less than the manageable number of 40—they are aggregated together in a Raw Table (Table 44). This is rapidly achieved using squared paper, with the species names written at the left-hand side of the table and each relevé assigned a single vertical column. It is usually convenient at this preliminary stage to group the species under the separate headings of trees, herbs, mosses, lichens, etc., or at least to phanerogams and cryptogams. The addition of more new species to the later relevés produces a characteristic tailing off to the right of the table. The only additional information included in the table at this stage is the relevé number and the number of species per relevé.

TABLE 44

Stage 1: Completed Raw Table

Relevé number	1	2	3	4	5	6	7	8	9	10	11	12	13	14	15	16	17	18	19	20
Phanerogams																				
Hippophae rhamnoides	5·1	5·1	5·1	5·1	5·1	3·2	2·2	4·3	2·2	5·1	3·2	5·1	3·3	4·3	3·2	5·1	5·1	5·1	5·1	5·1
Senecio jacobaea	1·1	+	1·1	+	+	+	+			1·1	+		+	+	1·1	+		+	1·1	1·1
Solanum dulcamara	2·1	2·1	+	+	+		+			+	+	+		+		+	+	+	+	+
Rubus fruticosus s.l	+	1·1		+	1·1		+			+		+		+		+			+·3	1·1
Urtica dioica	3·3	1·3	1·3															1·3	1·1	2·3
Rumex crispus	+	+	+			+				+	+	+	+	+		+		+		
Montia perfoliata				3·4	4·4															
Stellaria media				1·2	+		1·1	1·1												
Festuca rubra					+	1·1	2·3	1·1	3·3	4·5	2·3	2·3	3·3	1·3	1·3	4·5	2·3	1·3		
Agropyron repens						2·3	4·3	2·3	3·3		3·3	3·4	1·3	2·3	3·2	+	2·3	1·2	2·3	
Ammophila arenaria						+·3		+	+		1·1			3·3	4·3					
Sonchus arvensis						1·1								+	+					
Ononis repens						+	+		+					+						
Galium verum						+														
Calystegia soldanella						1·1		1·1	+·2		1·1		+·1	+·1	+·2			1·1		
Poa pratensis						+		+	+++	+	+++	+		++++++	++·2		++	+	+·3	+
Agrostis stolonifera					+								+		1·1					
Ranunculus bulbosus				+	+									+			+			
Plantago lanceolata														+			+			
Veronica chamaedrys														+			+			
Chamaenerion angustifolium														+			+		+	+
Cerastium vulgatum																				+·2
Sambucus nigra																		1·1		+
Cirsium vulgare																	1·1	1·1		
Heracleum sphondylium											1·1	1·1			1·1		+	+		+
Inula conyza											++	+		++			++++			
Cardamine hirsuta											+			+						
Hypochaeris radicata										+	+		+	+						
Arrhenatherum elatius										+				+					+·1	
Sonchus asper										+		+								+·2
Cryptogams																				
Eurynchium praelongum	1·3	+·3	1·3																	
Hypnum cupressiforme	+	+·2	+·2	+	+															
Brachythecium rutabulum		1·3	1·3	+	+														1·3	1·3
Geastrum fornicatum																				
Brachythecium albicans						+·3			+·3					+				+·3		
Bryum inclinatum						+·3														
Tortula ruraliformis								1·3		+	+			+·3						
Cladonia rangiformis							+·3	+·3			+3			+·3						
Bovista nigrescens													+	+						
Lophocolea heterophylla																				+
Species Number	8	9	8	8	9	11	7	10	14	12	12	10	10	22	10	11	12	12	12	13

TABLE 45
Stage 2: Annotated Raw Table

Relevé number		1	2	3	4	5	6	7	8	9	10	11	12	13	14	15	16	17	18	19	20
Phanerogams																					
Hippophae rhamnoides	A	5·1	5·1	5·1	5·1	5·1	3·2	2·2	4·3	2·2	5·1	3·2	5·1	3·3	4·3	3·2	5·1	5·1	5·1	5·1	5·1
Senecio jacobaea		1·1	+	1·1	+	+	+	+			1·1	+	+	+	+	1·1	+		+	1·1	+
Solanum dulcamara		2·1	+	+	+		+				1·1	+	+	+	+		1·1		+	+	1·1
Rubus fruticosus s.l		1·1		1·1		1·1					+			+	+		+	+		+·3	1·1
Urtica dioica		3·3		1·3					+	+				+					1·3	+·3	2·3
Rumex crispus	B	+	+	+	3·4	4·4					4·5		2·3				4·5	2·3	1·3		+
Montia perfoliata	B				1·2	+					+		3·4			1·3	+	2·3	1·2		
Stellaria media	CCC					+	1·1	+	1·1	+		2·3		3·3	1·3	1·3					+
Festuca rubra							2·3	2·3	+	3·3		3·3		+	2·3	+·2					1·1
Agropyron repens								4·3	2·3	+		1·1		1·3	+·3	4·3					2·3
Ammophila arenaria							+								+	+					
Sonchus arvensis																					
Ononis repens																					
Galium verum							+	+	1·1	+	+	1·1	+	1·1	+	+			+		+
Calystegia soldanella	C						1·1		+	+·2				+	+	+·2		+			+·2
Poa pratensis							+		+	+·3					+	+					
Agrostis stolonifera	CC									++		++			++	1·1	++			2·3	+
Ranunculus bulbosus																	++		1·1		+
Plantago lanceolata									+	+	+++		1·1		++		++	1·1	1·1	1·1	1·1
Veronica chamaedrys	B																				+
Chamaenerion angustifolium																					
Cerastium vulgatum																					
Sambucus nigra																		+++			
Cirsium vulgare																		+++	1·1		
Heracleum sphondylium																			1·1		
Inula conyza	B																	++++			+
Cardamine hirsuta																				1·1	
Hypochaeris radicata																					+
Arrhenatherum elatius																					
Sonchus asper															+						
Cryptogams																					
Eurynchium praelongum	A	1·3	1·3	1·3	+	+									+					1·3	1·3
Hypnum cupressiforme	B	+	+·2	+·2	+	1·1					+·3		++		1·1			++	+·3	1·1	
Brachythecium rutabulum		+	+	1·3	++	+				1·3	+	+·3	1·1		+·3						
Geastrum formicatum							+·3														
Brachythecium albicans							+·3		+·3	1·3				+	+·3						+
Bryum inclinatum															+						
Tortula ruraliformis																					
Cladonia rangiformis																					
Bovista nigrescens																					
Lophocolea heterophylla																					+
Species Number		8	9	8	8	9	11	7	10	14	12	12	10	10	22	10	11	12	12	12	13

In the finished raw table it soon becomes apparent, even to the unskilled eye, that while no two relevés are alike, certain combinations of species recur again and again and that these groups may be mutually exclusive. Species which have a restricted occurrence in the raw table and which occur together in the same group-combination are then boxed or outlined to give an annotated or outlined raw table (Table 45). For example, *Montia perfoliata* and *Stellaria media* show a striking correlation in relevés 4, 5, 10, 12, 16, 17 and 18 while a second group of species with related distribution is to be seen in *Festuca rubra, Agropyron repens, Ammophila arenaria*, and so on.

Further annotation follows with a review of the variation in species number per relevé in a search for fragmentary or heterogeneous stands. Relevé 14 stands out immediately with a high species number of 22 indicating heterogeneity when compared to the overall species average. Reference to the field notes indicates that there was a small footpath passing through the stand which led to trampling, a different set of ecological pressures and hence a higher species number. This relevé is discarded. Finally, the presence of each species in the data is noted to the left of the table matrix.

The next step is to make a partial table in which only the noted and boxed species are extracted and used to rearrange the table in both a vertical and horizontal manner. This procedure highlights the existence of mutually exclusive groups of species or in other words groups of differential species. The differentiating species are written on the left of the paper in groups and the relevés rearranged vertically into a new order to give a progression of species groups from left to right of the table. To simplify the rearrangement of the relevés dictation or transfer strips are used (Table 46). These are narrow strips of squared paper

TABLE 46

Dictation or transfer strips

Strip B	1	2	3	4	5	6	7	8	9	10	11	12	13	14	15	16	17	18	19	20
Strip A	1	2	3	6	7	14	13	15	16	8	17	9	18	X	19	10	11	12	4	5

of similar type to that used in the making of the tables. One of the strips (B) is numbered consecutively from 1–20 and is used in the making of the partial table. The other strip (A) also bears the numbers 1–20 but arranged in a new progression so that relevés with the same groups of presumed differential species are brought together. This latter strip is placed on the annotated raw table and the appropriate data transferred to the partial table. The finished partial table (Table 47) shows the effect of regrouping the relevés. Three groups emerge and

TABLE 47

Stage 3: Partial or Extract Table

New relevé order	1	2	3	19	20	4	5	10	12	16	17	18	6	7	8	9	11	13	15
Group A																			
Urtica dioica	3·3	1·3	3·3	+3	2·3														
Eurynchium praelongum	1·3	+3	1·3	1·3	1·3														
Group B																			
Montia perfoliata						3·4	4·4	4·5	2·3	4·5	2·3	1·3							
Stellaria media						1·2	+	++	3·4	+	2·3	1·2							
Cerastium vulgatum						-	-			++	1·1	1·1							
Cirsium vulgare											+++								
Cardamine hirsuta						+	+	-	+	-	+								
Geastrum fornicatum										-	+++	++							
Group C																			
Festuca rubra													1·1	2·3	1·1	+	2·3	3·3	1·3
Agropyron repens													+	4·3	+	3·3	3·3	+	+·2
Ammophila arenaria													2·3	+	2·3	++++	1·1	1·3	4·3
Poa pratensis													-		+		+	1·1	+
Ranunculus bulbosus													-					+	-
Plantago lanceolata													+·3					+·3	
Brachythecium albicans													-						1·1

- : Dashes are of no floristic significance and are usually added to aid alignment of columns, etc. They are usually omitted in the final association table.

Stage 4: Differentiated Table

Revised relevé order	1	2	3	19	20	4	5	10	12	16	17	18	7	6	8	9	11	13	15
Group A																			
Urtica dioica	3·3	1·3	3·3	+·3	2·3														
Eurynchium praelongum	1·3	+·3	1·3	1·3	1·3														
Group B																			
Montia perfoliata						3·4	4·4	4·5	4·5	4·5	2·3	1·3							
Stellaria media						1·2	+	+	3·4	+	2·3	1·2							
Geastrum fornicatum						+	+	·+	+	·+	·+	·+							
Cerastium vulgatum						·	-	+	+	+	1·1	1·1							
Cirsium vulgare										+	1·1	1·1							
Cardamine hirsuta						-			+	+	++	·+							
Group C																			
Festuca rubra							+						2·3	1·1	1·1	+	2·3	3·3	1·3
Agropyron repens													4·3	+·3	2·3	3·3	3·3	+·3	+·2
Ammophila arenaria														2·3	+·3	+++	3·3	1·3	4·3
Poa pratensis													-	-	+	+	+	+	+
Plantago lanceolata													-		-	+	+	1·1	1·1
Brachythecium albicans													-	+3		1·3	+·3	+·1	-
Ranunculus bulbosus													-		-	+	+·3	+	-
Group D																			
Hippophae rhamnoides	5·1	5·1	5·1	5·1	5·1	5·1	5·1	5·1	5·1	5·1	5·1	5·1	2·2	3·2	4·3	2·2	3·2	3·3	3·2
Senecio jacobaea	1·1	1·1	1·1	1·1	+	+	+	1·1	+	+·1	+	+	+	+	+	+	+	+	1·1
Solanum dulcamara	2·1	2·1	+	+	+	+	+	+	+	+	+	+	+	+	+	+	+	+	+
Rumex crispus	+	+·1	+	+·1	+	+	·+	+	+	+	+	+		+	+	+	+	+	+
Rubus fruticosus s.l	+	1·1	-	++	-	+	1·1	++	+	+	+	+		-	+	+	·+	+	+
Sonchus arvensis				++		-		+	+						+	++	-	+	+
Calystegia soldanella	+	+			++	+	1·1	+				+3	+	+	+·1	+		+	+·2
Brachythecium rutabulum												+3				·+			
Agrostis stolonifera		1·3			+2						+								
Chamaenerion angustifolium	+	+2	+2									1·1	+3	+		1·3	+3	1·1	1·1
Veronica chamaedrys			+	+									-	-	-	+3	+3	1·1	1·1
Inula conyza																			
Sambucus nigra																			
Hypochaeris radicata																		3·2	2·1
Arrhenatherum elatius																			
Sonchus asper																			
Lophocolea heterophylla							+			+									
Heracleum sphondylium											+								
Ononis repens														1·1					
Galium verum														+3					
Bryum inclinatum															+3				
Tortula ruraliformis																			
Cladonia rangiformis																		+	

the total cover of all species in each group indicates their value as differential species within the set of raw data available.

Finally, the species are scored for presence and the groups 'tidied up' by rearranging the species in a descending order of number of occurrences and the relevés rearranged vertically again if necessary. Species such as *Cirsium vulgare* and *Cardamine hirsuta*, although of restricted occurrence are included provisionally with the view that further data of the vegetation type will uphold or reject their status as differential species. From this rearrangement, a differentiated table is produced (Table 48) where the large Group (D) of companion species with a high or low presence value are appended, again in descending order of presence. At this stage, three sub-units with their diagnostic species stand out clearly and unequivocally at a glance and the structure of the whole unit is reflected by the first few species of Group D which are of constant occurrence. As Moore (1962) has pointed out, there may be objections that this pattern could be an artifact inherent in the handling of the data and that this pattern does not reflect the natural situation. But the differentiated table should only be a working hypothesis and should receive confirmation from further floristic data in the same and other localities, from mapping of the units distinguished and from the demonstration of a coincidence between the units and various environmental factors.

With reference to the latter approach, the sub-groups A and B occur under the older stands of *Hippophaë* where surface leaching has led to the formation of a redeposition zone of humus some 10 cm below the surface. There is some correlation between exposure and Group A, while all the representative stands of B are of colonies where female *Hippophaë* predominate. These areas are subject to high roosting and feeding pressures from migrants and the surface layers of the soil are consequently richer in nitrogen and phosphorus. Group C corresponds to areas where *Hippophaë* is invading the normal *Ammophila* dunes and where other dune grasses remain under a fairly open canopy. The sub-units thus appear to be ecologically viable in the broadest possible sense and to be valid units for the mapping of the vegetation of Spurn Point, the locality from which they were recorded. It is, in fact, this stage in the production of an association table which lends itself to the extraction of units for mapping on a small scale. There is no difficulty in donating names to the units, e.g. mature *Urtica-Hippophaë*, mature *Montia-Hippophaë*, advancing *Hippophaë-Ammophila* communities which will illustrate the pattern of the dune scrub in the locality (cf. Küchler, 1967, p. 254).

In the creation and delimitation of the association table, it must be shown that the pattern mirrored by the original data of the differentiated table is applicable in a number of geographical regions where

Hippophaë scrub occurs. In Table 48 it might be considered that each of the sub-units are associations in their own right. If the same or a similar pattern of the three sub-units crops up under *Hippophaë* in diverse geographical regions it becomes legitimate to infer that the same environmental conditions pertain and that these manifest themselves in three groups of differential species of similar ecological amplitude. One could thus conclude that there were three associations, *Urtico-Hippophaetum*, *Montio-Hippophaetum* and *Ammophilo-Hippophaetum*. If a different pattern emerges in a different region then it is likely that the data from the Spurn Point dune scrubs form the northern representative of a network of floristic variation within geographical region. The whole data would then be the association and the sub-units the sub-associations.

The dune scrub relevés were chosen as a relatively simple example of the so-called Braun-Blanquet table method (Küchler, 1967). Because of the inadequacies of the data in terms of numbers of relevés and geographical localities it is impossible to create and characterize an association table. More complete data, however, are available from Shimwell (1968) from calcareous grasslands in Britain, which because of the large species numbers involved were not used as an example of the table method. The example taken from this data is of the *Sesleria*-rich grasslands overlying the magnesian limestone soils and parent rock of County Durham. The initial stages of survey took the form of the collection of relevés from three of the main localities of calcareous grassland in the region and the completion of a differentiated table. Three main groups of differential species became apparent:

(*a*) A group from damper, predominantly north and north-west facing sites containing *Acrocladium cuspidatum* and *Carex pulicaris*.

(*b*) A group of grasses and herbs of rank growth which conferred a physiognomic appearance of a tall-herb grassland type on the communities. *Helictotrichon pubescens* was particularly instrumental in conveying this impression.

(*c*) A group comprising several acrocarpous mosses and phanerogams of open habitats which occurred in disturbed sites such as quarry spoil banks.

A fourth group of relevés was discernible in which there were no apparent differential species.

This pattern of variation was then tested for other sites in the region and was shown to be valid in that all relevés collected subsequently fitted one of the four groups. The next stage of definition of an association called for a study of grasslands of magnesian limestone derived soils further to the south and other *Sesleria*-dominated grasslands in the north of England, thereby covering any possible pedological, ecological and physiognomic relationships of the grasslands in question.

TABLE 49

Nomenclature of classificatory units

Rank	Ending	Example
Class	-etea	Festuco-Brometea
Order	-etalia	Brometalia erecti
Alliance	-ion	Mesobromion
Association	-etum	Cirsio-Brometum
Sub-association	-etosum	Cirsio-Brometum caricetosum
Variant	–	(specific names used)

Gradually, a picture of overall floristic relationships within calcareous grasslands is built up, and a detailed insight into the ecological amplitude of the component species attained. It becomes apparent that some species are of widespread occurrence in the calcareous grassland matrix and in fact do not show a pronounced calcicolous tendency. Others are widespread but are also more or less restricted to dry calcareous habitats, while a third group exhibit a pronounced geographical distribution to a particular western or sub-montane type of calcareous grassland. It is the latter two groups of species which become useful in the definition of the association.

The method used in all steps leading to the production of the final community table (in other words, a second differentiated table which contains all available data) is a purely comparative one demanding no knowledge of phytosociology or ecology. However, in the interpretation of the final association table and the positioning of the association in the hierarchical floristic system of Braun-Blanquet, the sociological rank and ecological amplitude of each species comes into question. It is because of a lack of basic phytosociological and field ecological knowledge that most opponents of the Z-M system fail to comprehend this next step. At this stage a slight digression is necessary.

The broad vegetation surveys of Braun-Blanquet and Tüxen which envelop all types of terrestrial vegetation have enabled a basic understanding of the ecological and phytosociological distribution of species and their values as indicator species. From this enormous matrix of data it has been possible to group species along a gradient of ecological amplitude the steps of which are rungs in a hierarchical classification system comprising alliances, orders and classes. These terms are an integral part of the hierarchical system of plant community classification proposed by Braun-Blanquet (1921) and now widely applied in Europe.

Specific endings are given to community names which indicate the rank of the community in the system. The examples in Table 49 illustrate this point where the rank ending is added to the genitive stem of the second genus in the name or to the first where there is only one genus in the name, as in *Brometalia*. If the generic name of the plant labelling the community is not sufficient for the naming of the community (e.g. there are more than one species of *Bromus*) the species name is included in the genitive as *Brometalia erecti*. When two species are used to describe a rank, e.g. *Sesleria* and *Helictotrichon*, the first name is modified by the addition of a connecting vowel to its genitive stem— hence *Seslerio-Helictotrichetum*. Further details on the naming of community ranks may be obtained by reference to Bach, Kuoch and Moor (1962) who have published a set of precise rules of which one of the basic is 'for the naming of an association or higher unit no more than two plant names should be employed'.

To return to the groups of species of different ecological amplitude, it becomes possible, partly by reference to copious Continental literature, partly by extensive fieldwork in a particular geographical region to define groups of species which are similar in this respect. Thus, for the British Isles, Shimwell (1968) was able to arrive at the results shown in Table 50. These then are considered the character species (Charakterarten, Kennarten) for the different levels of the hierarchy which describes British calcareous grasslands.

It is by these species which the association is in the first place characterized and pigeon-holed. The rewriting of the differentiated table whereby each species is given its correct sociological rank now gives the community new meaning. The Kennarten of the alliance, order and class can be extracted from the companion species and either grouped together at the top of the table or separated below the character and differential species of the sub-alliance, association and sub-associations (Table 51). The association is thus characterized by three groups of species: (*a*) the alliance and order character species in combination with; (*b*) the character and differential species of the sub-alliance and association; (*c*) the group of companion species of high presence or constancy which confer a certain structure upon the community.

At this stage, the environmental characteristics of the relevés are added at the top and the average species number per sub-unit calculated. Life-form symbols may be added before the species names. Finally, the association is labelled by the choice of a name. In this case, *Sesleria* is chosen because of its role in characterizing the community, and *Helictotrichon* because of the continuous presence of the order character species *H. pratense* and the occurrence of *H. pubescens* as a differential species, in the majority of stands of rank growth (Table 51). The table

TABLE 50

Character species for the Festuco-Brometea hierarchy in the British Isles

Festuco-Brometea and Brometalia character species

Acinos arvensis
Anthyllis vulneraria
Blackstonia perfoliata
Brachypodium pinnatum
Carex humilis
Carlina vulgaris
Centaurea scabiosa
Filipendula vulgaris

Gentinaella amarella
Helianthemum chamaecistus
Helictotrichon pratense
Koeleria cristata
Poterium sanguisorba
Scabiosa columbaria
Viola hirta ssp. calcarea

Bromion and Sub-Alliance Xerobromion character species

Helianthemum appenninum
Koeleria vallesiana
Trinia glauca

(? Bupleurum baldense
Seseli libanotis
Teucrium botrys?)

Mesobromion and Sub-Alliance Eu-Mesobromion character species

Anacamptis pyramidalis
Asperula cynanchica
Bromus erectus
Campanula glomerata
Cerastium pumilum
Cirsium acaulon
Helianthemum canum
Hippocrepis comosa

Onobrychis viciifolia
Orobanche elatior
Picris hieracioides
Polygala calcarea
Potentilla tabernaemontani
Pulsatilla vulgaris
Thymus pulegioides
Thalictrum minus

A number of species which are restricted geographically, but which are good Mesobromion character species in continental Europe:

Aceras anthropophorum
Carex ericetorum
Dianthus gratianopolitanus
Euphrasia pseudokerneri
Galium pumilum
Gentianella anglica
G. germanica
Herminium monorchis

Himantoglossum hircinum
Orchis ustulata
Phyteuma tenerum
Polygala austriaca
Salvia pratensis
Senecio integrifolius
Thesium humifusum
Veronica spicata

Sub-alliance Seslerio-Mesobromion character species

Euphrasia salisburgensis
Gentiana verna
Polygala amara

Sesleria caerulea
Viola rupestris

is reduced by the inclusion of incidental species of only one or two occurrences in a list at the base.

To enable a complete understanding of the nature of the association, its ecology and distribution, a description of the edaphic, climatic and structural vegetation features usually accompanies the association table. The approach used is essentially that of the synecosystem concept of Braun-Blanquet (1964) who states that these environmental factors are best reflected in the vegetation or vice versa. For the purposes of climatic reference, the *World Climate Diagram Atlas* of Walter and Lieth (1967) provides an excellent background and the background pedological data is usually available from Soil Survey Memoirs. The following example is taken from Shimwell (1968):

Association Seslerio-Helictotrichetum, Shimwell, 1968.
Synonymy: Magnesian Limestone Rough Pasture, Heslop-Harrison and Richardson (1953).

Habitat details

This association is peculiar to the magnesian limestone escarpment of eastern Durham and a few localities in the plateau region further to the east. Representative stands have been recorded from Boldon and Tunstall Hills in the north to Bishop Middleham and Cornforth in the south at altitudes varying between 200 and 500 ft (60–160 m).

On a broad regional basis the east Durham plateau represents an extension of the drier eastern climate of south England. However, Walter and Lieth (1967) subdivide Type IV so that the region in question is separated under IV_2. In relation to the Durham City region and the plateau region, the west-facing escarpment has a considerably higher rainfall and cooler temperatures. At elevations of 500–600 ft (160–190 m), the average annual rainfall is over 760 mm as opposed to 635–700 mm on the dip slopes. There is a longer snow-lie period and the most striking temperature differences are seen in the average minimum winter temperatures: 27·5°F ($-2·5$°C) compared to 32·5°F (0·3°C) (Durham University Observatory Records).

Over most of the eastern plateau the soils are never free from the influence of drift and consequently the deeper soils thus formed are mainly cultivated. The stands of the association are found only in isolated localities on steep slopes, around quarries and rock outcrops along the western escarpment and in local areas on the undulating plateau (see Figure 50). On the steeper western escarpment the soils are generally free from drift and the shallow rendziniform profiles of the Cornforth and Middleham Series (McKee, 1965) are developed from the parent friable, yellow, lower limestones. Both soil types are exceptionally thin, uniformly sandy loams directly over limestone with little or no drift influence and good drainage. The Middleham Series differ mainly in that they are brash-like throughout with diffuse boundaries between horizons and are developed over the upper beds of the lower limestone series around Bishop Middleham and Fishburn. The maximum profile depth in both soil types is

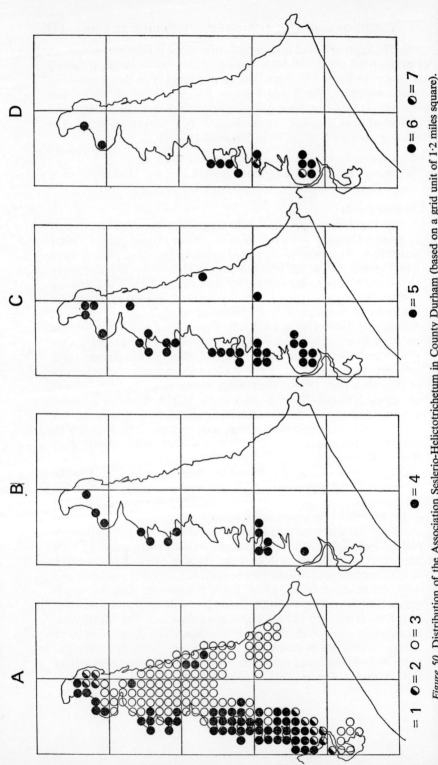

Figure 50. Distribution of the Association Seslerio-Helictotrichetum in County Durham (based on a grid unit of 1·2 miles square).

● = 1   ◑ = 2   ○ = 3   ● = 4   ● = 5   ● = 6   ◑ = 7

A   B   C   D

around 24 in. (60 cm) with a typical profile A : C/D, but a B horizon may appear as a shallow weathered drift. The pH range is between 7·2 and 7·8 and there is a relatively high sand content (50–60%) and low (3–20%) humus content.

Characteristics of the association

The association is characterized by the occurrence of the Seslerio-Mesobromion differential species *Sesleria caerulea* and *Epipactis atrorubens* in combination with such alliance and order character species as *Poterium sanguisorba*, *Helictotrichon pratense*, *Anthyllis vulneraria*, *Scabiosa columbaria*, *Koeleria cristata*, *Helianthemum chamaecistus* and others. The gregariousness and dominance of *Sesleria* usually reduces the cover and abundance of these species and also many of the companion species. However, species such as *Carex flacca*, *Thymus drucei*, *Carex caryophyllea*, *Festuca ovina*, *Lotus corniculatus* and *Plantago lanceolata* maintain a high constancy.

The Seslerio-Helictotrichetum is the most thermophilous association in the sub-alliance and forms an important link with the Eu-Mesobromion associations to the south and the damper upland associations of the Seslerio-Mesobromion to the west. These relationships are well illustrated in the sub-associations that have been delimited.

Sub-association *Typicum* occurs in areas where anthropogenic influences have been minimal and is probably the most natural group of grasslands in the association. The areas appear not to have been burned or disturbed and are generally free from species indicative of interference other than light grazing. The better areas are located at Thrislington Plantation, and on the escarpment above Pittington and the south side of Cassop Vale. *Epipactis atrorubens* is more or less restricted to this sub-association and *Betonica officinalis* and *Campanula rotundifolia* occur more frequently here.

Sub-association *Caricetosum pulicaris* forms an important link with the *Seslerio-Caricetum pulicaris* described from the mountain limestones of the northern Pennines. It generally occurs on north- and west-facing slopes with a more humid microclimate. *Carex pulicaris* occurs as co-dominant with *Sesleria* and in several places at Thrislington almost replaces the latter. Two distinct variants are recognizable: variant of *Selaginella* and *Pinguicula* is found only at Cassop Vale on the steep, damp, north-facing slopes. The main localities appear to be on quarry spoil-banks, but doubtless the community survived here in a more natural habitat before quarrying began. The abundance of bryophytes such as *Ctenidium molluscum*, *Acrocladium cuspidatum* and *Hylocomium splendens*, indicates the damper nature of the situation. Variant of *Antennaria dioica* has been recorded only at Thrislington and is differentiated by the extremely gregarious pattern exhibited by *Antennaria. Ctenidium molluscum, Acrocladium cuspidatum, Rhytisiadelphus squarrosus* and *Fissidens cristatus* also show a high constancy to this variant. Although *Primula farinosa* is reported from several localities in the magnesian limestone region of Durham, its only occurrence in this association was at Thrislington in a relevé of this variant.

Sub-association of *Helictotrichon pubescens* is fairly widespread along the escarpment and is differentiated by the presence of *Helictotrichon pubescens, Dactylis glomerata* and *Daucus carota.* The sub-association is found in small areas which have been disturbed in the past and are now grazed to varying extents. Indicator species which occur more commonly in this group are *Centaurea nigra, Plantago media* and *Trifolium repens.* These species and the differential species form a group which represents the first stage transition from a Mesobromion to an Arrhenatheratalia community. In general, the climate on the exposed hills such as Sherburn Hill is too dry and grazing too light to enhance further transition. On the damper slopes of Cassop Vale, *H. pubescens* assumes dominance and in combination with other tall grasses and herbs such as *Arrhenatherum elatius, Heracleum sphondylium* and *Filipendula ulmaria,* gives the characteristic community structure of an Arrhenatherion association.

A variant of *Rosa pimpinellifolia* is found in areas which are periodically burned, or where grazing pressures have been gradually lessened over a period of a few years.

A second variant of *Bromus erectus* is described from Tunstall Hills, and on roadsides at Catley Hill and near Bishop Middleham. This variant which forms an important link with the Eu-Mesobromion associations to the south of Co. Durham is characterized by the dominance of *Bromus erectus* and the greatly reduced cover of *Sesleria.*

Sub-association of *Encalypta* and *Plantago maritima* represents the second stage of colonization of quarry spoil-banks replacing a weed community which includes *Chamaenerion angustifolium, Hypericum* spp., *Erigeron acer* and *Equisetum arvense.* The sub-association was located in the quarries at Pittington, Sherburn Hill, Cassop and Bishop Middleham. The communities are typically open with large patches of *Encalypta vulgaris* often alternating with *E. streptocarpa* and occasionally *Ceratodon purpureus.* The areas in between the moss patches are occupied mainly by tufts of *Sesleria* and three moderately good differential species—*Hypochaeris radicata, Hypericum montanum* and *Plantago maritima.* It is puzzling whether this latter species forms a link with the montane *Seslerio-Caricetum pulicaris* or with the common maritime grasslands.

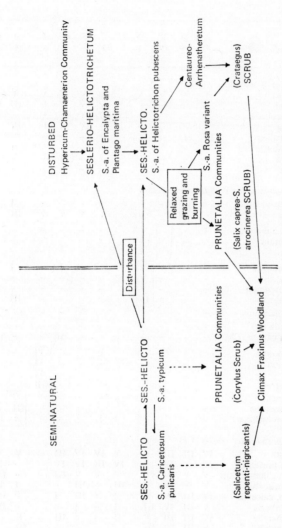

Figure 51. The seral relationships of the Association Seslerio–Helictotrichetum.

TABLE 52

The phytogeographical relationships of the associations of the sub-alliance Seslerio-Mesobromion in Britain

	A	B	C	D	E	F	G	H	J	K	L
Number of Aufnahmen	9	8	19	8	29	5	5	8	6	7	12
Differential species of sub-alliance (D)											
Sesleria caerulea	V	V	V	V	V	V	V	V	V	V	V
Galium sterneri	–	–	–	–	IV	V	V	V	III	IV	III
Cornicularia aculeata	–	–	–	–	III	IV	IV	V	–	–	–
Gentiana verna	–	–	–	–	I	IV	I	II	IV	–	II
Epipactis atrorubens	IV	I	r	–	r	–	r	–	–	I	II
Euphrasia salisburgensis	–	–	–	–	–	–	–	–	II	–	II
Polygala amara	–	–	–	–	I	–	I	–	–	–	–
Viola rupestris	–	–	–	–	I	–	–	–	–	–	–
Association and sub-association differential species											
Acrocladium cuspidatum	r	V	–	–	r	–	–	–	I	–	–
Carex pulicaris	–	V	–	–	V	V	V	V	III	–	III
Helictotrichon pubescens	I	–	–	I	–	–	–	–	–	–	–
Dactylis glomerata	–	–	V	I	r	–	–	–	–	–	–
Daucus carota	–	–	IV	I	–	–	–	–	–	–	–
Hypochaeris radicata	–	–	–	V	–	–	I	–	I	–	II
Encalypta vulgaris	–	–	–	V	–	–	–	–	–	I	–
Hypericum montanum	–	–	–	V	–	–	–	–	–	–	–
Plantago maritima	–	–	r	IV	r	–	–	–	III	I	IV
Kobresia simpliciuscula	–	–	–	–	–	V	–	–	–	–	–
Dryas octopetala (D)	–	–	–	–	–	–	–	–	–	–	V
Helianthemum canum (Ch)	–	–	–	–	–	–	II	–	–	IV	IV
Saxifraga hypnoides	–	–	–	–	r	–	–	V	–	–	–
Cochlearia alpina	–	–	–	–	–	–	–	V	–	–	–
Draba incana	–	–	–	–	–	–	–	V	–	–	–
Myosotis alpestris	–	–	–	–	–	–	–	III	–	–	–
Asperula cynanchica (Ch)	–	–	–	–	–	–	–	–	V	V	V
Breutelia chrysocoma	–	–	–	–	r	–	–	–	III	–	IV
Camptothecium lutescens	–	–	–	–	I	–	–	–	V	IV	III
Alliance and order character species (Ch)											
Koeleria cristata	IV	III	III	V	V	IV	IV	III	V	V	V
Helictotrichon pratense	IV	V	IV	V	IV	III	IV	IV	–	II	I
Helianthemum chamaecistus	V	V	IV	IV	V	V	V	III	–	IV	–
Poterium sanguisorba	V	V	V	V	III	–	III	–	I	II	I
Viola hirta ssp. calcarea	IV	III	IV	V	II	–	III	I	–	IV	–
Scabiosa columbaria	V	III	IV	III	II	–	I	–	–	IV	–
Gentianella amarella	IV	III	III	V	II	–	–	III	II	I	–
Anthyllis vulneraria	V	III	III	III	–	–	–	–	III	I	–
Carlina vulgaris	IV	I	I	–	II	–	II	–	III	III	IV
Centaurea scabiosa	I	I	IV	III	–	–	–	–	–	–	–
Hippocrepis comosa	–	–	–	–	II	–	III	–	–	IV	–

TABLE 52—*continued*

Number of Aufnahmen	A 9	B 8	C 19	D 8	E 29	F 5	G 5	H 8	J 6	K 7	L 12
Potentilla tabernaemontani	–	–	–	–	r	–	–	–	–	II	–
Carex ericetorum	–	–	–	–	r	–	II	–	–	II	–
Blackstonia perfoliata	–	–	–	–	–	–	–	–	II	–	–
Bromus erectus	–	I	–	–	–	–	–	–	–	–	–
Filipendula vulgaris	–	–	–	–	I	–	–	–	–	–	–
Companions (of constancy V and IV)											
Carex flacca	V	V	V	V	IV	V	V	III	IV	IV	V
Thymus drucei (x)	V	V	V	V	V	IV	V	V	V	V	V
Lotus corniculatus	IV	V	IV	III	V	–	IV	II	V	IV	IV
Plantago lanceolata	V	III	IV	III	III	II	II	IV	IV	III	V
				ETC.							

A. Seslerio-Helictotrichetum typicum
B.   ,,   ,,   caricetosum
C.   ,,   ,,   of *Helictotrichon*
D.   ,,   ,,   of *Encalypta* and *Plantago*
E. Seslerio-Caricetum typicum
F.   ,,   ,,   kobresietosum
G.   ,,   ,,   dryadetosum
H.   ,,   ,,   of *Saxifraga*
J. Asperulo-Seslerietum typicum
K.   ,,   ,,   typicum, *Helianthemum* var.
L.   ,,   ,,   dryadetosum

Geographical character and differential species for Continental associations

1. Vincetoxico-Seslerietum R. Tx. 1967
   Vincetoxicum officinale ⎫ occurring in isolated localities on the Jurassic limestone
   Hypericum perforatum ⎬ cliffs of Belgium and northern Germany, particularly in
   Origanum vulgare ⎭ the Wesergebirge.

2. Alvar vegetation of Öland and Gotland
   Brachypodium pinnatum ⎫
   Cirsium acaulon ⎬ an overlap with Eu-Mesobromion.
   Euphorbia cyparissias ⎭

3. Seslerio-Koelerietum (Kuhn, 1937) Oberd, 1957
   Carling acaulis ⎫
   Buphthalmum salicifolium ⎪
   Aster bellidiastrum ⎬ pre-alpine regions of southern Germany.
   Prunella grandiflora ⎪
   Carex montana ⎭

This sub-association is probably the seral stage below the sub-association of *Helictotrichon pubescens* which develops as a more compact turf is formed.

Zonation and succession

The variation in edaphic factors within the association *Seslerio-Helictotrichetum* and the parallel variation in vegetation enables several tentative suggestions to be made with regard to the mechanisms involved in some stages of the theoretical succession on the magnesian limestone as outlined in Figure 51. On disturbed sites the initial colonizers of quarry waste are ruderals such as *Hypericum* spp. and *Chamaenerion angustifolium*. Several mosses such as *Encalypta* spp. and *Ceratodon purpureus* colonize damper, more shaded habitats and form a patchy carpet which gradually adds more humus to the habitat. The subsequent colonization of the habitat by the coarse grass *Sesleria caerulea* and herbs which require semi-open, unstable habitats, gives the characteristic structure of the sub-association of *Encalypta* and *Plantago maritima*. On older disturbed habitats where there has been (*a*) considerable humus accumulation; (*b*) some nitrogenous enrichment via grazing; (*c*) reduction in pH and decrease in total carbonates due to leaching; (*d*) development of a deeper soil profile, coarse grasses, such as *Helictotrichon pubescens* and *Dactylis glomerata*, and mesophilous indicators of grazing, e.g. *Daucus carota* and *Briza media*, begin to replace the acrocarpous bryophytes and *Sesleria* to give the sub-association of *Helictotrichon pubescens* typical variant. In other areas, this stage in succession is prevented by the invasion of *Salix capraea* and *S. atrocinerea* bushes to produce a low hybrid scrub.

It cannot be suggested that there is a successive step from sub-association of *Helictotrichon pubescens* to sub-association *typicum*, because the community organization and edaphic factors controlling development are more or less the same. The decrease of floristic diversity in sub-association *typicum* and hence the appearance of a more stable community indicate the grasslands are older and perhaps semi-natural. For similar reasons, it is difficult to envisage the short-term development of this grassland type in recent times from the disturbed sub-association of *Encalypta* and *Plantago maritima*.

It appears that burning and the consequent destruction of much of the raw humus in the habitat causes invasion of *Rosa pimpinellifolia* and *Crataegus monogyna*, and observations over three years have indicated that this step may proceed when grazing pressures are relieved.

The communities of the sub-association of *Helictotrichon pubescens*, *Bromus erectus* variant, appear to have no place in the disturbed

TABLE 53

Constancy classes

r	Less than 1 % presence in relevés
I	1–20%
II	21–40%
III	41–60%
IV	61–80%
V	81–100%

succession sequence, in spite of the relationships suggested by their brash-like soils. In fact, *Bromus erectus* was observed growing on similar soil types over Jurassic limestone, throughout the Cotswold Hills and on the magnesian limestone of Yorkshire and Derbyshire.

If a knowledge of the general phytogeography of a region is a desirable outcome of the application of this method, associations which are related in terms of floristics, physiognomy, structure and ecology are frequently grouped together into a constancy table (Table 52). For this purpose, the presence is recorded for each species in each unit described. Thus if a species occurs in all ten relevés of a particular association it is given a constancy figure of five represented by the roman numeral V. The range of constancy groups is given in Table 53. The range of cover values for the species may also be represented thus V^{1-4}. Table 52 illustrates a constancy table for the associations of the sub-alliance *Seslerio-Mesobromion* in north-west Europe. The differences between units are obvious at a glance. Apart from the *Seslerio-Helictotrichetum*, the other associations of Britain, *Seslerio-Caricetum pulicaris* and *Asperulo-Seslerietum*, are neatly distinguished by species such as *Carex pulicaris* and *Asperula cyanchica*. The Continental associations such as those from Oland, the Jurassic limestones of the Wesergebirge, Germany, and the pre-alpine region of southern Germany, whilst maintaining the basic floristic and structural characteristics of the sub-alliance all have their own distinguishing species.

Because of the copious amount of sociological data in the form of associations it is frequently impossible to include all the data in a single constancy table. A syntaxonomic table of the class or even alliance is in some ways a replacement for this (Table 54) and is merely a table in which the hierarchy is summarized. Although this is perhaps the ultimate aim of some branches of the Z-M School, it is by no means the only aim. Classification for the sake of classification although a basic human characteristic is seldom the only outcome. The differentiated table provides a basic level for the actual mapping of vegetation;

the final association table provides an ultimate level for the representative mapping of vegetation similar to the *Atlas of the British Flora* for species and inevitably a broader and deeper understanding of phytogeography. This in turn provides some idea of the rarity of a community and of its biological diversity and in this repect enables an initial grasp of the need for conservation. Perhaps on a more practical front, the grassland surveys such as that of O'Sullivan (1965) and Jasnowski (1966) on the mires of Poland provide basic data on pasture composition relative to agriculture and land use and the distribution and type of peat resources respectively.

Fidelity: fact or friction?

Up to now no mention has been made of the concept of fidelity, a concept which 'is of prime importance for the Z-M School' (Becking, 1957). Reviews of the nature of fidelity have tended to suggest that it is over-emphasized (Poore, 1956) and that its use is questionable. Its bases are seen in the strong perference of certain species for particular communities. As Ellenberg (1954) points out every species has a potentially much wider ecological amplitude than that which it normally demonstrates in Nature. The amplitude is restricted by niche space competition with other species and other life forms so that the species can only survive in its optimal habitat conditions. This existence is manifest in the selective preference of certain species for a particular community so that ecological amplitude plus sociological niche performance is basically synonymous with sociological preference.

In a discussion of fidelity, it is necessary to invoke reference to the concept of presence (constancy). If a species occurs in a high percentage of a number of relevés it will show high constancy for the community represented by the relevés. If the same species is restricted to relevés of one vegetation unit, then it will show a high fidelity for the said vegetation unit, and will be a community-specific or faithful species. Fidelity can only be determined, however, if a broad spectrum of relevés representative of different plant communities has been sampled and compared. It must be stressed that fidelity and presence are not correlated in all cases. *Sesleria caerulea* is of high constancy in Table 51 and shows a high fidelity to the association *Seslerio-Helictotrichetum* and a higher relation to the sub-alliance *Seslerio-Mesobromion*. *Carex flacca* is also of high constancy buts its fidelity to either association or alliance is negligible, a fact revealed by reference to its occurrence in diverse habitats and vegetation types. On the other hand *Polygala amara*, which is of rare occurrence in the Craven and Teesdale stands of *Seslerio-Caricetum pulicaris* is faithful to this association in the British Isles.

Five classes of fidelity are commonly distinguished and these, with

TABLE 54

Syntaxonomy of the alliance *Mesobromion erecti* in the British Isles

ALLIANCE MESOBROMION ERECTI Br.-Bl. and Moor, 1938, emend. Oberd., 1949	Sub-alliance Eu-Mesobromion Oberd., 1957	1. CIRSIO-BROMETUM
		2. HELICTOTRICHO-CARICETUM FLACCAE
		3. HELIANTHEMO KOELERIETUM
		4 CARICETUM M ONTANAE
		5. ANTENNARIETUM HIBERNICAE Br.-Bl. and Tx., 1952
		6. CAMPTOTHECIO-ASPERULETUM CYNANCHICAE Br.-Bl. and Tx., 1952
	Sub-alliance Seslerio-Mesobromion Oberd., 1957	7. SESLERIO-HELICTOTRICHETUM
		8. SESLERIO-CARICETUM PULICARIS
		9. ASPERULO-SESLERIETUM (Br.-Bl. and Tx., 1952) emend.
		(Sesleria-Carex ornithopoda community)
		(Breutelia-Sesleria pavement community)
		(Empetrum-Epipactis Nodum Iv.–Ck. and Proct., 1966)

examples from the British Isles, are shown in Table 55. Originally, only three categories of diagnostic species were distinguished by Braun-Blanquet (1928)—Character species (Fidelity V, IV and III), Companions (Fidelity II) and Strangers, but gradually the concept of the character species had to be modified as more phyotsociological data became available from diverse geographical regions. Species which had been designated as 'Gesellschaft treue' (Fidelity V) became demoted to 'feste' so that the present knowlege of character species suggests that there are few which can now be regarded as exclusive to a single association throughout their complete distribution. Consequently, at this level of the hierarchy it has been recognized that the value of each character species is much more limited than was originally thought. However, at the alliance level, there is an infinitely greater number of species which show a fidelity of Class V and it is by such a group of characteristic species that an association may be recognized (e.g. *Seslerio-Helictotrichetum*). The actual features of the vegetation types

TABLE 55

Fidelity classes (with examples from the British Isles)

1. *Character species* (Charakterarten, Kennarten)

   Class V.   Species exclusively or almost exclusively restricted to certain vegetation units (German: Treue Arten). *Koeleria vallesiana* to the Poterio-Koelerietum vallesianae, Shimwell, 1968; *Cirsium acaulon* to the Mesobromion

   Class VI. Species with a strong preference for a specific vegetation unit but also found infrequently in other units (German feste). *Thlaspi alpestre* and *Minuartia verna* to the Minuartio-Thlaspetum, Shimwell, 1968; *Dryas octopetala* to the Elyno-Dryadetalia Br.-Bl., 1948, though occasionally in Seslerio-Mesobromion

   Class III. Species often occurring in other vegetation units but with their optimum definitely in one unit (German holde). *Bromus erectus* in the Cirsio-Brometum, Shimwell, 1968, though also in the Seslerio-Caricetum pulicaris at its regional limits

2. *Companions* (Begleiter)

   II. Species without a definite preference for certain vegetation units (German: vage)
   Carex flacca in Seslerio-Caricetum pulicaris
   Urtica dioica in Hippophaetum

3. *Strangers*

   I.   Species rare or accidental in the vegetation unit in question (German: fremde)
   Ononis repens to Hippophaetum as a relict from an earlier stage in succession
   Montia perfoliata in the Hippophaetum as an introduced species more or less peculiar to the Spurn Point *Hippophaë* scrub

which seem to distinguish between related associations are commonly constant species of low fidelity which may or may not confer a recognizable structure on the association, e.g. *Helictotrichon pubescens*, *Carex pulicaris* or *Dryas octopetala* for the three British associations of the *Seslerio-Mesobromion*.

Frequently, transgressive character species or species with an overlapping character value which are generally regarded as faithful species to an alliance crop up in a single association of a related alliance. *Asperula cynanchica* in Britain is such a species. It is fairly widely distributed in British associations of the *Eu-Mesobromion* but only occurs in one association, *Asperulo-Seslerietum*, in the *Seslerio-Mesobromion* (Shimwell, 1968) (Table 56). It is thus quite a widespread *Eu-Mesobromion* character species of high fidelity (IV) and an equally good character species for the single association within the *Seslerio-Mesobromion*. A transgressive species thus has character value for vegetation units of different syntaxonomic rank.

TABLE 56

Asperula cynanchica as a transgressive character species

	A	B	C	D	E	F	G	H	I	J	K	L	M	N
Number of relevés	70	22	30	33	5	7	7	11	19	34	44	47	13	10
Asperula cynanchica	IV	V	–	III	V	–	V	V	–	–	–	–	V	V

Eu-Mesobromion   Seslerio-Mesobromion

A. Cirsio-Brometum. Jurassic limestones of England
B. Cirsio-Brometum. Chalk of south-west England
C. Helianthemo-Koelerietum. North Wales
D.   ,,   ,,   South Wales
E. Caricetum montanae. South Wales
F, G. Antennarietum hibernicae. West Ireland
H. Camptothecio-Asperuletum. West Ireland and South Wales
I, J. Helictotricho-Caricetum flaccae. Derbyshire and Somerset
K. Seslerio-Helictotrichetum. East Durham
L. Seslerio-Caricetum pulicaris. Teesdale and Craven Pennines
M. Asperulo-Seslerietum. South Lake District and West Ireland
N.   ,,   ,,   West Ireland

In addition to absolute and transgressive character species, one frequently finds reference to geographical (regional) and local character species. The orginal definition of a regional character species supplied by Braun-Blanquet and Moor (1938) was a species which is definitely limited to the range of distribution of the association and never or seldom present outside that range. Under this definition they are more or less identical with absolute character species in the accepted modern meaning. On the other hand, a local character species is a faithful species to a single association for only a part of the range of the association, having limited geographical distribution within the association. This same species may, however, be a character species for other associations but in another region or it may serve as a differential species of a sub-association. Considerable confusion over the nomenclature of geographically restricted character species has arisen in the past. Meijer Drees (1951) attempted to overcome this confusion by re-defining the regional character species as a species characteristic of the whole range where vegetation unit and character species occur in common. Within this definition the species/unit relationship is defined as:

(a) conregional—range of species and vegetation unit coinciding.

H

(b) intraregional—range of species within the range of the vegetation unit.

(c) circumregional—vegetation unit range within the species range.

(d) pararegional—species range and vegetation unit range overlapping partly only with each other.

Examples of the four types of regional character species are shown in Figure 52 at the levels of two vegetation units, association and sub-alliance. As often happens, there are frequently no conregional and intraregional character species at the association level. Circumregional species are well represented and it is usually on this particular group of species that the association is characterized. An example of a pararegional character species is to be found in *Bromus erectus* which in the north of England is at its geographical limits and its distribution only just overlaps with that of the association. At the level of sub-alliance, *Sesleria caerulea* becomes a conregional species being more or less restricted to this sub-alliance in the British Isles while the local species *Polygala amara* is a good example of an intraregional character species.

The same categories exist for local character species or those characteristic for part of the total range where the vegetation unit and species occur in common. In this case, the ecological distribution of the plant is given more attention. Thus in the magnesian limestone region of Co. Durham (the distribution of *Seslerio-Helictotrichetum*), *Antennaria dioica* and *Epipactis atrorubens* are restricted to stands of this association and are intralocal character species. Similarly, *Plantago maritima* occurs on coastal cliffs and in disturbed magnesian limestone grasslands and is thus a circumlocal character species for *Seslerio-Helictotrichetum*. It should also be apparent that these local character species can serve as differential species for the delimitation of sub-associations. A similar situation prevails at the sub-alliance level.

Regional and local character species are of character value in a large regional for a number of allied associations or vicariants. In north-west Europe there are two types of vicariant series:

(a) Broad regional variation comprising four major vicariant zones (Figure 53): (i) *Cirsio-Brometum* type—perhaps the characteristic *Eu-Mesobromion* association type of north central Europe from Germany to southern England and southern Sweden; with four characteristic species present, *Bromus erectus, Brachypodium pinnatum, Cirsium acaulon* and *Asperula cynanchica;* rich in alliance character species; (ii) *Avenetum* type—lacking the four above species and with *Helictotrichon (Avena) pratense* and *Festuca ovina* the predominant grass species; fewer alliance character species of which *Helianthemum chamaecistus* and *Poterium sanguisorba* are common—west central

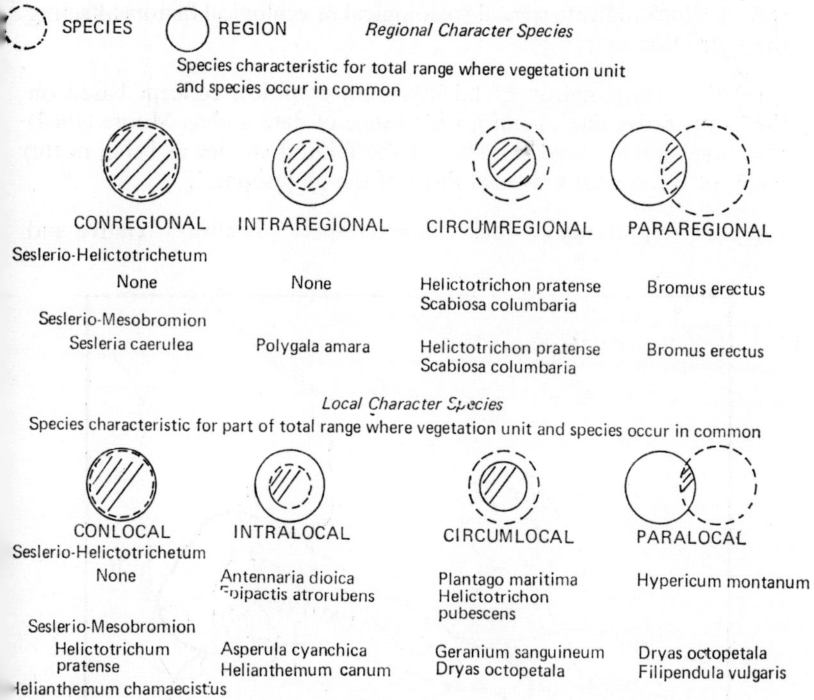

◯ SPECIES ◯ REGION *Regional Character Species*

Species characteristic for total range where vegetation unit
and species occur in common

CONREGIONAL	INTRAREGIONAL	CIRCUMREGIONAL	PARAREGIONAL

Seslerio-Helictotrichetum

| None | None | Helictotrichon pratense | Bromus erectus |
| | | Scabiosa columbaria | |

Seslerio-Mesobromion

| Sesleria caerulea | Polygala amara | Helictotrichon pratense | Bromus erectus |
| | | Scabiosa columbaria | |

Local Character Species

Species characteristic for part of total range where vegetation unit and species occur in common

CONLOCAL	INTRALOCAL	CIRCUMLOCAL	PARALOCAL

Seslerio-Helictotrichetum

| None | Antennaria dioica | Plantago maritima | Hypericum montanum |
| | Epipactis atrorubens | Helictotrichon pubescens | |

Seslerio-Mesobromion

| Helictotrichum pratense | Asperula cyanchica | Geranium sanguineum | Dryas octopetala |
| Helianthemum chamaecistus | Helianthemum canum | Dryas octopetala | Filipendula vulgaris |

Figure 52. Regional and local character species.

England and southern Scandinavia; (iii) *Western* type—lacking *Helictotrichon pratense* and with *Helianthemum chamaecistus* and *Poterium sanguisorba* of low constancy; poor in alliance character species; Wales and western Ireland; (iv) *De alpine* type—characteristic of the pre-alpine region of southern Germany where alpine species invade the *Cirsio-Brometum* type.

(b) Vicariance on a local regional scale within these main zones, e.g. variation between the *Cirsio-Brometa* of the Jurassic limestone and chalk downlands of England, manifest via presence of more alliance character species in chalk areas such as *Phyteuma tenerum, Thesium humifusum,* etc.

From the foregoing discussion with the examples quoted it should have become clear that:

(a) the term character species should be reserved for species used in the recognition and definition of vegetation units corresponding with their fidelity and that these are special cases of ordinary differential species which indicate special sociological or ecological factors affecting the vegetation unit;

(b) the determination of fidelity is an empirical concept based on the comparative analysis of a wide range of data and as Moore (1962) says 'one should avoid thinking of the faithful species as being in the realm of the eternal universal ideas of the philosopher';

(c) the importance of each character species is always relative and

Figure 53. The distribution of the major Association-groups of the Alliance Mesobromion in north-west Europe.

all degrees of fidelity exist, most character species being of reliable diagnostic value only within certain geographical limits;

(d) absolute association character species are rare, perhaps non-existent above the level of ecotype or intraspecific taxon of uncertain taxonomic status (e.g. *Thlaspi alpestre* var. *calaminare*); but that regional and local association character species are extremely valuable in defining and describing vegetation units;

(*e*) absolute character species, however, are relatively common at the alliance and association level.

(*f*) the overestimation of the value of fidelity in the establishment of vegetation units and the misconception that the determination of fidelity involves a circular argument (Poore, 1956) are criticisms which do not apply to the Z-M system as it is practised in Europe at the present time.

The final words on the subject are taken from Braun-Blanquet (1959) himself (as translated by Moore, 1962):

'It should not be concluded, as has sometimes been maintained, that the associations are based on fidelity. This is emphatically not the case. The association is an abstraction based on the totality of more or less homogeneous relevés which floristically correspond closely with one another; it is however, characterized not merely floristically, but also ecologically, dynamically and geographically (*chorologisch*). Nevertheless, in distinguishing from one another the associations which have been conceived on a floristic basis, far greater importance and more general significance is attributed to fidelity than to purely quantitative characteristics, especially when fidelity is combined with high constancy.'

The table and text concerning the association *Seslerio-Helictotriche-tum* epitomize the combined floristic, dynamic, ecological and chorological approach of the Z-M School. When it comes to the relation of the association to a reference point in a hierarchical classification, it is clearly advantageous to have the background knowledge on species distributions such as that provided in Table 50. But this type of result is only the product of a detailed phytogeographical and phytosociological knowledge based on extensive field work and derived from literature. In an attempt to provide this background, the major classificatory units of the Z-M system derived from Lohmeyer *et al.* (1962) are summarized in Appendix III.

Scandinavian Traditions

Scandinavian methods of vegetation description fall into three broad groups—the Uppsala Tradition which for a long time dominated Scandinavian phytosociological studies; the Norse–Scots School which in many ways combines the best approaches of the Uppsala and

Zurich–Montpellier Schools; and the Danish School represented mainly in the works of Raunkiaer. These three different approaches are considered in the subsequent sections of this chapter, each approach being illustrated by reference to a particular paper or papers which appear to be the most representative of the tradition. In this way, the difficulties of interpretation are minimized, since reference to copious works often produces ambiguities.

As field material to exemplify the methods of the various schools, data from five areas of soligenous mires dominated by *Molinia caerulea* and *Myrica gale* from the Glen Trool Forest Park in southern Scotland is used. This data is presented in different table forms to enable a grasp of the methods and, it is hoped, a guide to table interpretation for those unfamiliar with German or the Scandinavian languages. Vegetation of this type, relatively poor in species and with several prominent physiognomic features was chosen deliberately in an attempt to fully comprehend the arguments against the floristic characterization of vegetation put forward by Scandinavian workers on the grounds that their vegetation is comparatively uniform and species poor. The *Molinia-Myrica* vegetation which is particularly characteristic of the Glen Trool region of the southern uplands of Scotland is widespread on thin topogenous peats at the base of montane slopes which supply a continuous source of ground water, or on the sloping rands of ombrogenous peat bogs.

The Uppsala School

Reference to Chapter 2 and the appropriate figures will indicate that the concepts of the Uppsala School fall into three periods of investigation:

1. The early period up to 1928 when the term association was used to describe a real or 'concrete' vegetation unit observable in the field, which had a certain minimal area and which could be defined by constant species—those which occurred in all or 90% of the sample quadrats. Above the association level, the next unit, the association-complex, corresponded to a definitive topographic or ecological entity, whilst the next level, the formation, comprised a number of association-complexes of related physiognomy.

2. The period 1928–35 when the term sociation proposed by Rübel (1927) was accepted by Du Rietz (1930) and a complex nomenclature built up around the term, with greater emphasis placed on the homogeneity of the different layers of the vegetation.

3. The period 1935 onwards saw an increasing convergence towards Zurich–Montpellier methods and also a trend towards an ecological description and classification of vegetation as opposed to one which was

based solely on the characteristics of the vegetation itself—as had been the case in the second period.

There is clearly a need for amplification of the trends in those three periods, but in doing so there is a danger of misinterpretation of the methods of the Uppsala School as they are now practised. Clearly the methods of each period must be considered separately and interpreted in their own light to enable an understanding of the literature of the period. Some of this literature still ranks as classic: for example, there is seldom a paper on mire ecology which does not carry a reference to the work of Osvald (1923). But the two earlier periods need only a summary reference.

Du Rietz et al. (1920) erected four major 'laws' of vegetation which summarize the work of the first period.

(a) Every sociation (association in the literature of the period) requires a specific minimal area for its development. Each layer of the community has its own minimal area, the size of which depends upon (i) the size of the component plants; (ii) the number of different species present; (iii) the population sizes of the component species. To determine the minimal area of each stratum, a large sample plot of the sociation is subdivided into smaller areas until a single group of constants, or a single dominant characterizes the layer.

(b) Developing from this, every sociation has a number of dominant species, and one or more will be present in every sample stand of the sociation provided the sample is larger than the minimal area. These species are also known as constants.

(c) There are degrees of phytosociological affinity between constants of the same sociation.

(d) Sociations are divided from one another by distinct boundaries.

Analysis techniques of the period involved the division of the single-layer community into a number of small quadrats usually 0·25 m² and ten in number. After noting species occurrence and cover values according to the Hult-Sernander Scale (see Table 57) species constancy (referred to as frequency) was calculated along with '*durchschnittlicher Bedeckungsgrad*'—the arithmetic mean of the cover values. Thus Osvald (1923) recognized no less than 164 different sociations on a single peat bog of 40 square miles, which he placed into seven association-complexes. These sociations are concrete units readily studied in the field and with homogeneity in all layers. The type of unit described may be imagined with reference to a simple situation of *Molinia-Myrica* mire. Some areas have a canopy of *Myrica gale* over a ground layer of *Molinia caerulea*, whilst others lack *Myrica* and are composed of *Molinia* with a *Sphagnum* moss layer. In Osvald's terms these two types are two associations, a *Myrica-Molinia* Association and a *Molinia-Sphagnum* Association.

TABLE 57

The Hult-Sernander scale of cover representation

Degree of cover	Cover range	Middle of cover range
1	to 1/16	1/32
2	1/16–1/8	3/32
3	1/8–1/4	6/32
4	1/4–1/2	12/32
5	>1/2	24/32

The period 1928–35 was notable for three main features: (i) the adoption of the term sociation or microassociation for what had previously been termed the association; (ii) the development of the stratal aspects of community study; (iii) the development of a complex classification of both layers and communities.

The bases of these developments have already been reviewed in Chapter 2 and summarized in Figures 18 and 19. The methods were first published in a volume on plant sociology, *Abderhaldens Handbuch der biologischen Arbeitsmethoden* in 1930, and in a Scandinavian journal as 'a compromise between the systems of the leading schools of present-day ecology' (Du Rietz, 1931), in preparation for their proposed acceptance by the Fifth International Botanical Congress of 1930. The classificatory methods, however, were not accepted and at the next Congress were rejected in favour of the Zurich–Montpellier system. As such, the different levels of classification, the stratal communities, or synusiae, remain mainly as examples in primarily theoretical texts and appear only infrequently in detailed analyses.

The third period is distinguished by the development of the resolutions of the 1935 Congress and their application to Scandinavian vegetation. One of the first contributions to these resolutions is probably to be found in the work of Nordhagen (1936) on the sub-alpine and alpine vegetation of Norway, in which he related much of his previously amassed data to the alliance and order units developed by the Z-M School. It is from the work of Nordhagen that the later developments of the Norse–Scots School were made and a review of his work would be equally at home in that section to give full continuity of concept development. However, the use of the Hult-Sernander scale and the unit, sociation, requires its inclusion in the present section.

In most cases the sociations of Nordhagen (1936) were united directly into alliances as constancy figures:

Sociation	I	II	III	
Number of quadrats and stands	10 : 5	20 : 7	10 : 4	etc.
* Dryas octopetala	100^2	80^2	100^4	
* Salix reticulata	60^1	100^1	80^1	
etc.				

where the figures 100^2 denote a constancy in the sociation of 100%
with an average cover value of 2. The character species for the alliance
were denoted by an asterisk and other ecological indicator species
received an appropriate letter to denote their habitat relationships,
e.g. (K) = *Kalkliebende* or calcicolous.

In his next major treatise (1943) on the vegetation of the Sikilsdalen
region of Norway, he adopted a technique of recording and tabulation
shown in Table 58. Stands of vegetation are chosen and within each
stand a varying number of 1 or 4 m² quadrats (mainly 5 or 10) are
recorded for floristics and cover. The quadrats are then grouped into
sociations where there is a recognizable physiognomic dominant, e.g.
Myrica gale in Stands I and II, *Molinia caerulea* in Stands III, IV and V
in Table 58. The species are arranged in groups according to their
physiognomic life forms in the order shrubs, dwarf shrubs, grasses,
sedges and rushes, herbs, mosses, liverworts and lichens partly to give an
impression of layering of the vegetation types, partly as a method of
rapidly finding a particular species. Sociations thus recognized are
grouped together on physiognomy and floristic affinities mainly via
the constant species (species of constancy III and above). At the
extreme right of the table figures for the constancy of each species in
the table as a percentage figure and dominance as a mean value of the
total species occurrences with + for greater than the value, − for less
but more than half, are added either for each component sociation or
the association as a whole. At the base of the table the constancy values
are summarized in a constancy diagram where the species are recorded
in one of ten constancy class ranges. Associations are grouped into
alliances according to the relationships of their constant species using
Jaccard's (1902) coefficient of similarity $K = c/a+b-c \times 100$ where
a is the number of constants in the first association, b the number in
the second and c the number common to both. Applying this to the
two sociations in Table 58, the formula is represented as $6/10+7-6 =$
54·5% similarity. Nordhagen used a figure of 50% similarity as the
level necessary for two sociations to be united into an association. It is
in this aspect of the use of coefficients of similarity that the methods
of the Norse–Scots School are closely related to those of Nordhagen
(q.v.).

TABLE 58

The Association Myrico-Molinietum: Nordhagen's Table Method

Sociation	Myrica—Molinia *Sociation*									Molinia *Tussock Sociation*						
Nature of substrate	*Thin topogenous peat*															
Stand number	I		II		III			IV			V				C	D
Plot number (4 m²)	1	2	3	4	5	6	7	8	9	10	11	12	13	14	15	
Myrica gale	5	5	5	5	5	5	1	2	4	3	1	2	2	1	2	100 3+
Calluna vulgaris	1	–	–	–	1	1	1	1	1	–	–	1	2	1	1	67 1
Erica tetralix	1	1	1	1	1	–	1	1	1	1	1	–	–	1	2	80 1
Anthoxanthum odoratum	–	–	1	–	–	–	–	–	–	–	–	1	–	–	–	13 —
Carex nigra	–	–	1	–	–	–	–	1	–	–	–	–	–	1	1	26 1
Deschampsia flexuosa	–	–	–	–	–	–	–	–	–	1	1	–	–	–	–	13
Molinia caerulea	2	4	2	2	3	4	5	5	5	5	5	5	5	5	5	100 4
Trichophorum caespitosum	–	–	–	1	1	2	2	–	2	1	–	–	–	–	2	46 1+
Drosera rotundifolia	–	–	–	1	–	–	–	–	–	–	–	–	–	–	1	13 —
Dryopteris spinulosa	–	–	–	–	–	–	–	–	–	1	–	–	–	–	–	7 —
Galium saxatile	–	–	1	–	1	–	–	1	1	–	1	–	1	1	1	53 1
Pedicularis sylvatica	1	1	1	1	1	–	–	–	–	–	1	1	1	–	–	53 1
Polygala vulgaris	–	–	–	1	1	–	–	1	1	–	–	–	–	–	–	26 1
Potentilla erecta	–	1	1	–	–	1	1	1	1	–	–	1	1	1	–	59 1
Hypnum cupressiforme	1	–	1	1	–	–	–	–	–	4	1	–	–	1	–	33 2
Orthodontium lineare	–	–	–	–	–	–	–	–	–	–	1	–	–	–	–	7 —
Sphagnum papillosum	–	–	–	4	1	1	2	–	1	1	–	–	–	1	1	53 2+
,, rubellum	–	–	–	1	–	3	1	–	–	1	–	–	–	1	2	40 2+
,, subsecundum	–	–	–	1	4	1	–	–	–	–	–	–	–	1	1	33 2
Lepidozia setacea	–	–	–	–	1	–	–	–	–	–	1	–	–	–	–	13 —
Odontoschisma sphagni	–	–	–	–	–	1	1	–	–	–	–	–	–	1	1	26 1
Cladonia arbuscula	1	1	–	–	1	–	–	–	–	1	–	–	1	–	–	33 1
,, coccifera	–	–	–	–	–	–	–	–	–	1	1	–	–	–	–	13 —
,, uncialis	–	–	–	–	–	–	–	–	–	–	–	–	1	–	–	7 —
Phanerogam number	5	7	8	6	7	4	5	8	7	5	6	5	6	7	8	
Cryptogam number	2	1	1	4	4	4	3	0	1	5	4	1	1	5	4	
Total	7	8	9	10	11	8	8	8	8	10	10	6	7	12	12	

Average species number = 9

	0	10	20	30	40	50	60	70	80	90	100%
Constancy diagram	3	5	3	3	2	4	1	–	1	2	

Other authors, whilst accepting the classificatory levels of the Z-M system, preferred to erect their own systems within this framework. At first, the classification proceeded only to the level of the alliance (Du Rietz, 1949) but later full classifications were produced (Du Rietz, 1954). Much attention was focused on gradients of variation, particularly in mire vegetation and the plant communities were related to major environmental and vegetational gradients (Sjörs, 1948; Malmer, 1957, 1962). Recently Malmer (1968) has combined both approaches in a detailed review of mire types in southern Sweden, and it is via the work of this author that modern trends within the school are perhaps best exemplified.

The descriptive method begins with the siting of one or several small quadrats, 0·25 m² in each plant community *distinguished on the site*. This size is in general usage for the study of mire communities, but 1 m² quadrats have been used for other more homogeneous communities since the earliest period of the development of the school (Du Rietz, 1931, *vide* Brooks and Chipp, 1931, p. 175). Malmer, however, states that no two sociations are represented within a small quadrat, the familiar assessment of homogeneity. Within each quadrat all species are listed with their degree of cover derived from the Hult-Sernander scale (Table 57). To facilitate a qualitative comparison of quadrats, the small quadrat frequency (SQF, Du Rietz, 1957) and the characteristic degree of cover are calculated for those plant communities from which ten or more small quadrats are available. Note must be made here of the reference to plant communities as though they are concrete units recognizable in the field and not abstracted from the field data.

The SQF refers to the number of quadrats in which a species is found expressed as a percentage of the total number of squares—essentially the same thing as Osvald's '*Konstanzzahl*' or 'K'. The characteristic degree of cover as used by Malmer (1962) is calculated in the following manner. The cover values are transformed to the mean of the cover classes (Table 57), added, and divided by the number of quadrats in which the species is found. Two other related methods have been widely used. Du Rietz and all the investigators of the early period added the cover values and calculated the arithmetic mean, and this method is also found in several of the Norse–Scots School methods. Sjörs (1954), on the other hand, transforms the cover values to the upper limits of each class, i.e. for 1, 2, 3, 4 and 5, 2/32, 4/32, 8/32, 16/32 and 32/32 (Table 57). After division by the total number of quadrats the mean degree of cover is arrived at.

The small quadrat frequency values along with the characteristic degrees of cover as an exponent (SQF^c) are used for comparisons between different plant communities or as Malmer calls them small associations. These latter are delimited mainly according to their species composition but two kinds of species are regarded as important: *characteristic species*—'species more or less confined to one community'; *differential species*—'species more or less confined to one of two communities compared to each other'. Both these types can be divided into exclusive species and preferential species (the *Leitarten* and *Scheidearten* of Du Rietz, 1942). The quadrats are united into small association tables in the same manner as Table 58 with the species enumerated in their usual order according to layering, physiognomy, etc. The constancy and dominance calculations of Nordhagen are replaced with SQF^c at the right-hand side of the table. The different units in Table 58 would probably be regarded by Malmer as:

I. *Molinia-Myrica* small association.

II. *Molinia-Myrica* small association, *Trichophorum-Sphagnum* variant.

III, IV and V. *Molinia* tussock small association.

When further synthesis is required, Malmer (1962) groups the component mire species into seventeen distribution types, based on the overall distribution pattern of the species on the mire under study. These distribution types are then related to two major directions of variation, the mire expanse-mire margin direction of variation and the poor-rich direction of variation from ombrotrophic to minerotrophic mire type. Thus, the *Carex canescens* and *Potentilla erecta* types are characteristic species groups of mire margins, whilst the *Sphagnum imbricatum*, *Rhynchospora alba*, *Trichophorum caespitosum* types are characteristic of the central mire or mire expanse. Here, it is interesting to note that the character species refer to a species of a particular ecological or topographical distribution.

Each of the two groups of variation are then divided into series according to the variation of vegetation along the mud bottom-hummock gradient, six series being recognized in the mire expanse group and four in the mire margin. Each series is divided into smaller units along the poor-rich direction of variation and the divisions are made with the aid of the species groups or distribution types. Each unit is called a small association and because it is delimited only by species composition it is not really a sociation or even a micro-association of Rübel (1927) but rather a unit associated with the peculiar mosaic vegetation of mires. Malmer (1962) points out that as units they are valid only for the one mire, but in his later paper (1968) he extends their validity to a regional unit and for reasons which are not explicit in the paper proceeds to call them associations.

If one takes the *Molinia-Myrica* data as an example of a series in the sense of Malmer then the comparative series table would be represented in the manner of Table 59, where the species groups are enumerated down the left-hand side with their SQF^c opposite. Thus, the major species distribution groups of the solignous mires of the Galloway region are (i) the *Myrica-Molinia* group; (ii) the *Trichophorum-Sphagnum* group; (iii) the *Erica tetralix-Gallium saxatile* group; and

TABLE 59

Molinia-Myrica series table, after Malmer, 1962

A. *Molinia Myrica* small association typical variant
B. *Mol-Myr.* s-a Trichophorum-Sphagnum var.
C. *Molinia* tussock small ass.

	A	B	C
Quadrat numbers	16	16	48
Myrica gale	100[4]	100[4]	40[2]
Molinia caerulea	100[2]	100[2]	100[5]
Calluna vulgaris	–	–	40[2]
Trichophorum caespitosum	–	65[1]	15[1]
Sphagnum papillosum	–	90[3]	10[1]
S. rubellum	–	40[1]	10[1]
S. subsecundum	–	35[1]	–
Drosera rotundifolia	–	10[1]	–
Erica tetralix	50[1]	45[1]	30[1]
Galium saxatile	60[2]	–	15[1]
Potentilla erecta	15[1]	–	25[1]
Pedicularis sylvatica	40[1]	25[1]	10[1]
Carex nigra	20[1]	–	10[1]
Hypnum cupressiforme	45[1]	25[1]	30[2]
Cladonia arbuscula	–	–	25[1]
Polygala vulgaris	–	15[1]	5[1]
Deschampsia flexuosa	–	–	5[1]

Occasional species: Anthoxanthum odoratum, Cladonia coccifera.
(Samples 0·25 m² were taken in Plot numbers 3, 4, 8, 10 and 13, ref. Table 58.)

(iv) the *Hypnum-Cladonia* group. Such species groups as those enumerated by Malmer (1962) were later used by him to characterize associations (1968). His *Myrica-Molinia-Sphagnum* association thus contains representatives of some mire species groups of which the *Trichophorum-Sphagnum*, *Myrica-Andromeda* and *Molinia-Narthecium* groups are the strongest represented. He related his associations to the classifications of mires developed by Duvigneaud (1949) and Du Rietz (1954), adding several units of his own. The position of the *Molinia-Myrica-Sphagnum* association is problematical but Malmer places it in a hierarchical system in the following manner:

Class Vaginato-Sphagnetea Duvign. 1949
Order Trichophoro-Sphagnetalia Malmer 1965
Alliance Caricio-Sphagnion papillosi (Duvign. 1949) Malmer
Association *Myrica-Molinia-Sphagnum*

On the contrary Du Rietz (1954) would place the association in the following position:

Class Sphagno-Drepanocladetea
  Order Apiculetalia
  Alliance Euapiculation } from *Sphagnum apiculatum* agg.

This latter example provides a clear-cut indication of the lack of a uniform formal classification method in the Uppsala Tradition due mainly to the lack of a basic set of species types by which the communities are recognized, such as the character and differential species of the Z-M School. In his work on the late alpine snow beds of Norway, Gjaerevoll (1956) suggests that if a sociological classification were to be based exclusively on the principle of character and differential species, an obscurity of interrelationships would result. In place of these two concepts he uses groups of species which occur as dominants and have a narrow ecological amplitude. To these he attaches the 'very greatest sociological importance'. Malmer uses floristic groups of a certain ecological amplitude, Gjaerevoll, perhaps in the truer sense of the Uppsala Tradition, uses dominance groups of limited ecological amplitude. For his snow-bed community classification, Gjaerevoll bases the first major division into series of communities rich or poor in calciphiles—analogous to Malmer's poor-rich gradient of variation, but placing greater emphasis on vegetation than on environment. Within these two series two major sub-series, rich and poor in hygrophilous species due to differences of water supply from the melt water of the snow-beds are recognized. The sub-series contain alliances in the generally accepted sense of the term. These are assigned to a particular sub-series according to the presence or absence of a field layer, or in other words, whether the alliances are dominated by cryptogams or grasses and heath species.

It may, therefore, have become obvious that the modern trends within the Uppsala Tradition have been twofold: (i) The acceptance of the Z-M hierarchical units of classification, with attempts to relate Scandinavian communities to the system of classification already worked out for Central Europe, using character and differential species in the Z-M sense; (ii) the acceptance of the units of the Z-M hierarchical classification, but with the erection of a separate classification based on ecologically exclusive or preferential species of high constancy or dominance. These species have also been used to describe ecological gradients along which the communities are arranged in series.

The Norse–Scots School

The development of this school may be considered to begin with the work of Dahl and Hadač (1941, 1949), although several of its principles are drawn from the older works of Domin (1928) and Nordhagen (1936). The basic characteristics of the methods of this school are as follows:

(a) An eleven category scale of cover values—the Domin scale—is used to describe a homogeneous vegetation type.

(b) Plot samples are united into associations on the basis of overall floristic similarity, dominance and constancy. Generally speaking, the associations are characterized by a greater number of species of constancy class V than class IV; the concept of the character species is rejected.

(c) The Sørensen coefficient of similarity is used to indicate affinities between associations and for allocating them to a particular alliance.

(d) The general Z-M system of hierarchical units forms the basis of the classification and is supplemented with data from the works of Nordhagen (1936, 1942) and Dahl (1956).

(e) The vegetation is studied relative to major environmental gradients or in a general ecological framework from which groups of ecologically related species can be extracted.

The two early works of Dahl and Hadač (1941, 1949) apparently involved the assessment of homogeneity by eye or by the method of increasing the plot size to produce a flattening in the species number curve (see Chapter 1) which was taken as the homogeneous minimal area. The lists from each quadrat were united into 'homogeneous tables' with species arranged in order of decreasing constancy (Table 60). Three columns to the left of the table recorded constancy, the average cover value calculated as the total cover values over the number of occurrences, and the Raunkiaerian life form. In this way, a simple but fairly definitive association table was produced.

The later work of Dahl (1956) saw some alteration of the method of representation of tables in a reversion to the Uppsala method of separating the community components into their layers or taxonomic categories as shrubs, monocotyledons, herbs, liverworts, mosses and lichens. The field analysis differs from Uppsala methods in that only one quadrat is recorded in a particular vegetation stand, not several, and in consequence each stand is documented at the top of the table with notes such as slope, aspect, altitude, etc. To replace the use of character species in uniting associations into alliances, Dahl makes use of Sørensen (1948) coefficient $2c/a+b$ where c is the number of features in common with two units a and b. This method has been used by several exponents of the Z-M School, but it is only in the work of Dahl that it becomes the focal point of study. From his association tables, Dahl derives four properties for the calculation of correlations (K):

Ksp —the numbers of species present;

Ksc —using the total sum of constancy class numbers;

TABLE 60

Molinia-Myrica data: Dahl and Hadač type representation

Species number Quadrat size (m²)	1 7 4	2 8 4	3 8 4	4 10 4	5 11 4	6 8 4	7 8 4	8 8 4	9 8 4	10 10 4	11 10 4	12 6 4	13 7 4	14 12 4	15 12 4	C	A	L
Molinia caerulea	4	6	4	4	5	6	9	9	9	10	10	9	10	8	8	V	7·4	H
Myrica gale	8	9	8	8	9	9	3	4	6	5	3	4	4	3	4	V	5·8	N
Erica tetralix	2	2	2	3	1	–	2	1	1	+	2	–	–	3	4	IV	1·5	Ch
Calluna vulgaris	+	–	–	–	2	1	1	2	3	–	–	+	4	2	2	IV	1·1	C
Potentilla erecta	–	1	2	–	–	2	2	2	2	–	–	2	2	3	–	III	1·2	H
Galium saxatile	–	–	3	–	3	–	–	2	3	–	3	–	3	2	2	III	1·4	H
Pedicularis sylvatica	2	2	2	3	2	–	–	–	–	–	2	2	3	–	–	III	1·2	H
Trichophorum caespitosum	–	–	–	3	2	4	4	–	4	2	–	–	–	–	4	III	1·5	H
Carex nigra	–	–	1	–	–	–	–	1	–	–	–	–	–	2	1	II	0·3	Cr
Polygala vulgaris	–	–	–	1	2	–	–	3	2	–	–	–	–	–	–	II	0·5	H
Anthoxanthum odoratum	–	–	2	–	–	–	–	–	–	–	–	2	–	–	–	I	0·2	H
Deschampsia flexuosa	–	–	–	–	–	–	–	–	–	3	1	–	–	–	–	I	0·2	H
Drosera rotundifolia	–	–	–	2	–	–	–	–	–	–	–	–	–	–	2	I	0·2	H
Dryopteris spinulosa	–	–	–	–	–	–	–	–	–	1	–	–	–	–	–	I	0·1	H
Sphagnum papillosum	–	–	–	6	2	3	4	–	2	3	–	–	–	3	3	III	1·7	M
S. rubellum	–	–	–	2	–	5	2	–	–	2	–	–	–	2	4	II	1·1	M
Hypnum cupressiforme	3	–	–	3	–	–	–	–	–	6	3	–	–	1	–	II	0·8	M
Cladonia arbuscula	2	2	–	–	2	–	–	–	–	2	–	–	3	–	–	II	0·7	L
Sphagnum subsecundum	–	–	–	1	6	2	–	–	–	–	–	–	–	2	3	II	0·9	M
Odontoschisma sphagni	–	–	–	–	–	2	2	–	–	–	–	–	–	2	2	II	0·5	M
Lepidozia setacea	–	–	–	–	1	–	–	–	–	–	2	–	–	–	–	I	0·2	M
Cladonia coccifera	–	–	–	–	–	–	–	–	–	3	2	–	–	–	–	I	0·3	L
Orthodontium lineare	–	–	–	–	–	–	–	–	–	–	1	–	–	–	–	I	0·1	M
Cladonia uncialis	–	–	–	–	–	–	–	–	–	–	–	2	–	–	–	I	0·1	L

C—Constancy; A—Average cover value; L—Raunkiaerian life form.

Ksd —dominance, where each species is given a mean rating of the Domin value in the scale and any species with a value of over 3·0 is considered to be a dominant;

Kspc—combines presence with species of constancy III and above and scores 1 for such combinations.

Table 61 shows these values recorded for the *Molinia-Myrica* data of Table 60 divided for the purpose of the illustration to give two community units based on physiognomic appearance. The question arises as to whether the two units are two associations or merely representatives of the same association. The Sørensen calculations are as follows:

$$\text{Ksp} = \frac{2 \times 19}{19 + 24} \times 100 = 88 \cdot 3\% \quad \text{Ksd} = \frac{2 \times 21 \cdot 1}{28 \cdot 6 + 30 \cdot 1} \times 100 = 71 \cdot 8\%$$

$$\text{Ksc} = \frac{2 \times 45}{51 + 60} \times 100 = 81 \cdot 1\% \quad \text{Kspc} = \frac{14}{20} \times 100 = 70\%$$

In general, Ksp tends to give a higher index of similarity than Ksc and Ksd. When it comes to assessing the relationships of two communities, Dahl remarks that it is impossible to give a particular order of magnitude of indices required for distinguishing the association, alliance or order to which a community belongs, but offers two simple rules. First, any vegetation type within an alliance should have higher indices of similarity with some other type within the same alliance,

TABLE 61

The comparative table of Dahl (based on the data of Table 60)

	Molinia-Myrica				Molinia tussock				Shared			
	p	d	c	pc	p	d	c	pc	p	d	c	pc
Calluna vulgaris	1	0·6	III	1	1	1·6	IV	1	1	0·6	III	1
Erica tetralix	1	1·6	V	1	1	1·4	IV	1	1	1·4	IV	1
Myrica gale	1	8·5	V	1	1	4·0	V	1	1	4·0	V	1
Dryopteris spinulosa	–	–	–	–	1	0·1	I	–	–	–	–	–
Anthoxanthum odoratum	1	0·3	I	–	1	0·2	I	–	1	0·2	I	–
Deschampsia flexuosa	–	–	–	–	1	0·4	II	–	–	–	–	–
Molinia caerulea	1	5·0	V	1	1	9·3	V	1	1	5·0	V	1
Carex nigra	1	0·1	I	–	1	0·4	II	–	1	0·1	I	–
Trichophorum caespitosum	1	1·5	III	1	1	1·6	III	1	1	1·5	III	1
Drosera rotundifolia	1	0·1	I	–	1	0·2	I	–	1	0·1	I	–
Galium saxatile	1	1·0	II	–	1	1·6	IV	1	1	1·0	II	–
Pedicularis sylvatica	1	1·8	V	1	1	0·6	II	–	1	0·6	II	–
Polygala vulgaris	1	0·5	II	–	1	0·6	II	–	1	0·5	II	–
Potentilla erecta	1	0·8	III	1	1	1·4	IV	1	1	0·8	III	1
Hypnum cupressiforme	1	1·0	II	–	1	1·1	III	1	1	1·0	II	–
Orthodontium lineare	–	–	–	–	1	0·1	I	–	–	–	–	–
Sphagnum papillosum	1	1·8	III	1	1	1·6	III	1	1	1·6	III	1
S. rubellum	1	1·1	II	–	1	1·1	III	1	1	1·1	II	–
S. subsecundum	1	1·5	III	1	1	0·6	II	–	1	0·6	II	–
Lepidozia setacea	1	0·1	I	–	1	0·2	I	–	1	0·1	I	–
Odontoschisma sphagni	1	0·3	I	–	1	0·6	II	–	1	0·3	I	–
Cladonia arbuscula	1	1·0	III	1	1	0·6	II	–	1	0·6	II	–
C. coccifera	–	–	–	–	1	0·6	II	–	–	–	–	–
C. uncialis	–	–	–	–	1	0·2	I	–	–	–	–	–

p—presence; d—dominance; c—constancy;
pc—combined constancy and presence (III and above).

than within any vegetation type outside the alliance; and secondly, the same premise applies to the alliance-order relationship. At the basic level of association interrelationships, presence and constancy indices of over 70% are generally regarded as being levels which indicate a single vegetation type. It thus seems that Dahl would regard the data in Table 61 as being representative of the same association. Like Nordhagen, Dahl maintains the use of the suffixes of the Z-M terminology and relates all his associations and alliances to the higher units of the central European school with only slight modifications.

At the same time as the emergence of Dahl's monograph on the vegetation of Rondane, Poore (1955, 1956) reviewed the Z-M system of phytosociology and adopted a procedure of vegetation description and classification similar to that of Dahl. He considers that the systematic methods of phytosociology should consist of four stages; choosing uniform areas of vegetation; describing these areas; tabulating the lists obtained and segregating from the tables lists which are sufficiently alike to be considered to belong to the same vegetation unit; grouping these units according to their affinities. Taking each step in turn, he employs similar homogeneity tests to those described by Dahl and follows the basic Z-M method of description (see Chapter 6) except that the Domin scale is preferred and a standard plot size of 4 m² used. The lists are tabulated and vegetation units abstracted by what was later called a 'method of successive approximation'. The units abstracted are referred to as *noda*—or units of any category, and these are distinguished by the criteria of constancy and dominance. In general, he follows Dahl and Hadač (1949) in separating stands into noda where the number of species in constancy class V is greater than the number in class IV. The units so distinguished are grouped according to their floristic affinities using the Sørensen coefficient of similarity, with slight modification.

Poore and McVean (1957) suggest a new approach to Scottish mountain vegetation which is similar to that of Dahl (1956) in that the vegetation is studied in relation to the 'ecological framework' consisting of such major environmental gradients as altitudinal zonation, oceanicity, snow cover, base status and soil moisture. This work was later greatly elaborated by McVean and Ratcliffe (1962) who describe the vegetation of the Scottish Highlands in relation to the ecological background. Their methods of analysis are those of Poore and their tables follow the standard order of species groups, e.g. shrubs, grasses and sedges, herbs, etc. Species of constancy IV and V in a particular nodum or association are reproduced in heavy type and the combination of these species is referred to as the *association-element*. No calculations of indices of similarity are undertaken, but the associations described are related to the classification of Dahl by overall floristic similarities. The descriptive methods of Poore are widely used in modern British ecology but apart from the classification attempts by McVean and Ratcliffe, systems for other regions are generally lacking.

Finally, with regard to the classification of the *Molinia-Myrica* example, if the ideas of McVean and Ratcliffe are followed, the vegetation type would remain as a nodum of uncertain classificatory position. Otherwise, the closest floristic affinities, using a Sørensen coefficient calculation, suggest that the closest relationships shown are with the associations of the alliance *Oxycocco-Empetrion hermaphroditi*—an alliance of ombrogenous bog vegetation.

The Danish School

The concepts of the Danish School were outlined by Raunkiaer (1910, 1913, 1918, 1928) and since then have had only limited application by other Danish workers, mainly in Denmark and Danish colonies, e.g. Böcher (1933, 1940) in Greenland and in the Faeroe Isles, although Resvoll-Holmsen (1932) in her work on the Norwegian alpine snow beds and Hansen (1932) in Iceland provide notable exceptions.

As a result of much deliberation on the size of the quadrat used Raunkiaer (1928) based his analyses on 0·1 of a square metre, remarking that this size quadrat obtains realistic and constant comparative numbers which are as exact as any practical method allows. In the determination of the degree of frequency of an individual species (or the valency as Raunkiaer calls it), floristic analyses of a number of 0·1 m^2 quadrats are taken within a vegetation type at random. The homogeneity of the vegetation is apparently assessed by eye. In each quadrat species are recorded as present or absent and the scores added for all the quadrats thrown in a particular vegetation type in a single locality, in Table 62 where twenty quadrats were thrown in each of five localities, in the first locality *Myrica gale* was present in all twenty and is therefore recorded as 100% frequency. *Hypnum cupressiforme* was only present in two of twenty quadrats, giving it a 10% frequency value. Tabulation of data from other stands makes possible a rapid comparison of the frequencies of all component species. Further, the separation at the top of the table with species of high frequency in at least one stand (forty-five and above) indicates the major variation in predominant species, i.e. differentiates the stands in frequency terms. Four types of mire are thus recognizable—stand 1, a *Myrica-Molinia* mire with a 100 : 70 frequency relationship for the two major species; stands 3 and 5, a *Molinia* tussock grassland with a 3 : 2 *Molinia* : *Myrica* relationship; stand 2, a *Myrica-Molinia* mire with subsidiary *Sphagnum-Trichophorum*; stand 4, a drier *Molinia* tussock grassland indicated by the subsidiary frequencies of *Hypnum cupressiforme* and

TABLE 62

Molinia-Myrica data. I. Danish descriptive method

Stand number		1	2	3	4	5
Myrica gale	N	100	100	65	60	70
Molinia caerulea	H	70	85	100	100	90
Sphagnum papillosum	M	–	60	20	5	15
S. subsecundum	M	–	45	–	–	20
Trichophorum caespitosum	H	–	45	30	5	5
Hypnum cupressiforme	M	10	10	–	80	5
Cladonia coccifera	L	–	–	–	70	–
Erica tetralix	Ch	30	25	25	5	15
Calluna vulgaris	Ch	5	15	20	5	20
Galium saxatile	H	10	10	10	10	25
Potentilla erecta	H	10	5	15	10	15
Sphagnum rubellum	M	–	20	5	10	15
Cladonia arbuscula	L	10	5	–	5	5
Polygala vulgaris	H	–	10	15	–	–
Carex nigra	Cr	5	–	5	–	5
Anthoxanthum odoratum	H	5	–	–	10	–
Deschampsia flexuosa	H	–	–	–	15	–
Drosera rotundifolia	H	5	–	–	–	5
Lepidozia setacea	M	–	5	–	5	–
Orthodontium lineare	M	–	–	–	5	–
Cladonia uncialis	L	–	–	–	5	–
Pleurozia purpurea	M	–	–	–	–	5
Campylopus atrovirens	M	–	–	–	–	5
Carex pilulifera	H	–	–	–	–	5

N—Nanophanerophyte; Ch—Chamaephyte; H—Hemicryptophyte;
Cr—Cryptophyte; M—Moss and liverwort; L—Lichen.
1 Myrica-Molinia mire.
2 Myrica-Molinia-Sphagnum mire.
3 and 5 Molinia tussock grassland.
4 Molinia-Hypnum dry tussock grassland.

Cladonia coccifera on the tussocks. The table is completed by the addition of other species often in descending order of frequency importance, and a column after the species names with their life-form abbreviations.

Raunkiaer (1928) viewed this method as one which enabled the representation of a continuous quantitative change in species populations as a series of discontinuous units which can be treated as quantitative units in the classification of vegetation. The tables if so arranged can give a picture of the dominance area of a species or the regions throughout which the species occurs as a frequency dominant—

with a valency of 80 or over. This method is particularly instructive in autecological studies or studies on the interrelationships of co-dominants. The table of *Molinia-Myrica* stands could be re-written to illustrate this relationship, i.e.

Stand number	1	2	5	3	4
Myrica gale	100	100	70	65	60
Molinia caerulea	70	85	90	100	100

with the additional species appended to show any theoretical eco-sociological effects of the co-dominants on their frequencies. The species of high frequency above 80 % are further used to define and characterize formations (although Raunkiaer uses the nature of the habitat as a preliminary basis for defining formations—see Chapter 2) but percentage frequencies as low as 29 and 30 are also used for forma-tion characterization when they are combined with species with higher frequency values. For example, in Raunkiaer's investigations of the heaths of Aadum-Varde Bakkeø in West Jutland (1910) (see Table 63, columns A, B and C) the *Myrica* formation is characterized by a combination of moderate frequencies for *Myrica, Molinia, Calluna* and *Erica tetralix*, unlike the stands represented in Column D from Skagens Odde (1913) which have monodominant *Myrica*. The point emerging is that there is considerable variation in the frequencies of the dominants and co-dominants within each formation. So that all the data of Table 63 with the exception of columns E and 3, 4 and 5 would be classified by Raunkiaer as the *Myrica* formation. To the individual stand units the name facies was applied by Raunkiaer (1910), since he thought that the use of the term association was not only 'super-fluous' but also false (1934, p. 380). There is thus a *Myrica-Eriophorum vaginatum* facies in column B, a *Myrica-Trichophorum-Sphagnum* facies in column 2, etc; facies which are recognizable by one or more species of limited distribution within the formation which occur in one or more stands with a significant subsidiary percentage frequency. The remaining stands in Table 63 would be called the *Erica-Calluna* formation (column E) and the *Molinia*-formation, *Myrica* facies (columns 3, 4 5).

As Raunkiaer states (1934, p. 417) the use of species frequency values serves as a basis for the direct comparisons of formations which are floristically similar, for the detection of floristic homogeneity and the recognition of sub-units or facies. But Raunkiaer also requires that the formations should be characterized by their physiognomy and biological life forms. A physiognomic characterization merely involves the calculation of the proportion of the species present and expressing

TABLE 63

Variation in species frequencies in stands of the '*Myrica*-formation' (Raunkiaer, 1910) from Denmark and Scotland

	Danish data					Scottish data				
	A	B	C	D	E	1	2	3	4	5
Myrica gale	29	35	35	100	76	100	100	60	65	70
Molinia caerulea	43	13	19	–	–	70	85	100	100	90
Erica tetralix	49	41	50	40	100	30	25	5	25	15
Calluna vulgaris	30	33	35	24	100	5	15	5	20	20
Eriophorum vaginatum	2	35	–	–	–	–	–	–	–	–
Trichophorum caespitosum	27	2	39	–	–	–	45	5	30	5
Agrostis canina	–	–	–	60	–	–	–	–	–	–
Sieglingia decumbens	–	–	–	52	–	–	–	–	–	–
Carex nigra	1	5	14	44	16	5	–	–	5	5
Empetrum nigrum	–	26	2	36	20	–	–	–	–	–
Salix repens	2	–	5	24	72	–	–	–	–	–
Sphagnum papillosum	–	–	–	–	–	–	60	5	20	15
S. subsecundum	–	–	–	–	–	–	45	–	–	20
Hypnum cupressiforme			~	–	–	10	10	80	–	5
Cladonia coccifera	–	–	–	–	–	–	–	70	–	–

A, B, C—Columns 1, 2, 3 in Raunkiaer (1934), p. 267, Table 34. *Myrica*-formation from Aadum-Varde Bakkeø, West Jutland.

D, E   —Columns 3 and 4 in Raunkiaer (1934), p. 320, Table 5; 3, *Myrica*-formation; 4, *Calluna-Erica*-formation with *Myrica* and *Salix* from Skagens Odde, North Jutland.

the figures as a percentage of the total species records. The biological characterization of the formation similarly requires the assignation of each species to a particular life-form type and the calculation of the proportion of this life form expressed as a percentage of the total species records. Table 64 shows the biological characterization of the Scottish *Myrica-Molinia* data, where nanophanerophyte and chamaephyte, hemicryptophyte and cryptophyte are considered together. Stands 3 and 5 can be seen to be closely related biologically and structurally, whilst stand 2 differs only in its cryptogam percentage. Stands 1 and 4 are quite distinct, both in the nanophanerophyte and cryptogam percentages.

Raunkiaer uses the term 'formation' as 'species' is used in taxonomy, and just as species are grouped into genera and families, so formations can be grouped into groups and classes of formations. His 1910 conspectus of plant communities which, although he acknowledged was based on incomplete data, he thought would be applicable to

TABLE 64

Molinia-Myrica data: Biological characterization

	Species number	Species records	Life-form structure percentage		
			N and Ch	H and Cr	ML
1. Myrica—Facies	11	52	52	40·5	7·5
2. Myrica-Molinia—Facies	15	89	32·5	35	32·5
3. Molinia-tussock—Facies	12	63	35	43	22
5. Molinia-tussock-Facies	18	67	31·5	44·5	24
4. Molinia-Hypnum—Facies	17	81	17	37	46

Denmark at least and perhaps even further afield. But after enumerating the classes and groups he then stated that he would not keep to the classificatory order because of what seemed to be environmental gradients between classes and groups. Nevertheless, the classification is of interest as a broad ecological framework and as an illustration of Raunkiaer's 'vegetation reflects environment' ideas. The *Myrica* and *Molinia* formations from Scotland would find a classificatory position as follows, where the formation may be keyed out in the manner of a species in a dichotomous key:

I. Formations of fresh soil (Formation Branch)
II. Formation of salt soil
   A. Formations of aquatic habitats (Formation Class)
   B. Formations of terrestrial habitats
      1. Formation of clay (Formation Group)
      2. Formations of sand
      3. Formations of humus
         (*a*) Formation of dry heaths (Formation Series)
         (*b*) Formation of wet heaths and mires (peat)
            α *Myrica* wet heath formation
            β *Molinia* wet grass-heath formation

Perhaps the best-known applications of the Raunkiaerian frequency-dominant system of definition of plant communities is seen in the work of Böcher (1933, 1940, 1943, 1954) in his work on the heaths of the Faeroe Isles and north-west Europe and the vegetation of southern Greenland. In his early work (1933), he defines community types by frequency dominants calling them associations and grouping them into formation types similar to Rübel (1930). In later works, he uses the sociation as a unit below the level of the association, characterized by frequency dominance, physiognomy and ecology (1943). The nomenclature of Böcher compared to Raunkiaer is as follows:

Raunkiaer	Facies	Formation	Formation series
Böcher	Sociation	Association	Formation-type

Another aspect of the work of Böcher was the analysis of the geographical relationship of the species in the communities defined (1940, 1954) where the character species of Braun-Blanquet are rejected and four other types of diagnostic species preferred:

1. species found in similar situations and communities in different areas—*area-geographical differential species*;

2. species which indicate the affinities of the community in which they occur within a broad geographical and climatic region—*climatic species*;

3. species associated with particular edaphic or microclimatic conditions—*ecological differential species*;

4. species which combine properties relating to categories 2 and 3—*ecogeographical indicator species*.

Thinking on these lines, representatives of the four categories from the Scottish and Danish *Myrica*-heath types can be readily found. *Myrica* itself falls into category 1; *Erica tetralix*, a west European species, indicates the affinities of the communities with Atlantic heaths (category 2); category 3 has abundant representatives in *Eriophorum vaginatum*, *Salix repens*, *Sphagnum papillosum*, *Hypnum cupressiforme*, etc.; whilst *Calluna vulgaris* and *Trichophorum caespitosum* fulfil the requirements of category 4.

CHAPTER 7

VEGETATION GRADIENTS AND CONTINUA

As an alternative to the classification of vegetation, the analysis of gradients has found favour with a number of phytosociologists, particularly Americans working in a continent where the influence of man on the vegetation is negligible. Gradient analysis is simply a study of the spatial patterns exhibited by vegetation in relation to three types of variable, environmental factors, species population structure and composition and the characteristics of communities. Frequently, it may be demonstrated that an environmental gradient exists, along which many microclimatic or pedological factors change with a parallel variation in vegetation, due either to variations in species populations which show distinct eco-physiological responses or community composition and structure, or both. Brown and Curtis (1952) liken the method of gradient analysis to the spectrum in which the community types are the colours.

The analysis of gradients developed from the early papers of Gleason (1926) which showed a dissent from Clements's monoclimax, organismal theories on the nature of vegetation which was predominant in the ecology of the English-speaking countries. Much the same type of dissent was shown by Ramensky (1924, 1930) in Russia and people such as Lenoble (1927) in France. Implicit in all these works and clearly explicit in Gleason's were two central concepts:

1. Each species has a distribution which is related to a total range of environmental factors and which depends upon the limits of its own population dynamics, physiological properties and primarily its genotypic adaptability—*the principle of species individuality.*

2. There is a continuous intergradation of communities along environmental gradients with gradual changes in species populations and population interactions along the gradient—*the principle of community continuity.* Although the theories met with intense opposition, especially in North America, several major schools developed in the late 1940s—Whittaker's work in the Great Smoky Mountains, Tennessee (Whittaker, 1948, 1951), Curtis and his associates in Wisconsin (Curtis and McIntosh, 1951; Brown and Curtis, 1952), Ellenberg

235

(1950, 1952) and a development of Jenny (1941), Major (1951) and Perring (1958). Later developments in the subject are legion but the two major approaches which may be singled out are those of similarity treatment between samples and data representation as matrices and secondly, the statistical techniques known as factor analysis which are aimed at the analysis of certain underlying initially unknown parameters or factors which combine to determine the correlation between species or samples.

All these approaches have been subjected to a series of exchanges of ideas and an intermixture of concepts, but they may be grouped into two broad groups. In the first of the two approaches, the vegetation analyses or samples are arranged in positions along an environmental gradient. This approach, which may be termed *direct gradient analysis*, was originally applied by Whittaker (1951, 1953) in his criticism and consideration of climax concepts. The second approach, called *indirect gradient analysis*, involves the comparison of samples with one another usually in terms of species composition, and then their arrangement along axes or in a three-dimensional hyperspace based on these similarity measurements. The approach to environmental gradients is always abstract, inferred and indirect. Both methods emphasize the continuity of vegetation and seek to demonstrate a continuum rather than a classification.

Direct gradient analysis

Methods of direct gradient analysis can be conveniently divided into three main approaches (cf. Whittaker, 1967): (1) the simple, single gradient transect; (2) ordination of samples in (*a*) a composite transect, or (*b*) a hemispherical plot; (3) environmental pattern analyses arranged as areas of a composite chart delimited by isonomes.

Simple gradient transect

For the *simple gradient transect* method of vegetation analysis, samples are taken at equal intervals along an observable environmental gradient such as altitude. The samples are usually in the form of a series of sample quadrats at convenient intervals along the transect in which species coverage, frequency, basal area, biomass, etc., may be determined. As an example of the method Figure 54 illustrates the results of an altitudinal transect up the mountain Corserine in the Rhinns of Kells range in south-west Scotland with samples at 50-ft intervals derived from the combination of cover estimates for species in 10 m^2 quadrats at each point. The performance of selected species over the altitudinal gradient is shown clearly. *Molinia caerulea* has

its greater cover values at lower altitudes on the glacial dep\
the river basins, disappearing at 1,550 ft. *Nardus stricta* h\
maximum expression at the median altitudes of the transect as\
Festuca rubra whilst *Carex bigelowii* shows a marked increase in co./cr
values at higher altitudes. The dwarf willow *Salix herbacea* makes an
appearance at 2,250 ft and rapidly assumes dominance in terms of
cover. The results indicate clearly that species populations seldom have
sharp boundaries but a central peak of maximum cover, density or
performance decreasing either side of the mode. In this respect several
curves resemble the binomial or Gaussian model. In general, species
are not organized into groups showing parallel distribution patterns

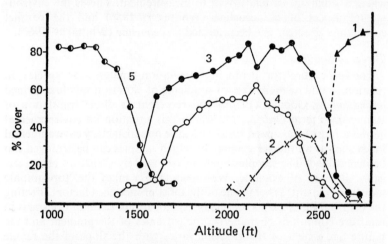

Figure 54. A single altitudinal gradient transect: Species 1, *Salix herbacea*; 2, *Carex bigelowii*; 3, *Festuca rubra*; 4, *Nardus stricta*; 5, *Molinia caerulea*.

(with possible exceptions in certain related species) and this difference
in distribution is related to the physiological properties of the species
or ecotype. Along the altitudinal gradient numerous and complex
interactions of environment produce physiological adaptations—
temperature, precipitation, humidity, length of growing season, wind
exposure, evaporation, etc.—all components contributing to a complex
climatic gradient for which altitude is a useful approximate gradient.
The evaluation of the role of individual climatic components in the
causation of the gradient is only possible by extensive and laborious
experiment and in general, the gradient relation observations must
remain as a generalization relative to altitudinal approximation.

As a direct result of the binomial forms of the species population
distributions, the composition of communities varies continuously

along the environmental gradient. The transect thus relates a community gradient to an environmental gradient and in the sense of Tansley (1939) the transect is a method of studying a gradient in ecosystems. For this study, Whittaker (1967) has adopted the term *ecocline* which has been otherwise restricted to the discipline of experimental taxonomy as a genetically based variation in a species population related to an environmental gradient. This present application of the term considerably extends this definition. Whittaker has introduced and defined two other situations which are of relevance in this respect. A gradient of environmental complexes where many factors vary together through space is termed a *complex-gradient*—as opposed to a *factor-gradient* where a single factor lends itself to measurement. This is the environmental aspect of the ecocline (Whittaker, 1956) and the parallel community gradient has been termed a *coenocline* (Whittaker, 1960).

Direct ordinations

The procedure for arranging vegetation analyses or species in relation to one or more axes or gradients of variation was first termed *ordination* by Goodall (1954b) and represents a direct translation of Ramensky's term *Ordnung* (1930). In this situation an environmental gradient must be assumed simply because no satisfactory environmental index such as altitude or geology by which samples can be arranged can be determined. The samples must be arranged by their own characteristics. By way of example, Whittaker (1967) takes the 'topographic moisture gradient' as being one of the most pronounced factors affecting the distribution of plants especially cryptogams in ravines. Contrasts in moisture conditions characterize the extremes of the gradient and the nature and extent of the gradient varies with the slope of the ravine sides as the stream or river passes over different geological strata or under a tree canopy. The gradient over the whole length of a ravine or stream course may be expressed in a composite transect in the following manner. A number of transects across the ravine are taken at regular or random intervals and the species recorded in each transect. A composite picture is built up of the distribution of each species for each transect and four broad groups are categorized:

(0) *Mesics*—species with population modes at or near the moisture extreme.
(1) *Submesics*—in less moist situations, but in the damper half of the transect.
(2) *Subxerics*—in the drier half of the transect.
(3) *Xerics*—species populations at or near the dry extremes.

The numbers preceding the four categories are applied as weights to data on the composition of each transect. Thus a bryophyte streamside

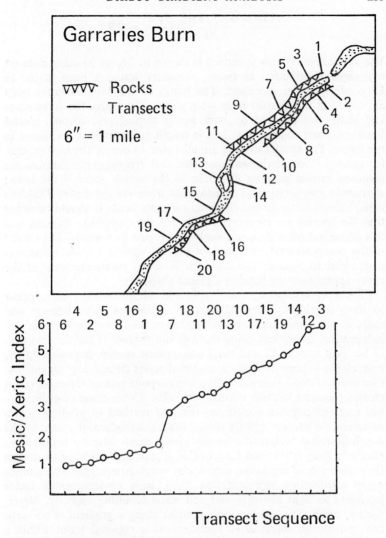

Figure 55. A composite moisture gradient transect of the Garraries Burn, Kirkud-brightshire, based on twenty bryophyte sample transects.

transect with thirty species may include fifteen mesic species, six submesics, six subxerics and three xerics. From this a weighted average of species composition is calculated where the species are multiplied by their weights, summed and divided by the unweighted number, thus:

$$\frac{(15 \times 0, \, 6 \times 1, \, 6 \times 2, \, 3 \times 3)}{30} = 0.9$$

The application of this approach is shown in Figure 55 using data on bryophyte distribution in twenty transects across a burn ravine in Kirkcudbrightshire, Scotland. The banks of the Mid-Garraries burn are varied in topography from steep rocky sides where the greywackes and shales are exposed to more gently sloping peat-covered glacial boulders. Each transect was 2 m in length and numbered as shown in the figure. The species present are allocated to one of the four groups, for example: *Rhacomitrium aquaticum* and *Hygrohypnum luridum* are common mesics growing on rocks in the splash zone of the burn; *Bartramia pomiformis* and *Solenostoma triste* are submesic; *Thuidium tamariscinum* subxeric and *Grimmia pulvinata* xeric. It should be noted that the species are categorized only for the composite transect and this allocation often does not apply to a region as a whole. The results of the compositional transects are plotted against transect sequence, rearranged to produce the gradient. In many cases, the form of the curve approaches the familiar sigmoid shape.

Ellenberg, Whittaker, and Curtis and McIntosh (loc. cit.) appear to have developed weighted average techniques independently and more or less simultaneously. Whittaker in his forest studies uses two independent weightings using these as the characters for the two axes of the plot instead of one weighting against species sequence. In the case of the bryophyte data, a weighted series comprising (0) thallose liverworts, (1) leafy liverworts, (2) pleurocarpous mosses, (3) acrocarpous mosses produces suitably correlated results. This method of ordination has received popular acclaim as the best method of gradient representation. Whittaker (1960) tested other methods and came to the conclusion that 'weighted averages give a more effective ordination'. Similarly Bray (1961) and Loucks (1962) produced evidence to suggest the ordination of vegetation samples by weighted averages gave a much more satisfactory representation than such environmental factor gradients as light intensity and soil water-retaining capacity. Nevertheless, ordination of vegetation samples along a gradient of soil type can provide an informative summary of a regional trend within a particular vegetation type. This is especially true where a climosequence of soil development is apparent. Bryan (1967) has described such a climosequence of development in soils over calcareous parent materials in the Peak District of Derbyshire, commencing with the primitive rendzinas and brown rendzinas through brown calcareous and brown podsolic soils to iron and humus-iron podsols. The sequence is characterized broadly by a progressive increase in soil depth, in humus content and a decrease in slope, pH and percentage of carbonates. To this sequence, the variations in vegetation types can be applied (Figure 56) (data from Shimwell, 1968a, b). For example, the species rich 'limestone grassland' or the association *Helictotricho-Caricetum flaccae* is more or less restricted to the rendziniform soils and becomes replaced by a species rich *Agrosto-Festucetum* on brown calcareous

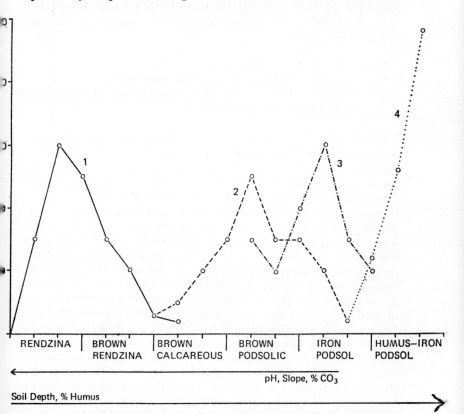

Figure 56. A composite vegetation-soil type gradient from the Peak District of Derbyshire. 1, Helictotricho-Caricetum flaccae (60 stands); 2, Agrosto-Festucetum (24 stands); 3, Nardo-Galietum (20 stands); 4, Trichophoro-Callunetum (12 stands).

and podsolic soils. With the development of an iron pan, ericaceous grass heath develops in which *Calluna vulgaris*, *Nardus stricta* and *Galium saxatile* are prominent (*Nardo-Galietum*). The final humus-iron podsols which are located mainly in plateau regions are characterized by a *Calluna-Molinia* or *Trichophorum* heath.

A second type of ordination technique is known as the hemispherical plot, a representation which is described by Perring (1959) in his investigations into variation in the composition of chalk grassland communities. His theoretical approach to the study is based on the suggestions of Jenny (1941) and Major (1951) that soil development and hence vegetation are determined by five independent factors— the regional climate (cl), the parent material from which the soil originated (p), topography (r), the biotic factor (o) and time (t). Thus Jenny showed that any soil type (S) may be expressed as a function of these five variables:

$S = f$ (cl, p, r, o, t ...) and so may vegetation:
$V = f$ (cl, p, r, o, t) where V is a plant community or
$v = f$ (cl, p, r, o, t) where v is some aspect of the vegetation which may be measured quantitatively.

Crocker (1952) considers that V and S are inseparable and that the five independent variables define the ecosystem of Tansley (1935). The climatic factor involved in the formula is that part of the climate which is not affected by any of the other variables and in this respect may be defined by the regional climate recorded at weather stations. Perring (1958) considers that the precipitation-saturation deficit ratio (P-SD) a useful representation of the regional climate. The P-SD ratio takes into consideration temperature, precipitation and relative humidity and may be calculated by dividing the mean annual precipitation by the absolute saturation deficit of air in millimetres of mercury. The second function, soil parent material is the initial source from which the soil was derived and at the best this categorization is obscure because of glacial or periglacial complications in many regions. The third function, topography, includes all features of relief which are responsible for the modification of the regional climate to produce a local or microclimate. Chalk and limestone grasslands provide an excellent example for the study of topographic variations in vegetation since the downlands and dale grasslands provide a wide range of slopes of varying exposition and angles.

The fourth function, the biotic factor is divided by Jenny (1941) according to the effects produced by four groups of organisms— micro-organisms O_m, vegetation O_v, animals O_a and man O_h. Jenny considered that micro-organisms could be ignored as independent soil forming factors, partly on the grounds that because of their widespread occurrence they could be regarded as a dependent factor, but also because of lack of evidence and the problems associated with their study. O_v in the case of the study of vegetation via the five variables may be referred to as the flora factor for which Perring considers there is no satisfactory general definition. To begin with there is great difficulty in the separation of the features of vegetation which may be looked upon as independent. The flora of the world or its regions are dependent upon climate and history and the only way to get around the dependence angle is to follow the suggestion of Crocker (1952) that the regional flora is that of an area at time zero of soil development. But the present vegetation selected from that particular flora has been dependent on the other four factors. There is thus no flora factor which is completely independent but there are degrees of independence, three of which are cited by Perring:

O_V *between areas with totally dissimilar floras:* comparisons of two regions with similar independent variables, but with completely different floras mean that V will be dependent upon flora.

O_V *between areas with partially dissimilar floras:* in this case the independent part of the flora is that which is not common to both areas, a phenomenon which can probably be explained by historical factors influencing migration patterns.

O_V *within the same area:* variations are due mainly to the reproductive biology of species of the flora which is limited or affected by climate. According to Perring, the importance of this O_V factor is not likely to be understood until either the distribution and biology of all the species of a flora have been studied or until climosequences have been investigated.

O_a, the fauna factor shows similar dependent and independent aspects as does the flora factor. One of the most important grazing animals, the rabbit, has a local distribution which depends largely upon topography. A similar situation applies to the distribution of snails and slugs whose winter distribution is primarily affected by temperature while topography determines microclimate and affects local distribution.

O_h the human factor, as Major (1951) suggests is the most important independent animal factor influencing vegetation, and the major developments in ecology should be aimed at a greater understanding of man's effects. All the processes which are described as management come under the head of the human factor, together with other disturbance factors such as mining and quarrying. The practices of deforestation, burning and the pasturing of grazing animals who add trampling, eating and manuring, have been the major ones determining the historical development of vegetation. This leads on to the last variable function—vegetation changes with time. The changes in ecosystems may be divided into five grades of time, seasonal, annual, successional, historical (geological) and evolution. Perring points out that the phytosociologist is primarily concerned with the successional time factor since the first two time types are merely cyclic variations in general development and the latter two involve a magnitude of time which for the most part is beyond the limits of developmental study. He

I

defines time as the number of years since a particular independent variable started to act.

From these basic premises Perring (1959, 1960) goes on to study climatic and edaphic gradients of chalk grassland. The main feature of the approach is to study changes in soil and vegetation values in relation to changes in one independent variable whilst keeping all others constant. His paper on topographical gradients (1959) is of particular interest since it gives a realistic summation of variations within chalk grassland vegetation relative to slope and aspect. To attempt the study Perring assumes ideal conditions as follows:

Isolated hemispherical hills in different humidity regions made of similar parent materials, undisturbed by ploughing, burning and grazed at the same intensity; isolated so that no major variations in local climate may upset the picture and with strata horizontal so that drainage patterns are similar.

Most of Perring's work on toposequences was centred on the downlands of Dorset where a network of sites was built up incorporating data from several hundred sites. For each site the exposition, angle of slope, soil colour, pH, organic carbon, exchangeable phosphate, exchangeable potassium, sodium and calcium, total carbonates, mechanical analysis plus vegetation analyses were recorded. The vegetation analyses consisted of twenty 1 dm^2 random quadrats with species cover estimates on a five category scale corresponding to cover percentages in 20% units. Soil and vegetation values were then plotted on a diagram representing an aerial view of an ideal hemispherical hill (Figure 57a) where the lines are the eight points of the compass which intersect at the highest point. Proceeding from the top, the angle of slope increases evenly in all directions up to 50° so that point X may be referred to as SW 40° and point Y N 15°. On these hemispherical plots individual site records may be located and all equal points or those within a certain range linked by isolines. Thus, Perring could come to such conclusions as pH is related to slope but not to aspect and that exchangeable calcium values are highest on northerly slopes between 10° and 18°. Similarly, vegetation cover values for each species may be plotted. This procedure inevitably gives rise to certain groups of species which have closely related topographical distribution—or maximum cover patterns which in turn may be related to edaphic features. This type of result is illustrated in Figure 57 b, c, d using data from Shimwell (1968a) from the limestone grasslands of the Derbyshire Dales. Figure 57b summarizes the aspect and angle distribution of 1 m^2 stands analysed for cover using the Domin scale—the data presented in the appendix table to Shimwell's work. It may be seen immediately that there is a preponderance of data from west- and north-facing slopes with angles between 40° and 45°. This is not due to oversampling

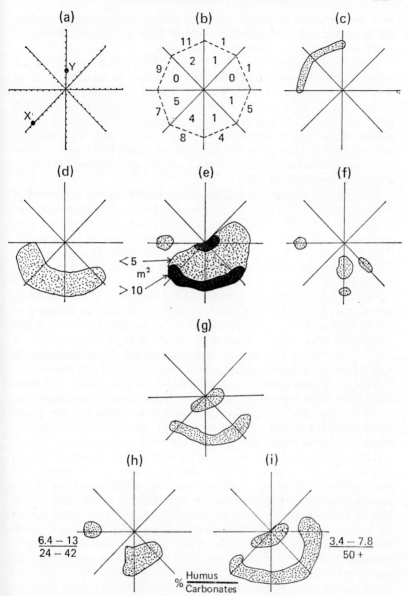

Figure 57. The hemispherical plot method. (*a*) representation; (*b*) distribution of stands of Helictotricho-Caricetum in Peak District, on slopes of greater and lesser than 40°; (*c*) *Carex pulicaris* group; (*d*) *Cirsium acaulon* group; (*e*) population density in *Verbascum nigrum*; (*f*) natural populations; (*g*) disturbed site populations; (*h*) and (*i*) humus/carbonate in soils of (*f*) and (*g*).

on these slopes but simply reflects the distribution pattern of the particular grassland type. Within this pattern there are certain groups of species which show interesting distributions:

Group a—Carex pulicaris group (Figure 57c) a group comprising this sedge, *Parnassia palustris*, *Potentilla erecta* and an increase in cover values for *Carex panicea*, *Viola riviniana* and *Acrocladium cuspidatum* occurring only on slopes of 30–40° W to N at altitudes between 800 and 1,200 ft. This is a group of species indicative of a special microclimate which reflects an overall climatic trend towards the north and west of Britain. Thus the composition of the stands involved here show a composition related to that of the limestone grassland of the Craven Pennines further north.

Group b—Cirsium acaulon group (Figure 57d) a group composed of *Cirsium acaulon*, *Filipendula vulgaris*, *Potentilla tabernaemontani* and *Hippocrepis comosa* which show a distribution restricted to south- and west-facing slopes at a variety of angles between 10° and 45°. This picture is further supported by data not included in the table from Staffordshire and the data of Pigott (1968) which is incorporated into the figure.

All this data comprises the study of a toposequence. Further developments into diverse limestone regions such as the Craven Pennines or North Wales enables a climosequence to be developed, where soil and vegetation values may be studied in relation to increasing humidity or the P-SD ratio—in this way, the behaviour of individual species or the type of species group described above may be studied and compared.

The study may also be approached in reverse, where a known vegetation type, community or species population is investigated in terms of the five variables in an attempt to determine the major function or functions which are affecting the distribution patterns. Here V or some aspect of V is taken as stable or equal at the time of sampling. This approach may be applied readily to a study of species populations and Figure 57 illustrates the results of an analysis of populations of *Verbascum nigrum* near the north-west limit of the species British distribution in Derbyshire. Fifteen populations of this species with numbers ranging between 25 and 420 plants occur mainly on south- and east-facing slopes with angles of slope varying between 0° and 45°. Figure 57e shows the density of the *V. nigrum* with variations in topography (r). The climatic variable is constant and no pronounced microclimatic factors are observable. The parent material (p) is either natural calcareous protorendzinas, lead mine spoil or quarry spoil and this as will be seen later is dependent upon the time factor. The major component of the O factor is O_h—anthropogenic influences which is inseparable from time. The species was first reported in the area in 1789, the *locus classicus* being a population of sixty individuals on protorendzinas. Assuming this to be a natural population little affected by O_h several other populations on similar sites fall into this category—all small populations on steep slopes between 15° and 40° (Figure 57f). A second group of populations on the spoil thrown out of lead mines abandoned by 1850 are of intermediate size and density as is a population on the spoil of a quarry opened in the 1920s. The highest population sizes and densities are to be found on recently (1960 onwards) disturbed spoil heaps of lead mines and quarries (Figure 57g). There is also a related variation of several soil factors. For example, the smaller populations in more or less undisturbed habitats on soils with high humus/low carbonate content when compared to the recently disturbed habitats of the larger populations (Figure 57 h, i). It may thus be seen that it is the factor O_h which coupled with time of development are the major factors controlling the population size of V. nigrum. The older the population the smaller it is; the more recent the disturbance the larger the population. Thus to summarize the time gradient or chrono-sequence in population variation:

O_ht 1790 = cl(k); r 15–40° W-E; p = rendzina; V_d 1·0–3·0.
O_ht 1850 = cl(k); r 0–5° S-SE; p lead spoil; V_d 3·3–5·0.
O_ht 1960 = cl(k); r 0–45° S-E; p lead, quarry spoil; $V_d = 8·3–11·6$.

If, via these methods, the major factors affecting the distribution and existence of rarer and precarious species can be demonstrated, some positive contributions towards an understanding of the methods of conservation can be made. In the example cited above, the future existence of the continental Verbascum nigrum at the north-western limits of its range seems to be assured as long as man as a disturbing factor remains active.

Environmental pattern analyses

Environmental pattern analyses, so called by Whittaker (1956, 1967) and exemplified neatly by Whittaker and Niering (1965), are methods of analysis of communities in relationship to two or more gradients and as such they fall far above the type of situation envisaged by Kershaw (1964) in his reference to environmental pattern. Much of Whittaker's work on this subject is aimed at the representation of observed variation in montane vegetation relative to two complex gradients, elevation and the topographical moisture gradient. The two gradients may be used as the axes of a chart on which the background vegetation types may be plotted. For example, the ordinate axes of Whittaker's (1956) study represents elevation, whilst the abscissa are represented as a mesic to xeric trend in which there are

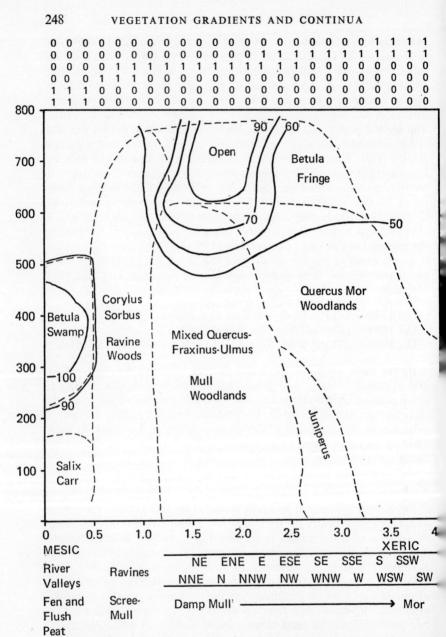

Figure 58. Mosaic and population charts for woodland types in the Galloway region, south-west Scotland. 1, *Acrocladium giganteum*; 2, *Sphagnum palustre*; 3, *Bartramia pomiformis*; 4, *Eurynchium praelongum*; 5, *Rhytidiadelphus loreus*; 6, *Leucobryum glaucum*.

topographic categories such as coves and canyons; flats, draws and ravines; sheltered slopes; open slopes; and ridges and peaks. On to this background pattern is superimposed data on populations of various important species, for which Whittaker uses the measure of percentage of tree stems of 1 cm diameter at breast height in composite samples approximating 1,000 stems. Plotted in this manner, the two-dimensional representations become binomial solids linked by isolines. As with previous gradient plots in one dimension no two species have population characteristics which are the same and the modes of the characteristics are distributed over the range of communities and environments represented by the chart. Figure 58 provides an example of a pattern analysis chart of woodland types in the Galloway regions. Data from sixty 10 m² sample plots are used to produce the data for the horizontal axis whilst the vertical axis represents elevation, The former axis uses the parallel gradients:

(i) A *mesic to xeric series* with indices values calculated on bryophyte composition recording scores of 0, 1, 2 and 3 for mesic, submesic, subxeric and xeric species as described previously. Examples of the categories are given at the top of the diagram and as before the choice of the category to which a particular moss belongs is based upon an overall regional appraisal.

(ii) *Broad topographical features* such as river valleys, ravines and open slopes of diverse aspect are used in the same sense as Whittaker (1956).

(iii) *Edaphic features* represented as overall soil type are included in a series which passes from the mesic peaty soils of flushes and fens, through damp mull-screes to upland acidic moor humus soil types.

The sixty sample plots from eleven localities are arranged upon the chart according to their coordinates and isolines constructed to enclose related types. Based on structure and canopy composition seven basic woodland types are recognizable each of which has a fairly distinct distribution pattern. Thus, the woodlands dominated by *Quercus* occur mainly on the open slopes with a southerly inclination and on acidic moor humus soils. Other woodland types have a more restricted distribution on the mesic-xeric scale. The *Corylus-Sorbus* woodlands have a mesic bryophyte component and consequently a low weighting which in this case ranges between 0·5 and 1·2. Similarly, the *Juniperus* woodlands are restricted to a narrow range, but this is more by virtue of the paucity of woodlands of this type in the region, than the constancy of a mesic-xeric index.

On to this background, data on the variation in populations of important tree species may be superimposed to give some comprehension of woodland structure. In Figure 58, the numbers *Betula* stems over 1 cm diameter are plotted as percentages of the total stems

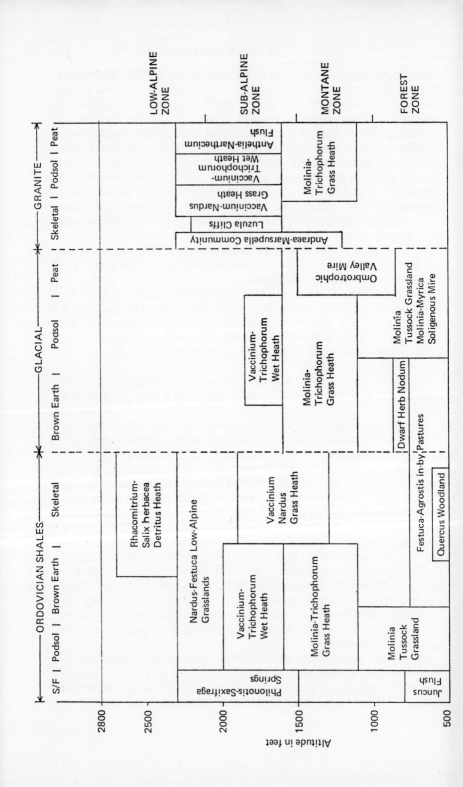

of all species per 10 m^2. Again isolines can be drawn to include all points within a particular percentage range to produce a population chart which indicates modal situations for species populations at zero time. On its own the background chart is often referred to as a *mosaic chart*. Mosaic charts have been widely used as vegetation representations by a number of workers. McVean and Ratcliffe (1962) combine classification and ordination in their diagrams of the distribution of Scottish Highland vegetation on different geological formations. In many ways, this is an extremely satisfactory end point to a study and provides a summary ready reference for future investigations and comparisons from different regions. Figure 59 provides such a set of comparative data for the communities of the montane-low alpine region of Galloway which includes the mountains of Merrick (2,770 ft), Corserine (2,669 ft) and Meikle Millyea (2,448 ft). The mosaic chart is divided primarily into three geological parent materials and these in turn are divided into three or more soil types. This type of representation of Scottish mountain vegetation was initiated by Poore and McVean (1957) where the communities or noda are fitted into an ecological framework based upon such characters as the interaction between altitude and exposure, oceanicity, snow cover, base status of the soils and soil moisture. Their diagrams are essentially the same as those of Whittaker but the horizontal axis uses increasing degrees of shelter and snow cover to the left, increasing exposure to the right, from an arbitrary central point on the axis. Plotted against elevation the chart method enables a rapid comparison between vegetation types and ecology in geographically separated regions. Figure 59 employs the same altitudinal delimitations of forest, montane, sub-alpine and low alpine zones as those of McVean and Ratcliffe (1962). The following comparisons between this data and that of the above authors (Figures 31–36, p. 396–407) reveals an approximately similar altitudinal distribution in the two regions. Several communities have similar altitudinal ranges, e.g.:

(i) *Molinia-Myrica* soligenous mire; *Molinia* tussock grassland
(*Molinia-Myrica Nodum*; Molinieto-Callunetum)

Ben Lui-Ben Heasgarnich District	100–1,900 ft siliceous soils
Cairngorms	1,000–2,100 ft granite
Kintail-Glen Affric District	100–2,000 ft granulite and schists
Ben More Assynt-Foinaven District	100–1,400 ft quartzite and gneiss
Galloway District	500–1,500 ft glacial, granites and shales.

(ii) *Philonotis-Saxifraga stellaris* springs
(*Philonoto-Saxifragetum stellaris*)

Ben Lui-Ben Heasgarnich District	1,500–3,300 ft siliceous
Clova-Caenlochan District	1,500–3,100 ft siliceous
Cairngorms	1,900–4,000 ft granitic
Kintail-Glen Affric District	1,200–3,600 ft granulite and schists
Ben Wyvis District	1,500–3,000 ft schists
Ben More Assynt-Foinaven District	1,600–3,000 ft quartzite and gneiss
Galloway District	1,500–2,300 ft shales.

By way of conclusion the major concepts of gradient analysis can be seen to present three types of transformation in the change from a single dimension to a multi-dimensional relationship (Whittaker, 1967).

(*a*) The ecosystem gradient or ecocline is transformed to an *ecosystemic* or *landscape pattern*, whilst the complex gradient of one dimension becomes an *environmental pattern* in several dimensions. Similarly, the coenocline may be interpreted as a community pattern.

(*b*) The normal binomial curves of population characteristics becomes *binomial solids* related to background ecosystemic patterns.

(*c*) The compositional gradient becomes a *complex population continuum*, envisaged as many binomial solids superimposed upon one another in the same ecosystemic pattern.

Indirect gradient analysis

As was pointed out in the introduction to this chapter indirect gradient analysis requires the comparison of samples of vegetation with one another in terms of some characteristic they possess so that gradient relationships result from an otherwise heterogeneous set of data. Simple measurements of the relative similarity of sample composition will usually give rise to a situation which enables the arrangement of the samples along axes which may correspond to underlying environmental gradients. Whittaker (1967) has classified the techniques of indirect gradient analysis into six main approaches, but for the purpose of this discussion, they may be readily considered under three main headings:

(1) Similarity measurements and their representation as matrices, plexuses or constellations of overall similarity.

(2) The geometric comparative ordination techniques of the Wisconsin School.

(3) Factor analysis where samples are points in an n-dimensional hyperspace projected on to an axis of maximum variance and a second axis at right angles along which the remaining variance is greatest.

Coefficients, matrices and plexuses

Perhaps the simplest measurement of similarity is the so-called *coefficient of community* first represented by Jaccard (1902) as K = c/(a+b−c) in which c is the number of species which two samples with a and b species respectively have in common. Various modifications of this formula are to be found in the literature, but by far the commonest is that of Sørensen (1948) K = 2c/ a+b . 100.

Several *percentage similarity* calculations have been derived and used for gradient analysis. Odum (1950) in similarity analyses of animal communities has used PS = $100 - 0.5\Sigma[a-b]$ = Σ min. (a, b) where a and b are, for given species, the percentages of importance values (e.g. density or quadrat frequency) in samples A and B which that species comprises. The interpretation may be easily modified for samples instead of species, but it is in the former fashion that it has found its greatest application to samples of animal communities. Recently, Orloci (1966) has suggested the corrected form PS = $\sqrt{[\Sigma(a-b)^2]}$ as a formula readily utilized in computation.

In central European phytosociology, a number of importance values have been used in various formulae for the calculation of *affinity indices* (A), the reciprocal of which are viewed as a measure of eco-sociological distance (D) between samples. Barkman (1958) uses five types of A and hence D calculations in his study on cryptogamic epiphyte vegetation. D_1 is a more or less direct modification of Jaccard's formula and is represented by

$$A_1 = \frac{c}{\sqrt{a \times b}}, \quad D_1 = \frac{\sqrt{a \times b}}{c},$$

being based upon species composition where a and b are the numbers of species in sample A but not B and in sample B but not A respectively, with c the number of species common to both. A_2 and D_2 involve the use of cover values for each species of two communities (Table 65), in the formulae $A_2 = \Sigma\gamma/\sqrt{\Sigma\alpha . \Sigma\beta}$ and $D_2 = 1/A_2$.

Comparing the total cover values for each species (T.C.V.), α and β are the differences in T.C.V. and γ the amount of T.C.V. common to both. For species 3, $TCV_A - TCV_B$ is positive and is referred to α. Alternatively, species 1 gives a negative value and is referred to β. Hence, from the values in Table 65, $A_2 = 30/\sqrt{70 \times 20} = 0.8$, $D_2 = 1.25$. D_3 is represented as $D_1 \times D_2$ but as this involves considerable calculation, D_4 was introduced. This is essentially the same as D_2 but uses the cover-abundance notation of the Braun-Blanquet scale instead of direct percentages. D_5, the final index used, derives its measurements from importance values represented as degrees of presence (constancy) on a five category scale.

Not to be confused with the distance measures listed below is the

TABLE 65

Example of the Affinity Index calculation (Barkman, 1958)

	TCV Community A	TCV Community B	α	β	γ
Species 1	10	15	–	5	10
2	15	15	–	–	15
3	20	–	20	–	–
4	50	5	50	–	5
5	–	15		15	
			= 70	= 20	= 30

statistic D^2 also known as Mahlanobis's D^2 as a meaure of the generalized distance between samples. Unlike the above formulae, the calculation of D^2 is complex and involves the computation of the means, variances and covariances of the various properties of replicate samples in a multivariate analysis. The greater the value of D^2 the larger the difference between the samples in terms of the quantities measured. The D^2 method is not widely used in similarity measurements but examples of its use are to be found in Hughes and Lindley (1955) who used the method to show trends in soil development and gradients in vegetation types in Snowdonia.

The computation of similarity values between all pairs of samples produces a large set of data which may be represented in the form of either a *matrix* or if the comparisons are relatively few in number, as a *plexus*. Figure 60 provides an example of a disorderly and an orderly matrix. To begin with the similarities between samples (in this case percentage similarity derived from a Sørensen coefficient) are located by their sample number co-ordinates. Ordering then proceeds either mechanically or by eye to give clusters of high percentage similarity. A sample which may be regarded as extreme for the data, for example, number 7 with its consistently low similarities with all other samples except number 10 is chosen as one end point for the ordination. The sample which shows the closest similarity to the first is placed next to it and so on until many or most of the highest similarity values have been manoeuvred down the oblique axis and definite clusters are visible. Identical techniques of similarity measurements and matrix representations can be used for both classification and ordination of stands and samples. In consequence, the methods have been widely used in several schools of phytosociology. The formula of Sørensen (1948) has been used by Dahl (1956) in the classification of montane vegetation, although he decided that it was impossible to specify what values of similarity were generally required for the definition of the different

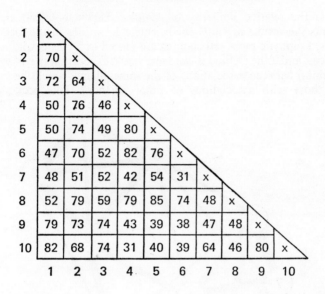

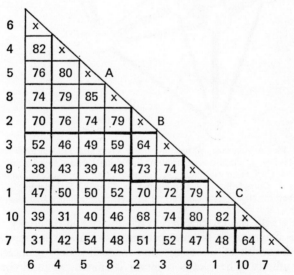

Figure 60. Matrices of Sørenson coefficients.

levels of the hierarchy. Similarly, Groenewoud (1965) has difficulty in defining the bounds of sample clusters in his use of Mahlanobis' D^2.

This latter measure of distance is more commonly represented as a plexus, a diagram in which lengths or widths of inter-connecting lines express the relative similarity of samples. Barkman (1958) also represents the results of his D-coefficients calculations in the form of a plexus. Epiphytic moss associations are placed in a plexus within their alliances, and if the method is valid one would expect that the differences (D-values) between associations of the same alliance would be smaller than those with associations of other alliances. This occurs most

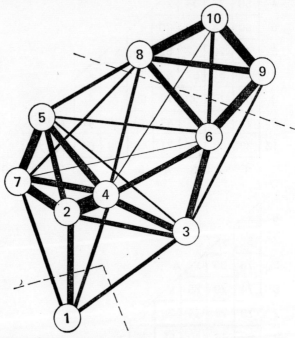

Figure 61. Plexus diagram of ten limestone grassland associations (data from Shimwell, 1968b). 1, Poterio-Koelerietum; 2, Helianthemo-Koelerietum; 3, Cirsio-Brometum; 4, Caricetum montanae; 5, Antennarietum hibernicae; 6, Helictotricho-Caricetum; 7, Camptothecio-Asperuletum; 8, Asperulo-Seslerietum; 9, Seslerio-Helictotrichetum; 10, Seslerio-Caricetum. Broken lines, alliance boundaries.

completely with the D_3 calculation of affinity and difference. Thus in the plexus diagram the distance between associations is represented as D_3 and the width of the connection made inversely proportional to D_3—the affinity. The plexus diagram is illustrated in Figure 61 in which the ten major associations of the alliances of the order *Brometalia* in Britain are depicted (data from Shimwell, 1968b). The D_3 calculation of Barkman is used, of which the components D_1 and D_2 are represented by total species composition and total cover values for character species and species of constancy V and VI respectively, using a standard number of representative stands of each association.

An analogous approach to the calculation of an eco-sociological distance is to be found in studies on the *distributional similarity of species* or interspecific correlation (association). Details of interrelationships can be calculated from presence and absence data in the form of a 2×2 contingency table with the appropriate χ^2 test, as outlined in Chapter 1. The major object of the method is to so subdivide the set of data that all correlations between species disappear and the resultant groups may be regarded as homogeneous. The early work on the subject, that of Goodall (1953a), suggested four ways in which quadrats with correlation between species could be categorized into homogeneous groups of uniform species composition, i.e. to regard as homogeneous groups: (i) all quadrats containing one of two species; (ii) all those lacking one of two species; (iii) all those with both species; (iv) all those containing neither species. Goodall concluded that the first method was the most successful in eliminating correlations and was also the least time-consuming. In the method, the most frequent species showing correlation is extracted first, and the number of quadrats (A) in which the species occurs recorded. These quadrats (A) are then re-examined and again the most frequent species showing correlation extracted, with its occurrence in (B) quadrats recorded. The remaining quadrats are pooled and the procedure is repeated for each group until the terminal groups are reached where no correlations are to be found. The method thus relies upon the elimination of all correlation and also takes into account positive correlations, negative correlations being ignored. Other authors, such as Agnew (1961) have used both negative and positive correlations, but rely solely upon positive values for data representation. In their analyses of quadrat and presence and absence data, widely known as *association-analysis*, Williams and Lambert (1959, 1960) also use positive and negative correlations for the division of data. They demonstrate that there are a number of alternative subdivisions which will produce the method's objective, and for the purpose introduce the concept of efficient subdivision, i.e. a divisive method based upon the species which produces the smallest total number of significant correlations in each of the dichotomous pairs of division. For normal amounts of data likely to be collected in field surveys, this is only realistic with the help of an electronic computer. A measure of correlation (C) is then required and Williams and Lambert show that division of the data on the species with the maximum ΣC will tend to reduce the residual ΣC in the terminal groups to a minimum—or will satisfy the concept of efficient subdivision. Several measures of correlation have been used, but mainly χ^2 or some derivative of it since χ^2 needs to be claculated for a test of significance Williams and Lambert selected $\sqrt{\chi^2/N}$ where N is the number of stands being analysed. Unlike the method of Goodall where all the residual stands are pooled for re-examination after a homogeneous terminal group has been delimited, Williams and Lambert prefer to use a dichotomous or hierarchical procedure where both groups resulting from a division are retained. This procedure posed a problem which did not arise in Goodall's method—except in subdivision termination—that of significance. It follows from a

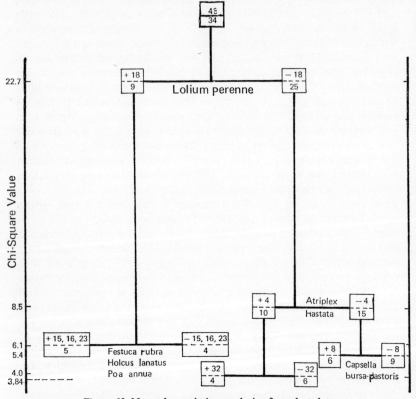

Figure 62. Normal association-analysis of quadrat data.

dichotomous divisive method that the relative importance of the different subdivisions at the same level may vary considerably. For example, one subdivision may reduce the residual heterogeneity by a large amount. A parallel division on the other side of the hierarchy may cause only a small reduction. It therefore becomes necessary to adopt some measure of heterogeneity which will record this disparity and the fall in heterogeneity with progressive subdivisions. For this purpose, Williams and Lambert have used the highest individual χ^2

value within the data as the *criterion of hererogeneity*. The level of χ^2 is also used as the criterion for termination of subdivision, i.e. subdivision continues until no individual χ^2 value exceeds 3·84, a figure which corresponds to a probability of 0·05.

The major points of the analysis are best reiterated with reference to an association-analysis diagram (Figure 62). A range of electronic computers is available for the handling of the basic programme and most of them have matrices which will readily handle up to seventy species and as many quadrats as desired. Others with high-speed stores provide larger matrices and can handle larger numbers of species. The punching of the cards for the computation of the data used for Figure 62 took just ten minutes, while the total computer time involved in the production of the output data was as little as twenty-three seconds. From the figure, the axis on the left-hand side is the scale of heterogeneity χ^2. The data in the box at the top of the hierarchy indicates that forty-three species and thirty-four quadrat samples form the matrix. The computer begins by stating the maximum χ^2 value in this matrix and then states the species with the highest value of $\sqrt{\chi^2/N}$ on which the first subdivision of the data is made. In this case, the species is number 18, *Lolium perenne*, and following this the number of individuals or quadrats in which the species occurs is divulged. These two figures are represented in the box to the left of the hierarchy. These nine quadrats are then searched for the next highest $\sqrt{\chi^2/N}$ and so on until subdivision is terminated. The cell which provides the other half of the first subdivision—minus *Lolium perenne*—is then returned to and the right-hand side of the hierarchy subdivided. There remain a series of terminal groups of quadrats which are fairly homogeneous with respect to floristic composition and which may have related ecological distributional ranges.

Perhaps one of the better applications of this method is seen when samples are taken in the form of a contiguous grid or transect across an observable series of gradients. The terminal groups of the association-analysis may then be recorded on the grid and it is often an enlightening exercise to compare this with a subjective estimate of quadrat grouping. Figure 63 illustrates such a comparison on a set of contiguous quadrats on an area of wasteland with an observable disturbance gradient from undisturbed grassland (approximately A1 quadrats) through disturbed revegetated areas (B, C, D, 3, 4, 5), recently disturbed open soil (E6) to a ditchbank (F2). Three of these areas receive approximately the same categorization, e.g. A1, E6, F2, but the revegetated areas are not quite so comparable to one another. On to the basic grid of groups may be superimposed environmental data, in this case pH values and the percentage of nitrogen in the surface soil determined by the Kjeldahl method. Other types of edaphic features may be super-

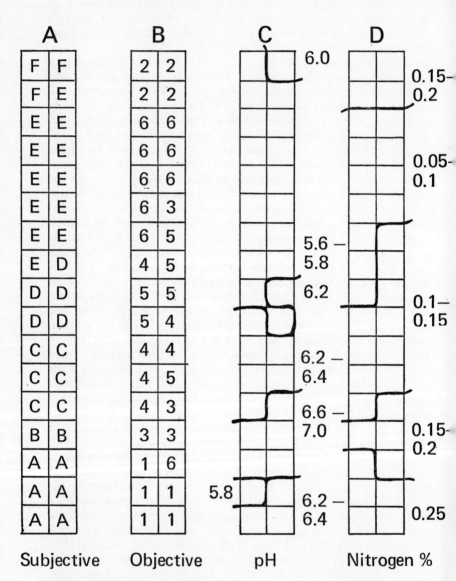

Figure 63. Base map of quadrat classification where A, subjective assessment; B, normal association analysis; C and D, superimposed environmental variables, pH and total nitrogen in percentage.

imposed until a detailed environmental complex of gradients is related to the vegetation gradient.

Two other aspects of association-analysis may now be referred to. As an alternative to an ordination interpretation, the data of Figure 62 may be conceived in the form of a dichotomous key for the recognition of quadrat types, similar to the keys used for the identification of species and genera in plant taxonomy. Thus the recognition of a quadrat of the group-type 1 would require the following characters to be present—*Lolium perenne*, *Festuca rubra*, *Holcus lanatus* and *Poa annua*—and so on for the other groups until group-type 6 requires negative records for all species used for subdivision. The interrelationships between stands can be further accentuated by inverting the association-analysis process so that the quadrats become the attributes and the species the individuals. The subdivisions are made on the quadrats with the highest ΣC and the resultant terminal groups are groups of species with a related distribution pattern in the data from which may be inferred eco-sociological relationship. This process of inverse association analysis (Williams and Lambert, 1961a) is identical with that of the so-called normal analysis outlined above, with the exception that the hierarchical representation is such that the χ^2 axis and subdivisions are horizontal instead of vertical. If the two analyses are brought together the resultant two-way table produces a matrix in which some assessment of how far a species grouping is linked with a particular stand-grouping. In the two-way table (Table 66) the normal groups of stands are arranged along the top of the table and are delimited by vertical boundaries, while the inverse species groups from the left-hand margin of the table, being separated by horizontal boundaries. The species-site data are recorded as + and the species and sites on which a subdivision was made recorded as an open circle in the appropriate cells. This in many ways is a satisfactory end point to the analysis except that the data in the cells might be tidied up according to species presence values. For example, a rearrangement of the species of cell E1 where *Agrostis tenuis* is placed below *Trifolium repens* and *Plantago major* relegated to the bottom of the group gives a more orderly representation of the species-in-habitat relationships of the data. A further development which seems to be merely an elaboration of the matrix and which is a questionable improvement is the technique of *nodal analysis* (Lambert and Williams, 1962) which according to the authors is devised 'to explore the extent of coincidence between the two forms of analysis'. Each group of quadrats is subjected to inverse analysis and each species group to normal analysis and the sites on which division is made in each case regarded as the coincidence parameters for the cells. If the cell is defined by a parameter in each direction the resultant cell or part of it is termed a *nodum* (analogous

TABLE 66

Two-way table derived from normal and inverse association analysis

Group	#	Species	1	2*	3	4	5	31	32	33*	34*	7	8*	10	24
A	2	Agrostis stolonifera	−	−	+	−	+	+	+	+	+	+	+	+	+
A	30	Senecio vulgaris	−	−	−	−	+	+	+	−	−	+	+	+	+
A	32	SONCHUS ASPER	−	−	−	−	+	−	−	−	−	O	O	O	O
B	5	Atriplex patula	−	−	−	−	−					−	−	−	−
B	8	CAPSELLA BURSA-PASTORIS	−	−	−	−	−					−	+	+	−
B	19	Matricaria matricarioides	−	−	−	−	−					−	+	+	+
B	28	Ranunculus repens	+	+	+	+	−					−	+	+	+
B	33	Sonchus oleraceus	−	−	−	−	+					−	−	+	−
C	12	Cirsium arvense	−	O	+	+	+					−	−	+	+
C	17	Lamium album	−	O	−	+	−					−	+	+	−
D	10	Chamaenerion angustifolium	−	O	−	−	−					−	−	−	−
D	23	POA ANNUA	O	O	O	O	O	−	−	−	−	+	−	+	+
E	3	Agrostis tenuis	+	O	+	+	−					−	−	−	−
E	15	FESTUCA RUBRA	O	O	O	O	O	−	−	−	−	−	−	−	−
E	16	HOLCUS LANATUS	O	O	O	O	O	−	−	−	−	−	+	+	−
E	18	LOLIUM PERENNE	O	O	O	O	O	O	O	O	O	−	−	−	−
E	22	Plantago major	+	O	−	−	−					+	−	−	−
E	38	Trifolium repens	+	O	+	+	+	−	+	−	−	+	−	−	−
E	39	Rumex acetosa	+	O	−	+	−					−	−	+	−
F	4	ATRIPLEX HASTATA	−	−	−	−	−					O	O	O	O
F	6	Barbarea vulgaris	−	−	−	−	+					+	O	−	−
F	29	Rumex obtusifolius	−	−	+	−	−					+	O	−	−
F	43	Stellaria media	−	−	−	−	−					+	O	+	−
G	1	Agropyron repens	+	−	+	−	−	−	−	O	O	−	−	+	−
G	7	Calystegia sepium	−	−	−	−	−	+	−	O	O				
G	21	Phragmites communis	−	−	−	−	−	+	−	O	O				
G	37	Urtica dioica	−	−	−	−	−	−	−	O	O				
H	9	Ceratodon purpureus	−	−	+	−	−	−	+	−	−	−	−	−	−
H	34	Taraxacum officinale	−	−	−	−	−	−	+	−	−	−	−	+	−
H	41	Veronica persica	−	−	−	−	−	−	−	−	−	−	−	−	−

I   plus 13 *other species*

O = quadrat and species on which subdivisions were made.

TABLE 66—*continued*

	4					5						6								
9	11	13	14	16	19	12	15	17	18	20*	22*	6	21	23	25	26	27*	28	29*	30*
+	+	+	−	+	+	−	+	+	+	O	+	+	−	−	+	+	−	−	−	O
−	−	−	+	+	−	−	+	+	+	O	+	+	+	+	+	+	+	+	+	O
−	−	−	−	−	+	−	−	−	−	O	−	+	+	−	+	−	+	−	+	O
−	−	−	−	−	−	−	+	+	−	O	+	−	+	+	+	−	+	+	+	−
+	+	+	+	+	−	O	O	O	O	O	O	−	−	−	−	−	−	−	−	−
+	+	−	+	+	−	+	−	−	−	O	−	−	−	+	−	+	−	−	−	−
−	−	+	+	+	−	+	+	+	+	O	+	+	−	+	+	+	+	+	+	−
−	+	+	−	+	−	−	−	−	−	O	−	−	−	+	+	+	+	+	+	−
+	+	+	+	+	−	+	−	−	−	−	O	+	+	+	−	+	−	−	−	−
−	−	+	−	−	+	−	+	−	−	−	O	−	−	−	−	−	−	−	−	−
−	+	+	−	−	−	−	−	−	−	−	−	−	−	−	+	−	O	+	−	−
+	−	−	−	−	−	−	−	−	−	−	−	+	−	+	−	+	O	−	+	+
−	−	−	−	−	−	−	−	−	−	−	−	−	−	−	−	−	−	−	−	−
−	−	−	−	−	−	−	−	−	−	−	−	+	−	−	−	−	−	−	−	−
−	−	−	−	−	−	−	−	−	−	−	−	+	−	−	−	−	−	−	−	−
−	−	−	−	−	−	−	−	−	−	−	−	−	−	−	−	−	−	−	−	−
+	−	−	−	−	−	+	−	−	−	−	−	−	−	−	−	−	−	−	−	−
+	−	−	−	−	−	−	−	−	−	−	−	+	−	−	−	−	−	−	+	−
+	+	−	−	−	−	−	−	−	−	−	−	−	−	−	−	−	−	−	−	−
O	O	O	O	O	O	−	−	−	−	−	−	−	−	−	−	−	−	−	−	−
+	−	−	−	+	−	+	−	−	−	−	−	+	−	+	−	−	−	−	−	−
−	−	−	−	−	−	+	−	−	−	−	−	−	−	−	+	−	−	+	−	+
−	−	−	+	−	−	−	−	−	−	−	−	−	−	−	−	−	−	−	−	−
−	−	−	−	−	−	+	−	+	+	−	−	−	−	−	−	−	−	−	−	−
−	−	−	−	−	−	−	−	−	−	−	−	−	−	−	−	−	−	−	−	−
−	−	−	−	−	−	−	−	−	−	−	−	−	−	−	−	−	−	−	−	−
−	−	−	−	−	−	−	+	+	+	−	−	−	−	−	−	−	−	−	−	−
−	−	−	−	−	−	−	−	−	−	−	−	−	−	−	−	−	−	−	O	−
−	−	−	+	−	−	−	−	−	−	−	+	−	−	−	−	−	−	−	O	−
−	+	−	−	−	−	+	−	−	−	−	−	−	−	−	−	−	−	−	O	+

cells in Table 66 are B5, D1, E1 and E3). Where only a single parameter defines a group (e.g. A3, A5) the term *sub-nodum* is used. As
Ivimey-Cook and Proctor (1966) point out the usage of this term nodum
must not be conceived in the original sense in which it was introduced
(Poore, 1955), but in a specialized sense related to the association-
analysis technique only. But to return to the use of nodal analysis, it
seems that this further refinement of the data, apart from the fact that
the coincidence parameters are determined objectively, has little
advantage over the mechanical reordering of data within the cells.
From the cells of any two-way table, be it of the simple kind or the
nodal kind, it is possible to see at a glance the groups of species which
are characteristic of a group of quadrats, to assess the interrelationships
of the groups and to pick out unsatisfactory classifications produced
by a single normal or inverse analysis. At a glance there are two types
of community labelled by *Lolium perenne* in Table 66, represented by
rows 1 and 2. One has a composition of three major components—E1,
D1 and C1, the other has a single component, G2. The normal analysis
grouped the four quadrats of row 2 together but the two-way table
reveals heterogeneity especially with respect to quadrats 31 and 32
which lack the parameters of the inverse analysis. Ivimey-Cook and
Proctor (1966) indicate that nodal analysis will remove such misfits
as quadrats 31 and 32 by creating a subnodum. They also make it
clear that some species and quadrats may have clear relationships with
other cells in the table, but that the process of nodal analysis cannot
help to place these in their 'best' location. Thus, it seems that the
'stark clarity' by which the nodal analysis table depicts the major
trends and groups in the data is equally well represented by a
mechanically arranged two-way table. One of the final pertinent
conclusions of Ivimey-Cook and Proctor is that association analysis is
extremely well suited to perform an initial breakdown of a set of
heterogeneous data and that other ordination methods such as factor
analysis can be used to explore relationships within or between a few
selected groups which are produced by association analysis.

The ordination techniques of the Wisconsin School

The techniques of this school of plant sociology fall naturally into
two types and two periods: the early gradient analyses beginning with
Curtis and McIntosh (1951) and culminating in Curtis's comprehensive
work *The Vegetation of Wisconsin* (1959); the comparative ordination
period beginning with Bray and Curtis (1957) with their geometric
construction of samples relative to compositional axes in a method
which is an approximation to factor analysis.

The early studies were of upland hardwood forests using the four
plotless sampling techniques outlined in Chapter 1 (Figure 6) but

initially the random-pairs method. They examined ninety-five stands of forest which were considered to be large enough, of natural origin and relatively free from interference. From the data collected by the random-pairs method of sampling an *importance value* (Chapter 1) was calculated for the leading tree species. To reiterate, this value is computed as the sum of (*a*) relative frequency (species frequency expressed as a percentage of frequency values of all species); (*b*) relative density; and (*c*) relative dominance (species basal area as a percentage of total basal area for all species). The higher the importance value, the greater the dominance of a species and in eighty of the ninety-five samples four species, *Quercus velutina, Q. alba, Q. rubra* and *Acer saccharum* proved to be the leading dominants. Dividing the samples into these four dominance-groups and calculating the average importance value for each dominant in each group, the species were shown to have a definite pattern of dominance behaviour. Similarly, subsidiary trees or understorey shrubs were shown to be related to a particular canopy dominant. A set of comparable data collected according to the method of Bray and Curtis (1957) from twenty Peak District woodlands is set down in Table 67. Three dominance types are recognizable, those of *Fraxinus excelsior, Ulmus glabra* and *Acer pseudoplatanus*, each of which have a maximum average importance value. Associated species also have maximum values in association with a particular canopy dominant, e.g. *Thelycranea* with *Fraxinus, Crataegus* with *Acer*. From the different patterns of importance value distributions for each species it becomes possible to donate to each species what Bray and Curtis refer to as a climax adaptation number. This is rather an arbitrary method based on overall importance values and for the purpose of the present example the three dominant species may be taken as end- and mid-points of a sequence which reflects an increasing anthropogenic influence on the woodland type. Based on their relationships to these three points subsidiary species may be given an 'adaptation value', but if the survey data is not complete then these values can only be regarded as tentative. Thus in the data of Table 67, species of *Tilia* are restricted to the *Fraxinus* dominance type, not occurring in others. They occur only in some stands, not in all, and are therefore considered to fall beyond the *Fraxinus* end-point. Both *Acer campestre* and *Thelycranea* occur outside the *Fraxinus* dominance type, but only sparsely and are placed next in the sequence; and so on. From this weighting of species the position of each of the stands is determined by a value known as the *continuum index*. For each stand, the component species are weighted by the multiplication of their importance values by their adaptation numbers and then added, i.e. Continuum Index (CI) = IV.A1, IV.A2, IV.A3, ... Because the greatest possible importance value is 300 (% frequency + % density + %

TABLE 67

Average importance values for some tree species in twenty
Peak District woodlands

| | Leading dominant in stand | | |
	Fraxinus	Ulmus	Acer
Dominant			
Fraxinus excelsior	126	48	21
Ulmus glabra	44	102	25
Acer pseudoplatanus	10	54	146
Subsidiary			
Tilia spp.	14	–	–
Acer campestre	8	1	–
Corylus avellana	22	10	1
Crataegus monogyna	2	5	8
Thelycranea sanguinea	11	2	–

Tentative climax adaptation (anthropogenic adaptation) numbers
for tree species

1. Tilia spp., Prunus padus, Quercus petraea, Taxus baccata, Sorbus aria
2. Fraxinus excelsior
3. Acer campestre, Thelycranea sanguinea
4. Ligustrum vulgare, Corylus avellana, Ilex aquifolium
5. Sorbus aucuparia, Malus sylvestris, Viburnum opulus
6. Ulmus glabra
7. Rosa canina, Prunus spinosa
8. Crataegus monogyna
9. Sambucus nigra, Salix capraea
10. Acer pseudoplatanus, Fagus sylvatica

dominance), the continuum index could theoretically range between 300 for a monodominant stand of *Fraxinus*, to 3,000 with a similar situation for *Acer pseudoplatanus*. The twenty stands are then plotted as a graph of performance of dominance types (importance value) against continuum index (Figure 64) and similar to the patterns inherent in some of Whittaker's direct gradient analyses, bimodal distributional species importance curves are manifest. This approach has been used not only for the major tree species of woodlands but for practically every component of many vegetation types, and related soil types. Gilbert and Curtis (1953) studied woodland understorey species, Hale (1955) epiphytic mosses and lichens, Bond (1957) investigated bird populations, etc. A slight variation of the representation was used by

Curtis (1959) in the study of a prairie continuum. From a study of sixty-five remnant prairies it became obvious that certain species were of restricted distribution with respect to a particular soil type. With a soil catena as a background it was possible to select groups of indicator species reaching optimum performance within a certain narrow range on the catena which was divided into five units—wet, wet-mesic, mesic, mesic-dry and dry, weighted 1 to 5 respectively. Ten indicator species were chosen for each site and the sixty-five stands were analysed for presence or absence of these fifty indicator species. The components

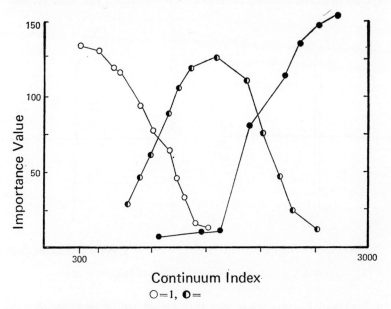

Figure 64. Continuum index ordination of twenty Peak District woodlands with either (1) *Fraxinus* (2) *Ulmus* or (3) *Acer pseudoplatanus* dominance.

of the five groups were weighted according to their 1 to 5 score and the derived total regarded as a *compositional index*. Stands containing only peak wetness indicators produce values of 100 whiie those of dry sites produce values around 500. Using the compositional index as the abscissa of a graph plot and species presence as the ordinate, the behaviour of individual species in relation to the moisture-based gradient becomes possible.

The work of Bray and Curtis (1957) produced a completely different approach to ordination. They calculated the coefficient of similarity for all samples using a coefficient which is often attributed to Czekanowski (1914), but which is essentially the same as that of Sørensen described previously; $C = 2w/a+b$ where a is the sum of quantitative measures of species in one stand, b is the similar sum for a second stand and w is the sum of the lesser value for species common to both stands. The resultant coefficients are placed in a matrix and then the least similar stands are chosen as end-points of a gradient on which other stands are placed according to their relative similarities. Such a comparative ordination has two main advantages over other types. First, samples may be ordinated without the previous assumption of an environmental gradient; and secondly, it provides a method for the arrangement of samples in relation to two or three axes of variation. The basic premise

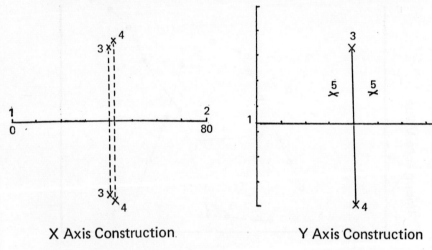

X Axis Construction.   Y Axis Construction

Figure 65. Wisconsin ordination; method of axis construction.

of Bray and Curtis's study was that the degree of phytosociological relationship between stands may be used to show the distance by which they are separated within a spatial ordination. Thus, given a matrix of values of distance between points in Euclidean space (wrongly assumed so, *q.v.*) it becomes possible to construct their spatial loci. It was considered that the distance between two stands could be translated from their degree of similarity by an inverse relationship, e.g. $1-C$. Using groups of stands which were least similar as the end-points of an axis, the relative positions of other stands were located along this axis by a geometric technique of arc projection, where the radii of the arcs are the inverted coefficients of similarity. The method is best described practically (Figure 65, a parallel example to that featured by Bray and Curtis). Because coefficients calculated from field data will inevitably be subject to sampling error, Bray and Curtis,

TABLE 68

Wisconsin ordination, a matrix of coefficients

	1	2	3	4	5
1	–	1	30	28	40
2	79	–	30	30	50
3	50	50	–	8	60
4	52	50	72	–	35
5	40	30	20	45	–

after replicate samples in two stands found a mean coefficient of 82 with each of seven replicates. They therefore considered that a value of 80 was the maximum coefficient and hence coefficients were subtracted from 80 instead of 100 to obtain interstand distance. Use is made of the $C_{max} - C$ calculation in Table 68, where the upper right portion of the table shows indices of similarity, the lower left portion interstand distances. Axis construction begins with the selection of the two stands with the largest interstand distance, which from the data is between stands 1 and 2, a figure of 79 units. These are therefore placed at opposite ends of the first (x) axis. Stand 3 is 50 units from both 1 and 2 and is therefore located at the intersection of arcs with radii 50 units and centres at points 1 and 2. Two intersections are possible and these are projected perpendicularly on to the axis to give an x axis location of 40 units. Similarly, stand 4 may also be located along this x axis at approximately 42 units, and so on. After the projection of all points on to the x axis, a second axis (y) may be constructed by the same method, using two stands which are close together in their projection on to the x axis. In the example stands 3 and 4, being only 2 units apart, are used. Reference to the matrix shows that they are a relatively large distance apart (72 units) and therefore on opposite sides of the x axis. Which point is selected above and which below the x axis is decided arbitrarily. When it comes to a decision on which is the correct position for stand 5, the intersection arcs for this point are drawn with reference to points 3 and 4, i.e. 20 units from 3 and 45 from 4. Of the two possible positions for stand 5, the correct one is determined by reference to its position relative to points 1 and 2. The distance from point 1 is 40 units and from point 2, 30 units, and therefore the position of 5 is that to the right of the y axis. This simple geometric technique can be extended for all points to give the spatial locations of all stands in a two-dimensional system, which may be readily extended to a three-dimensional 'lollipop' effect by the construction of a z axis.

The finished product of such a geometric process is illustrated in Figure 66, from Shimwell (1968b) using data from the magnesian

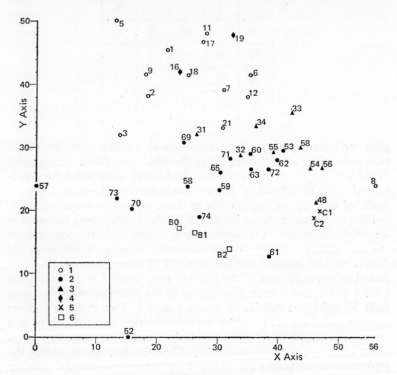

Figure 66. Ordination of stands of the Association Seslerio-Helictotrichetum (from Shimwell, 1968b). 1, S-ass. of *Helictotrichon*; 2, S-ass. typicum; 3, S-ass. of *Encalypta* and *Plantago*; 4, S-ass. of *Helictotrichon, Rosa* variant; 5, S-ass. caricetosum; 6, S-ass. of *Helictotrichon, Bromus* variant.

limestone grasslands of Co. Durham. Coefficients of similarity are based on species composition and the greatest $C_{max} - C$ value calculated as 56 units between stands 57 and 8, which were used as reference points for the x axis. The projection of all stands on to this axis revealed that stands 52 and 5 were two units apart but separated by a large interstand distance of 50 units. These two points were consequently used for the y axis and the position of all stands relative to these two axes plotted. If it is so desired, the predetermined classificatory units may be depicted as different symbols in an attempt to compare the methods of classification and ordination. For the most part the classificatory units form localized areas in the spatial pattern, e.g. the unit, *Seslerio-Helictotrichetum* sub-ass. of *Helictotrichon* is more or less restricted to the upper regions of the ordination.

Ordinations of this type were common in the early work of the Wisconsin School. On to the basic stand ordination, features such as

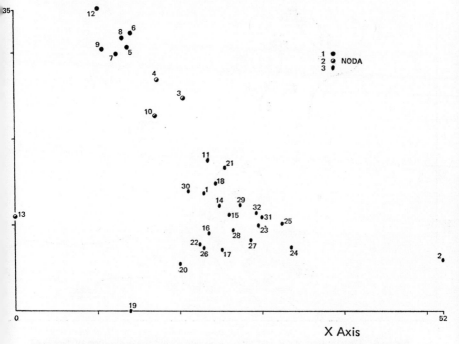

Figure 67. Ordination of stands of vegetation along an environmental gradient (from Shimwell, 1968b). 1, Helictotricho-Caricetum flaccae; 2, Mixed scrub/grassland stands; 3, Seslerio-Helictotrichetum.

the dominance distribution of leading species were plotted and delimited by isolines. The application of the method in British phytosociology has been of a more restricted nature. Gittins (1965 a, b) produced two types of ordination from different types of limestone grassland in North Wales in a locality where there was some observable gradient of species composition in the field. His second type of ordination, a direct species ordination, was completed by transposing the data matrix and estimating the similarity between species on the basis of their different frequency representation in the stands examined.

The former method was also used by Shimwell (1968b) on a topographical gradient in limestone grassland along which there was some apparent compositional variation. Thirty-two 1 m² quadrats were taken in the form of a regularly spaced transect along a gradient in which three limestone grassland communities were represented. These samples were then ordinated according to total floristic similarities. The maximum interstand distance was 52 units between stands 13 and 2

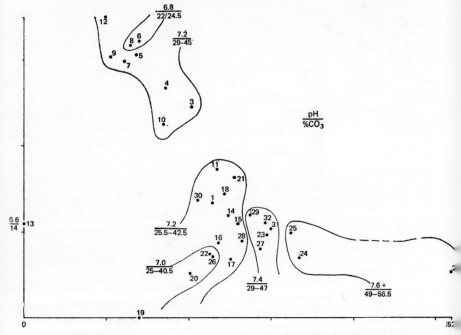

Figure 68. Distribution of pH and total carbonate within the ordination of *Figure 67* (from Shimwell, 1968b).

and these were used for the *x* axis, whilst stands 19 and 12 provided the end-points of the *y* axis. The spatial pattern of the ordination (Figure 67) is quite dissimilar to the type of 'scatter diagram' ordination described previously. Instead, there is a definite south-east north-west gradient of composition along which three noda may be recognized. This pattern emerges from similar small-scale studies where the ordination is aimed at observable field differences and used to provide a background representation of vegetation types for a particular nature reserve or site of special scientific interest. On to this background, data on environmental variables or species performance may be superimposed. Shimwell demonstrated that the correlated distribution of pH and per cent carbonates and calcium with magnesium could be readily related to segments of the gradient (Figure 68). Similarly, the dominance behaviour of grasses and other species may be plotted on the gradient.

There have been several modifications and criticisms of the Bray and Curtis ordination method. Beals (1960) who subtracted similarities from an upper limit of 85% has pointed out that the distance along the first axis at which a stand is placed may be obtained by calculation of the Pythagorean theorem $x = (L^2 + D_1^2 - D_2^2)/2L$, where D_1 and D_2 are the distances of the sample from end-point samples and L the distance of end-point samples from each other. Distance of the sample from the x axis is estimated from $d = \sqrt{(D_1^2 - x^2)}$. Gittins (1965a) allows 10% similarity deduction for the inevitable sampling error and uses a C_{max} value of 90, while McIntosh and Hurley (1964) use as C_{max} the highest value found in the data matrix. Finally, in a series of comparative tests Bannister (1968) concludes that the use of $1 - C$ as a measure of distance gives no less efficient ordinations than those using $C_{max} - C$, adding that there seems little point in the continuation of the use of the latter.

A major criticism of the Bray and Curtis method is found in the work of Austin and Orloci (1966) who after comparison of the $C_{max} - C$ distance with a known Euclidean distance (D) calculated from the Pythagorean theorem, indicate that $C_{max} - C$ is not in fact a Euclidean distance because: (*a*) negative distances can arise; (*b*) two stands may have zero distance even though they are not identical; (*c*) the sum of two sides of a triangle formed by three stands may be less than the third side, suggesting that the sides are not straight lines. Orloci (1966) further introduces a new method of stand ordination based on the Euclidean interstand distance (D). Here the distance between two stands (j and h) is derived from the species scores for the two stands X_{ij} and X_{ih} from the formula:

$$D_{jh} = \left[\sum_{i=1}^{n} (X_{ij} - X_{ih})^2 \right]$$

or in its expanded form (Bannister, 1968)

$$D_{jh}^2 = \sum_{i=1}^{n} (X_j^2 + X_{ih} - 2X_{ij}X_{ih})$$

i.e. D_h is calculated by subtracting twice the product of species common to both stands from the total sum of squares of species in each stand. Orloci (1966) also defines a related technique of perpendicular axis construction as opposed to the oblique axis method of Bray and Curtis, which Austin and Orloci (1966) consider to overcome the distortion of stand relationships experienced with the oblique method. They also submit that the use of extreme stands as reference points for the axes results in a loss of ordination efficiency, but also that the Bray and Curtis method exaggerates the appearance of a continuum. Bannister (1968) reviews the two approaches and arrives at two major conclusions: (a) the use of the Euclidean distance (D) results in a 10–20% increase in the efficiency of interpretation of interstand distances; (b) the techniques of Bray and Curtis and Orloci of simple ordination may be regarded as complementary; the use of the D-perpendicular axis method appears more efficient in vegetation types of relatively low species diversity and abundance; the Bray and Curtis method may produce a more readily interpreted picture in other situations. But neither method is considered better in terms of axis construction than principal components analysis.

The same sentiments are echoed by van der Maarel (1969), who, because electronic computers are not generally available to phytosociologists, develops a close approximation to principal component analysis (P.C.A.) which may be completed by hand. This method is known as *principal axes ordination* (P.A.O.). As a basis, van der Maarel begins with the Sørensen coefficient written as $C_s = 2P_{jk}/P_j + P_k$ where P_j and P_k are the sums of species representation values in two stands j and k and P_{jk} is the sum of lesser representation values. He goes on to conclude that a formula in which both similarity and dissimilarity are considered is more appropriate and from the Sørensen coefficient derives this value, C_m as equal to $2C_{s-1}$. The corresponding distance coefficient is $Dm = 1 - Cm$ or from Sørensen $Dm = 2 - 4P_{jk}/P_j + P_k = 2(1 - C_s)$. This may be used for both quantitative and presence-absence data.

The ordination begins in the usual manner with a matrix of Cm values and a choice of the lowest or another low value. The position of stands should reflect a maximal part of the total variation and van der Maarel suggests that the best criterion for finding this maximal variation is the negative correlation between the two columns which contribute the lowest C value. This negative correlation tendency (N.C.T.) is determined by adding the differences (ignoring the sign) between corresponding cells in the same column. The higher the N.C.T. the stronger the negative correlation and the better the axis in terms of maximal variation. Once the reference points for the x axis are determined the first ordination is carried out in the usual method of Bray and Curtis (1957) or the calculation method of Beals (1965).

$$XJ = \frac{(D_{AJ} + D_{BJ})(D_{AJ} - D_{BJ})}{2 D_{AB}}$$

where D_{AJ} and D_{BJ} are the distances between stand J and reference stand A and stand B and D_{AB} that between the two reference stands. The y axis may then be selected after the calculation of $D_{AJ}+D_{BJ}$ and $D_{AJ}-D_{BJ}$ for each stand and the comparison of stand pairs selected for the high sum $(D_{AJ}+D_{BJ})$, low difference $(D_{AJ}-D_{BJ})$ values, e.g. for stands V and W

$$(D_{AV}+D_{BV}+D_{AW}+D_{BW}) \quad - \quad [D_{AV}-D_{BV}-(D_{AW}-D_{BW})]$$

The highest species pair value for this formula or if there are two with the same value, the highest pair N.C.T. value are used as y axis reference stands. The calculation method of Beals (1965) may again be used for the location of stands on the y axis.

The method may seem complex at first sight but with a set of practical material and a desk calculator, its value soon becomes apparent. Van der Maarel (1969) applies the method to four examples; the ordination of structural characters, of stands, of successional relationships and of predetermined vegetation types, all of which give results comparable with the results of principal components analysis.

Principal components analysis

In ecological investigations, principal components analysis is generally considered to be synonymous with the technique of factor analysis, although a statistical interpretation such as that of Lawley and Maxwell (1963) considers the two processes distinct in that P.C.A. is variance-orientated and factor analysis covariance-orientated. The simple ordination methods outlined above are an approximation to P.C.A. of which the principal axes ordination of van der Maarel is the closest. P.C.A. may be considered to be concerned basically with the distribution of stands or species relative to the axes of greatest variation in the data, and if there are a few strong directions of variation in this data they should be reflected in a P.C.A. The actual statistical interpretations of the technique of axis selection are reviewed by Lawley and Maxwell (1963) and Seal (1964) but may be considered as three basic steps. Correlations may be shown between either attributes (the species in an R-type analysis) or between individuals (the quadrats of a Q-type analysis, cf. association analysis). Orloci (1966) states that when there are more individuals than attributes then an R-type analysis is possible and in the alternative case the Q-type analysis is applicable. After the compilation of a matrix of similarities of species or quadrat data (1), a correlation or loading matrix is calculated for the overall distributional similarities of species in quadrats (2). Whittaker (1967) represents the coefficients of correlation as

$$r_{jk} = a_{j1}a_{k1}+a_{j2}a_{k2}+a_{j3}a_{k3} \ldots +a_{jm}a_{km}$$

K

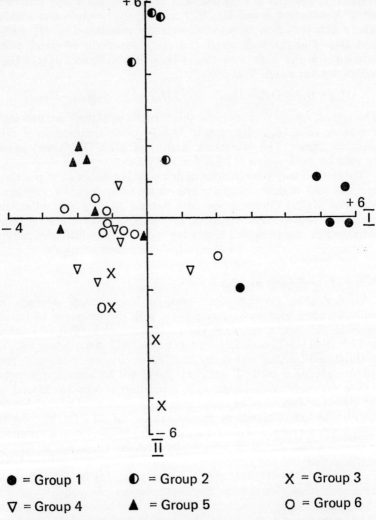

Figure 69. Principal components analysis (data and groups as for *Table 66*).

where r_{jk} is the coefficient of correlation for the distributional similarity of stands j and k according to their species loadings a_1, a_2, etc. The matrix is analysed to find linear combinations of variables (components) with large variance (3). The first principal component is the combination with the maximum variance, the second is a linear combination uncorrelated with the first with as large variance as possible, and so on.

From the correlation or covariance matrix, the variances of the principal components or eigenvalues and the principal components or eigenvectors are calculated using the QR alogrithm described by Bowdler *et al.* (1968). The eigenvalues are listed in order of descending variance, and in general, if there is a strong underlying factor affecting the species distribution then the first value will extract a large amount of the variance—usually in the region of 40–60%. If the data is fairly homogeneous or if no strong gradients are present the components will extract a small amount of variance. For example, the data used for Figure 69 produced eigenvalues of only 15·45 and 13·80 for the first two components with a total accumulated value of only 59·30 after six components. This tends to indicate the lack of strong gradients, but, nevertheless, the component axes erected are accurate in reflecting the major variance of the data.

Figure 69 shows a P.C.A. of the same data used for the association analysis (see Table 66), where the positions of the thirty-four stands are sited according to their component loadings on axes I and II. The symbols refer to the six groups delimited by association analysis and it may be seen that the grouping remains much the same. Anomalies such as quadrat 32 which was placed in group 2 by the normal analysis are picked out by their distances. Orloci (1966) has stated that interstand distances calculated from presence-absence data reflect differences in the species richness between stands under comparison, and the two axes do in fact represent variation in species richness. Group I at the positive end of axis I has a mean species number of 10 while those at the negative end of the axis have a mean of 6. The same applies to axis II. For similar reasons the variation along both axes represents a gradient from stable to unstable and high soil nitrogen to low soil nitrogen.

As Whittaker (1967) points out the techniques of P.C.A. are often laborious in relation to their research awards. The extracted factors may be considered as compositional gradients and may often produce gradients which are ecologically unidentifiable. As a method of stand ordination, research has shown that it is the most accurate one, but the divorced nature of the technique often leads to difficulties of gradient interpretation and misinterpretation. Moreover, the need for a knowledge of computation technique and, in some cases, the constant attention of a mathematician, gives the method little advantage over either the Bray and Curtis or van der Maarel methods in terms of the construction of a theorectical hyperspace into which vegetation stands and site data may be superimposed to depict a series or a single vegetation gradient.

CHAPTER 8

COMPARISONS AND CONCLUSIONS

Is there a 'best method' for classifying vegetation?

The past decade has witnessed a proliferation of approaches to the organization of phytosociological data and the subsequent classification or ordination of data. Adding to the earlier exchanges of concepts and criticisms of the European schools of phytosociology, this development places the average ecologist in a quandary when it comes to the choice of a method for describing and cataloguing the data pertinent to his particular problem. Should he choose a classificatory or ordination method? What characters of the vegetation should receive the greater attention and greater weight? How will he unite his site descriptions into units, by simple geometric methods or successive approximations or by the use of an electronic computer? Which is the best method for his particular problem? It may be answered that from a phytosociological point of view the best method may be taken to be that one which enables the maximum comprehension of the structural complexity of vegetation relative to the background environmental variables, in turn relative to the amount of time input. Several investigators have become aware of these problems and have attempted to compare and contrast various methods and it is with these studies that the first part of this chapter is concerned.

Speaking of the rift between the British dynamogenetic approach to vegetation description and the Continental floristic approach, Tansley (1922) said: 'The increasing recognition of the genetic principle on the Continent is a great gain on the one hand, and increased attention in England and America to the Continental methods of analysing communities will also contribute in an important degree, to mutual understanding and eventual accord.' But Tansley attempted to apply Continental techniques (Tansley and Adamson, 1926) and was unable to form an opinion as to their usefulness. The process towards 'mutual understanding and eventual accord' received its first setback. On the Continent, the war of criticism between the Uppsala and Zurich–Montpellier traditions built up in the late 1920s but finally subsided away from feuding level after the 1935 Botanical Congress only later

278

at the Stockholm Congress of 1950 to be reduced 'après une discussion très animée' to a predominance of Z-M methods and classificatory system. Latterly, within this tradition, there has been a growing awareness of the diverse trends of classification and a proliferation of association names. Lohmeyer et al. (1961) in a synopsis of the hierarchical classification of the vegetation types of north-west Europe have partially remedied the former problem, whilst the rules for the naming of an association outlined by Moravec (1965) are now accepted within this school of formal vegetation classification.

Westhoff (1967) has recognized the increasing emphasis of structure in the methods of this school and erects a structural classification which encompasses the usual floristic groups of the hierarchy. From the other side, Dansereau and Arros (1959) have applied Dansereau's structural representation of vegetation to several European associations in the sense of the Z-M School. This contribution is offered in an attempt to develop a universal system of graphic representation of structural and functional aspects of vegetation irrespective of composition, in other words to open the eyes of Z-M phytosociologists to another dimension of the vegetation units. As such it failed, but it probably contributed to an awakening of dormant views on vegetation physiognomy and structure.

The development of association analysis as a method for classifying vegetation into either groups of quadrats or to species groups with a related distribution in the data probably represents the major contribution to the phytosociological investigations of the past decade. There are few modern phytosociological research projects which do not make use of it in one of its component forms, usually normal analysis. After its establishment, the inevitable, but nevertheless important, comparisons of its methods with those of traditional classification and ordination procedures followed. In their application of the analysis to various types of phytosociological data, Ivimey-Cook and Proctor (1966) concluded that not only did it emulate the methods of traditional phytosociological techniques in recognizing species equivalent to the *Kennarten* or characteristic species, but it also produced divisions relative to directions of variation which the traditionalist tended to overlook. In a second direction, association analysis came to be compared with the methods of ordination used by the Wisconsin School (Gittins, 1965). He concludes that stand ordination and normal association analysis 'lead to substantially identical interpretations of the data', but that the former results in a more informative and ecologically more satisfying model of stand interrelationships.

This leads to a problem which has loomed large in recent ecological literature—classification or continuum? The relative merits of ordination and classification and their individual contribution towards a demonstration of the nature of vegetation have been discussed at length by McIntosh (1967). He submits that for most schools of phytosociology, classification is an end in itself, the aim being a taxonomy of vegetation. He asks why we attempt to classify vegetation, and if a preference for classification is developed in childhood as Goodall (1954a) suggests, or if the biologist is so indoctrinated at a later date with the hierarchical classification of organisms that he finds it difficult to think in other ways? (Webb 1954). He likens proponents of traditional phytosociological methods in their view towards modern numerical and geometric approaches to the Duchess in *Alice in Wonderland* with her 'Oh don't bother me; I never could abide figures' attitude. On a slightly more serious note, McIntosh also indicates that methods of classification and ordination have often been regarded as mutually exclusive, but that they have also been referred to as interrelated. Both Greig-Smith (1964) and Lambert and Dale (1964) are in agreement that the two processes are not necessarily incompatible and the work of Curtis (1959), Groenwoud (1965) and Shimwell (1968b), *inter alia* have shown that classification may lead to ordination, that an ordination may be classified arbitrarily into areas of high similarity, and that the two processes can be complementary in depicting variation in environment within the hyperspace of a classificatory unit. We may, therefore, delete the word 'controversy' from Anderson's (1965) question on the relationships between classification and ordination—'Controversy over a non-existent problem?'—and bury the horse long-since dead.

Within the traditions of ordination of vegetation, there has been a progressive refinement of both the calculation of coefficients of similarity and distance and the methods of axis construction. The reviews of Orloci (1966) and Austin and Orloci (1966) have set the scene on the modern approaches to such problems, but the basic method of Bray and Curtis (1957) has not been without its advocates also (Bannister, 1968). Specialization in this line of approach has undoubtedly been due to the development of computer programming for ordination methods or the adoption of statistical techniques such as factor analysis for the use in the extraction of the major trends of variation from a data matrix as the axes of an ordination. Recognizing that not all phytosociologists have access to computing facilities, van der Maarel (1969) has developed an approximation to principal components analysis called principal axis ordination which may be solved using a desk calculator. This paper is of interest not only for its description of the approximate hand principal components technique, but is also of immense value in its detailed interpretation of ordination models in terms of the successional relationships of vegetation and the traditional phytosociological approaches of the Z-M School. Using the diagnostic species for the levels of the hierarchical classification system of this latter school, van der Maarel ordinates the various alliances on his model and indicates relationships by lines called isocenes to show the percentage contribution of each species group to a particular area of the model. This, he concludes, leads to regular, easily interpreted synecological patterns. Moreover, he suggests that because the two methods of ordination and sociological group classification in combination are consistent, then it is extremely improbable that both techniques are wrong—some further evidence for the complementary nature of classification-ordination techniques.

In a similar vein to van der Maarel, Moore *et al.* (1970) have compared the relative merits of a computerized form of the Braun-Blanquet classification technique with an agglomerative cluster analysis technique of Sokal and Sneath (1963), normal and inverse association analysis and a principal components analysis. These four methods lead to the production of an association table, a dendrogram, a hierarchical subdivision of data and an ordination, respectively. The conclusion reached by Moore *et al.* is that the Braun-Blanquet method combines most of the advantages of representation and interpretation of both the dendrogram and the ordination diagram, and that it is the most 'efficient' for a general ecological survey. This 'efficiency' refers to the fact that the method is the most economical in terms of both computer and human time whilst yielding results which are at least, if not more, as informative as the other methods. The synthetic stage of this table method is viewed as a polythetic subdivisive classification of *relevés* which immediately has an advantage over the monothetic process of association analysis. After this, the reordering of species and quadrats to produce a neat table is merely a simple mechanical ordination based on both interstand similarities in terms of floristics and also ecological gradients. Thus, it is common to find that many terrestrial associations have a xeric sub-type, a typical sub-type and a mesic sub-type, relative to environmental moisture. Not only does the reordering of *relevés* approximate to a single dimensional indirect gradient analysis based on composition, but can also reflect the direct gradient analyses which are so strongly advocated by Whittaker (1967). As an intermediate method between the systems which rely upon dominance and physiognomy for characterization and strictly quantitative descriptions, the methods of the Z-M system provide a happy mean, and the present author would readily endorse these suggestions of Moore *et al.* However, the implication is here that this method is the best for all purposes and all vegetation types. Its applicability has not really been proved with respect to tropical rain-forest vegetation, although there has been a general lack of permanent exponents of the system in these regions.

But things are happening in the tropics. A team based in Australia has begun a series of numerical analyses of rain forest based primarily on floristic data rather than the other features of this vegetation type listed by Richards, Tansley and Watt (1939). Webb *et al.* (1967) review four numerical methods which they consider reflect known environmental variables 'with great precision' and also demonstrate that basic floristic units responsible for forest structure are integrated vertically and are not synusial. Later, Webb *et al.* (1970) examine and compare the properties of floristic and physiognomic-structural data and their use in rain forest description and classification. Their results indicate that a classification based 'on a limited number of physiognomic features noted by untrained observers can prove as useful as a floristic classification'. For this purpose, they devise a questionnaire consisting of twenty-four physiognomic-structural observations to be answered by inexperienced observers. Using both agglomerative polythetic (information analysis) and divisive monothetic (association analysis) methods they demonstrate that both floristic and structural data reflect environmental information and that there is no support for the suggestion that structural characters are less efficient at lower levels of a hierarchical classification. Finally, Austin and Greig-Smith (1968) in their assessment of some of the methodological problems of rain-forest data include amongst the conclusions the remark that '. . . qualitative data are likely to be more satisfactory than quantitative'.

By way of a conclusion for this particular section, four points are of significance:

1. The period of maximum diversification of ideas on the methods of the description and classification of vegetation is past.

2. The growing awareness of this diversification has led to a series of comparative studies, and though they mainly take the form of academic exercises their value has been immense in assessing the relative merits of the various methods.

3. The widespread proliferation and elaboration of computational techniques for vegetation description needs to be terminated, with the maintenance of a central core of divisive, agglomerative and ordination techniques maintained for general usage.

4. The results of comparative studies and studies on the 'best' method for a particular problem have shown: (*a*) that there is no general 'best method'; and (*b*) with respect to time involved and information obtained the methods of some of the so-called traditional schools of phytosociology present a better overall understanding of the complexity of the nature of vegetation.

Vegetation, education and conservation

In a book which has attempted to be practical throughout it seems that there should be some section reserved for some of the more theoretical aspects of vegetation description and classification. Two main discussion points may be extracted from an otherwise disparate mass of debating material: on the academic front—what role in the education of students of biological and environmental sciences should a study of classificatory and descriptive methods occupy; and on an applied front—towards what possible applications should the classification and description of vegetation be aimed?

Taking the first question—although many would claim that the two questions are too closely interrelated to discuss separately—the problems of education fall at three basic levels; school (and workers), undergraduate and graduate. Although the book is aimed mainly at the latter two levels, the extension of many of the methods of vegetation analysis and comprehension may be readily made with respect to the sixth form level. At the moment most G.C.E. 'A' level syllabuses in either botany or biology pay only perfunctory notice to ecology as such, with such directional statements as 'some first-hand observational work should be carried out in the field to illustrate the concept of the plant community and plant succession'. The scanty references to this aspect often lead the teacher to place greater emphasis on the heavily weighted plant-groups study and either relegate the ecological aspects to a place of minor importance or to delegate to each pupil a project to be completed with little or no direction. Other sixth forms receive more tuition on the ecological aspects of the syllabus through fieldwork at recognized field stations. Here it seems that it is not so much the theoretical studies of the plant community and succession which should form the major line of investigation, but some knowledge of the ecological amplitude of selected species and their contributions to vegetation in different sites. For this purpose the transect studies, whether based on percentage cover, performance or frequency, and the simple gradient analyses along an observable physiographic or altitudinal gradient are perhaps the most rewarding lines of observation. Only in the classroom afterwards should ideas about the nature of a plant community and succession be encouraged, based on field observation of combined species distributions or study of zonations around lakes.

At the undergraduate level the concept of ecological amplitude may be developed to a study of vegetation types of different soils and substrata. In the second and third years there is normally room in the timetable for at least two courses aimed at the collection of quantitative data and such considerations as the detection of non-randomness and pattern in vegetation. Field courses provide ample opportunity for the application of the overall survey method of the type used by the Z-M School and Poore (1955) and also an elaboration of the continuum concepts of the Wisconsin School. Evidence of the practicability of field courses as teaching platforms may be derived from the numerous figures in the preceding text which use data from the Galloway region of southern Scotland, data which is largely the result of three week-long field courses in the region. With a basic ecological background provided in the first two years, the integration of vegetation study with courses on plant-soil relationships and genecology becomes possible. Techniques such as association analysis may be taught on a general biological platform for the ordering of data rather than a process peculiar to vegetation analysis. These lines of proposed study are made with reference to a botany honours course only, and if opportunity arises for such a course as joint botany and geography in the university curriculum, then the proportion of vegetation study within the course is greatly enlarged.

It is at the graduate level that our two questions become most closely contingent. The comparative studies in the past, while teaching the researcher an understanding of diverse vegetation analysis methods, have unfortunately become enshrouded in a preoccupation with technique. Further, there are now reviews of the relative merits of the various methods, all of which more or less demand that the type of project undertaken by the postgraduate student should not merely be an academic exercise. A problem needs to be chosen where the results of the study will be of some application in a wider field than simple vegetation syntaxonomy or static hyperspace ordination. Analyses of the type of ecological problems which need to be answered suggest that the major ones are those which concern the effects of man on the ecosystem—the 'fitting of man back in to the ecosystem' cliché which has been bandied about recently. The central features associated with this problem may be regarded as twofold: (1) the careful monitoring of the effects of general pollution on the environment; (2) the maintenance of maximum biological diversity in all remaining ecosystems—synonymous with conservation.

The solution to pollution is in the first instance, political, and it is really only as a secondary, peripheral factor that the monitoring of the levels of pollution by a study of plant and animal communities can be contributory towards a biological solution. Organisms and vegetation can act as biometers or living measures of the levels of environmental variables, a fact which should have emerged from the foregoing text. The time is ripe for the widespread application of analysis methods towards the new environmental variables introduced by man. An excellent example of the type of application required is that found in the work of Hawksworth and Rose (1970). In a survey of epiphytic lichen communities in England and Wales they recognize a series of ten noda which they consider to form a continuum developed along an air pollution gradient. From the ten noda they erect ten zones of atmospheric purity or contamination, recognizable by communities, or (because lichen communities are often fragmentary for other reasons than air pollution) species-groups found on deciduous trees with rough bark. To the zones they relate measurements of mean winter atmospheric SO_2 in μ g/m^3 derived from rurally situated recording stations of the Ministry of Technology. Thus, communities rich in the large foliose lichens of the genus *Lobaria* and other sensitive species such as *Teloschistes flavicans* represent the zone at the 'pure' end of the atmospheric pollution gradient. Other species with greater tolerance of sulphur dioxide replace these latter as the concentration of the gas in the atmosphere increases, until at about 150 μ g/m^3 SO_2 only *Lecanora conizaeoides* remains as the sole representative of the epiphytic lichen flora.

Although air pollution was postulated as the causal factor behind the paucity of lichen floras near industrialized areas as long ago as 1866, it is only recently that attention has been focused on the possibility of the existence of an environmental scalar for reflecting the magnitude of air pollution. It seems that this type of study may be readily adapted to other aspects of pollution and used as a background monitoring technique for the basis of deeper investigations into the cause, extent and control of widespread environmental pollution.

Leading on directly from this applied aspect of study, the maintenance of biological diversity holds a position of equal priority. Many studies of vegetation are aimed directly at various aspects of nature conservation. The third aim of the Z-M school of plant sociology stated by Tüxen (1937) requires the study of the life conditions and developmental possibilities of the vegetation types which are recognized. Conservation and landscape architecture were not words which were in fashion at the time. There was not a great pressure on the environment and as yet there was no need for landscape architecture. But these two aspects, plus features such as the agricultural, educational and recreational possibilities of the vegetation and landscape types were the type of aims he had in mind. In later papers (Tüxen 1967, 1969), the economic and social aspects of the study are more apparent. Several authors have followed this lead in undertaking an applied line of study, that of O'Sullivan (1965) on Irish lowland grasslands providing an excellent example from the British Isles. Much of the present author's work has been carried out in the Peak District National Park, since it is in regions such as this, with one-third of the population of Britain sitting on its borders that a rapid assessment of developmental and management methods relative to recreational pursuits needs to be assessed in terms of pressures on the vegetation and environment.

Against this 'management for the masses' aspect of vegetation study, the 'management for the grasses' approach must be balanced, for conservation studies, to which this latter maxim applies, rank high on the list of applied vegetation study projects. Taking the maxim literally, the advent of myxomatosis completely removed rabbit populations from whole areas of chalk downland where they reigned supreme as the major grazing animal. The predominance of a short sward turf enabled the development of large populations of comparatively rare herb species, many with basal rosettes closely adpressed to the soil surface and resisting grazing. With the loss of the grazers, the coarse grass components of the sward which are fairly sensitive to grazing, e.g. *Bromus erectus* and *Brachypodium pinnatum* formed extensive rank growth, thus placing the small, precarious species, under severe threat of extinction. Thus, to maintain maximum biological diversity, it is the job of the conservationist to describe quantitatively the changes which occur following cessation of grazing, to simulate populations of rabbits or other grazing animals and to assess the best treatment for the answer to the problem. This approach is one of the basic problems of the Lowland Grassland and Grass Heath Division of the Nature Conservancy at Monks Wood Research Station in Huntingdon.

Moving away from the grasses, myxomatosis has been shown to have had unpredictable effects on many species and vegetation types. Perhaps the most interesting and the one where the maximum effect is visible is to be found with reference to the expansion of *Hippophaë rhamnoides* on the Spurn Peninsula in Yorkshire (Figure 70 from Dargie, unpubl.). The use of sequential aerial photographs for 1959, 1962, 1966 and 1971 enables an interpretation of the spread of the plant which reproduces by rhizomes. Before 1959, the buckthorn had been kept in check by a large population of rabbits, which in spite of the high tannin content of the leaves, grazed any new shoots produced by the rhizomes. Since this time, the buckthorn has progressively invaded open *Ammophila*-dune communities and the short sward dune grasslands rich in coastal species at the northern limits of their British distribution. But on the other hand, the peninsula is gradually moving southwards into the Humber estuary due to erosion at its northern end and accretion to the south. As this occurs the *Hippophaë* begins to stabilize the dunes and prevent further erosion. The conservationist is thus placed in a cleft stick with regard to the management of the dune scrub. Should he manage for the maintenance of biological diversity and attempt to eradicate the scrub from areas where this diversity occurs? In doing so will he initiate widespread erosion? The problems can only be answered by a detailed study on the biology, performance and management of the species in question, and this falls to the lot of the student of vegetation.

But conservation does not only revolve around the study of manage-

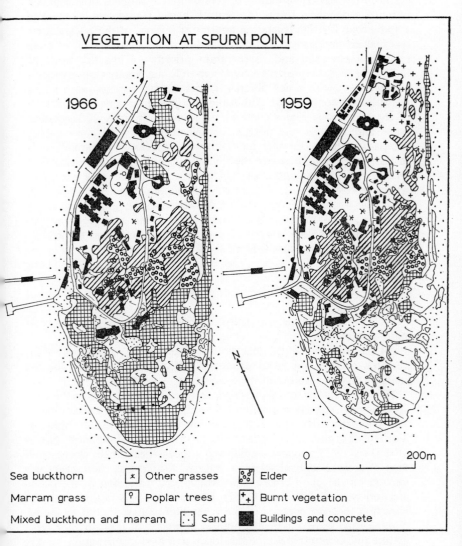

Figure 70. The expansion of *Hippophaë rhamnoides* at Spurn Point, Yorkshire, in the years following myxomatosis (from Dargie, unpublished).

ment practices for the maintenance or control of single species. Rather, the need for the conservation of representative samples of the major types of ecosystems forms the central theme. Consequently, there is ample scope for the vegetation analyst in reviews of these vegetation types, their comparison and classification with an assessment of their relative importance and conservation priorities. Thus the work of Shimwell (1968a) represents a contribution to the Nature Conservancy's Reserve Review in which different habitat teams were set up to assess priorities for conservation within the major environment types. Which of the Derbyshire Dales contains the greatest diversity of vegetation types and how do the floristic complements of these communities compare with one another and with communities of calcareous soils in other regions? These are the type of questions that the vegetation analyst needs to ask and answer before conservation can be put on an objective, scientific basis, rather than an aesthetic, sentimental approach.

On the wider scale than the nature conservation groups in single countries, the International Biological Programme (I.B.P.)—a worldwide plan of research concerned with 'the biological basis of productivity and human welfare'—has a component section referred to as the *Conservation of Terrestrial Communities* Section (C.T.). The major objects of this group are:

(1) to provide a scientific basis for the conservation of areas and of species;

(2) to assure, for present and future needs, an adequate supply of natural habitats;

(3) to provide unique and rapidly disappearing opportunities for research on the biological processes which are responsible for the maintenance of stability in the ecosystems;

(4) in the fulfilment of the previous three objects, to fill in the gaps in biological and environmental knowledge by means of a co-ordinated and comprehensive approach according to a standardized approach.

For the purpose of standardization the I.B.P. Handbook No. 4 (Peterken, 1967) provides a guide and a check-sheet for the documentation of areas of special scientific interest. Within this check-sheet is a section on the plant communities catalogued according to the classification system of Fosberg (1967). It seems that it is in this particular direction that the energies of the vegetation analyst should be aimed. Failing this, some phytogeographical appraisal of the rarity or relict nature of communities should form a major contribution to the international conservation programme. The type of study undertaken by Shimwell (1968b) shows that a grassland type such as the *Seslerio-Helictotrichetum* of the magnesian limestone of County Durham is not only a rarity on the ground, but also unique with respect to other grassland types in north-west Europe.

We may conclude this section with the following two summary points:

(1) Education in vegetation analysis is best started with simple gradient analyses, transects and quadrats to illustrate coincidence between species and between species and their environment. At the undergraduate level there is scope for courses on different vegetation types and the application of quantitative techniques with field courses designed to integrate these methods with a broader vegetation survey. Postgraduate studies should be aimed at a particular ecological problem and as far as possible should avoid methodological preoccupations.

(2) The main applications of vegetation analysis are threefold. First, as the form of a study to monitor the effects of the progressive pollution of the environment; secondly, as a sequential study on the effects of man upon the environment in terms of recreation and agriculture within areas of great aesthetic landscape beauty such as National Parks; and thirdly, to provide a scientific phytosociology, phyto-geographical and ecological basis for nature conservation.

Margalef (1968) viewing Nature as a total biological channel of information depicts life in three idealized secondary channels relative to their development in time. The genetic channel on the one hand has continued to develop gradually through time while on the other hand, the cultural information channel has exploded with development of the higher vertebrates. The central ecological channel has enlarged negligibly through time, and with the advent of mechanical man it will begin to decrease unless something is done to prevent the loss of ecological information. It is at this stage that man needs to step in to prevent this diminution of information and by conservation, maintain the central level stable, whilst expanding the information laterally via genecology to the genetic channel and by human eco-sociology to the cultural channel. As Margalef states 'a proper subject of ecology is the study of the increase and preservation of organization at the eco-system level'. Vegetation is a major component of the ecosystem and the subjects reviewed in this book provide pointers to the available methods for the study and preservation of its organization.

BIBLIOGRAPHY

AGNEW, A. D. Q. (1961). The ecology of *Juncus effusus* L. in North Wales. *J. Ecol.*, **49**, 83–102.

ANDERSON, D. J. (1965). Classification and ordination in vegetation science: controversy over a non-existent problem? *J. Ecol.*, **53**, 521–26.

ANDERSON, D. J. (1967a). Studies on structure in plant communities. III. Data on pattern in colonizing species. *J. Ecol.*, **55**, 397–404.

ANDERSON, D. J. (1967b). Studies on structure in plant communities. IV. Cyclical succession in *Dryas* communities from north-west Iceland. *J. Ecol.*, **55**, 629–36.

ANDERSON, E. (1953). Introgressive hybridization. *Biol. Rev.*, **28**, 280–307.

ARCHIBALD, E. E. A. (1948). Plant populations. I. A new application of Neyman's contagious distribution. *Ann. Bot. Lond.*, N.S. **12**, 221–35.

ARCHIBALD, E. E. A. (1949). The specific character of plant communities. II. A quantitative approach. *J. Ecol.*, **37**, 274–88.

ASHBY, E., and PIDGEON, I. M. (1942). A new quantitative method of analysis of plant communities. *Aust. J. Sci.*, **5**, 19.

AUSTIN, M. P. (1968). An ordination study of a chalk grassland community. *J. Ecol.*, **56**, 739–58.

AUSTIN, M. P., and GRIEG-SMITH, P. (1968). The application of quantitative methods to vegetation survey. II. Some methodological problems of data from rain forest. *J. Ecol.*, **56**, 827–44.

AUSTIN, M. P. and ORLOCI, L. (1966). Geometric models in ecology. II. An evaluation of some ordination techniques. *J. Ecol.*, **54**, 217–27.

BACH, R., KUOCH, R., and MOOR, M. (1962). Die Nomenklatur der Pflanzengesellschaften. *Mitt. flor.-soz. Arbeitsgem.*, N.F. **9**, 301–8.

BAKKER, D. (1966). On the life forms of hapaxants in the Dutch flora. *Wentia*, **15**, 13–24.

BANNISTER, P. (1968). An evaluation of some procedures used in simple ordinations. *J. Ecol.*, **56**, 27–34.

BARCLAY-ESTRUP, P. (1970). The description and interpretation of cyclical processes in a heath community. II. Changes in biomass and shoot production during the *Calluna* cycle. *J. Ecol.*, **58**, 243–50.

BARCLAY-ESTRUP, P., and GIMMIMGHAM, C. H. (1969). The description and interpretation of cyclical processes in a heath community. I. Vegetational change in relation to the *Calluna* cycle. *J. Ecol.*, **57**, 737–58.

BARKMAN, J. J. (1958). *Phytosociology and ecology of cryptogamic epiphytes.* Assen.

BARNES, H., and STANBURY, F. A. (1952). A statistical study of plant distribution during the colonization and early development of vegetation on china clay residues, *J. Ecol.*, **39**, 171–81.

BARROW, M. D., COSTIN, A. B., and LAKE, P. (1968). Cyclical changes in an Australian fjaeldmark community. *J. Ecol.*, **56**, 89–96.

BEADLE, N. C. W. (1953). The edaphic factor in plant ecology. *Ecology*, **34**, 426–8.

BEALS, E. W. (1960). Forest bird communities in the Apostle Islands of Wisconsin. *Wilson Bull.*, **72**, 156–81.

BEALS, E. W. (1965). Species patterns in a Lebanese Poterietum. *Vegetatio*, **13**, 69–87.

BEARD, J. S. (1944). Climax vegetation in tropical America. *Ecology*, **25**, 127–58.

BEARD, J. S. (1955). The classification of tropical American vegetation-types. *Ecology*, **36**, 89–100.

BECKING, R. W. (1957). The Zürich–Montpellier School of phytosociology. *Bot. Rev.*, **23**, 411–88.

BEEFTINK, W. G. (1965). De zoutvegetatie van ZW-Nederland, beschouwd in Europees veband. *Meded. Landbouw. Wageningen*, **65** (1); 1–167.

BEWS, J. W. (1917). The plant ecology of the Drakensberg Range. *Ann. Natal Mus.*, **3**, 511–65.

BEWS, J. W. (1920). The plant ecology of the coast belt of Natal. *Ann. Natal Mus.*, **4**, 367–469.

BILLINGS, W. D., and MOONEY, H. A. (1959). An apparent frost hummock-sorted polygon cycle in the alpine tundra of Wyoming. *Ecology*, **40**, 16–20.

BLACKMAN, G. E. (1935). A study by statistical methods of the distribution of species in grassland associations. *Ann. Bot. Lond.*, **49**, 749–78.

BLYTT, A. (1876). *Essay on the Immigration of the Norwegian Flora during Alternating Rainy and Dry Periods*. Oslo.

BOATMAN, D. J., and ARMSTRONG, W. A. (1968). A bog type in north-west Sutherland. *J. Ecol.*, **56**, 129–41.

BÖCHER, T. W. (1933). Phytogeographical studies of the Greenland flora. *Meddel. om Grønland*, **104**, 1–56.

BÖCHER, T. W. (1940). Studies on the plant geography of the north-atlantic heath formation. I. The heaths of the Faeroes. *K. danske. Vidensk. Selsk., Biol. Skr.*, **15**, 1–64.

BÖCHER, T. W. (1943). Studies on the plant geography of the north-atlantic heath formation. II. Danish dwarf shrub communities in relation to those of northern Europe. *K. danske Vidensk. Selsk., Biol. Skr.*, **1** (3).

BÖCHER, T. W. (1954). Oceanic and continental vegetational complexes in south-west Greenland. *Meddel. om Grønland*, **148**, 1–336.

BOND, R. R. (1957). The ecological distribution of breeding birds in the upland forests of southern Wisconsin. *Ecol. Monogr.*, **27**, 351–84.

BOWDLER, M., *et al.* (1968). The QR and QL alogrithm for symmetric matrices. *Springer Handbook on Linear Algebra*, No. 2, Vol. II.

BRAUN, E. L. (1947). Development of the deciduous forests of eastern North America. *Ecol. Monogr.*, **17**, 211–19.

BRAUN, E. L. (1950). *Deciduous Forests of eastern North America*. Philadelphia; Blaikston.

BRAUN-BLANQUET, J. (1913). Die vegetationsverhältnisse der Schneestufe in den Rätisch-Lepontischen Alpen. *Schweiz. Naturf. Gesell., N.D.*, **48**, 1–347.

BRAUN-BLANQUET, J. (1915). Les Cèvennes meridionales. Étude phyto-géographique. *Biol. Univ. Arch. Sci. Phys. Nat. Ser. 4*, 39.

BRAUN-BLANQUET, J. (1921). Prinzipen einer Systematik der Pflanzengesellschaften auf floristischer Grundlage. *St. Gall. Naturwiss. Gesell. Jabrb.*, **57**, 305–51.

BRAUN-BLANQUET, J. (1928). *Pflanzensoziologie. Grundzüge der Vegetationskunde* Springer, Berlin.

BRAUN-BLANQUET, J. (1932). *Plant Sociology; the study of plant communities* (English translation). McGraw-Hill, New York.

BRAUN-BLANQUET, J. (1959). Grundfragen und Aufgaben der Pflanzensoziologie. *Vistas in Botany*, pp. 145–71.

L

BRAUN-BLANQUET, J. (1969). Die Pflanzengesellschaften der rätischen Alpen im Rahmen ihrer Gesamtverbreitung. *S.I.G.M.A. Comm. No. 185*, 99 pp.

BRAUN-BLANQUET, J., and MARIE, R. (1924). Études sur la végétation er la flore Marocaines. *Mém. Soc. Sci. Nat. de Maroc*, **8**, 1–244.

BRAUN-BLANQUET, J., and JENNY, H. (1926). Vegetations-Entwicklung und Bodenbildung in der alpinen Stufe der Zentralalpen. *Schweiz Naturf. Gesell.*, *Denskschr.*, **63**, 181–349.

BRAUN-BLANQUET, J., and MOOR, M. (1938). Verband des Bromion erecti. *Prodromus der Pflanzenges.*, No. 5, 64 pp.

BRAUN-BLANQUET, J., and PAVILLARD, J. (1922). *Vocabulaire de Sociologie Végétale*. Montpellier.

BRAUN-BLANQUET, J., and TÜXEN, R. (1943). Ubersicht der höheren Vegetationseinheiten Mitteleuropas. *S.I.G.M.A. Comm.* **84**, 1–11.

BRAY, J. R. (1961). A test for estimating the relative informativeness of vegetation gradients. *J. Ecol.*, **49**, 631–42.

BRAY, J. R., and CURTIS, J. T. (1957). An ordination of the upland forest communities of southern Wisconsin, *Ecol. Monogr.*, **27**, 325–49.

BROCKMANN-JEROSCH, H. (1907). *Die Pflanzengesellschaften der Schweizeralpen. I. Die Flora des Puschlav und ihre Pflanzengesellschaften.* Engelmann, Leipzig.

BROCKMANN-JEROSCH, H., and RÜBEL, E. (1912). *Die Einteilung der Pflanzengesellschaften nach ökologisch-physiognomischen Gesichtspunkten.* Leipzig.

BROOKS, F. T., and CHIPP, T. F. (ed.) (1931). Fifth International Botanical Congress: Report of Proceedings. Cambridge.

BROWN, R. T., and CURTIS, J. T. (1952). The upland conifer-hardwood forests of northern Wisconsin. *Ecol. Monogr.*, **22**, 217–34.

BRYAN, R. B. (1967). Climosequences of soil development in the Peak District of Derbyshire. *E. Midl. Geog.*, **4**, 251–61.

BURGES, A. (1951). The ecology of the Cairngorms. III. The *Empetrum-Vaccinium* zone. *J. Ecol.*, **39**, 271–84.

BURTT-DAVY, J. (1938). The classification of tropical woody vegetation-types. *Oxf. Univ. Imp. For. Inst.*, *Paper 13*, 1–85.

CAIN, S. A. (1932). Concerning certain phytosociological concepts. *Ecol. Monogr.*, **2**, 475–508.

CAIN, S. A. (1934). A comparison of quadrat sizes in a quantitative phytosociology study of Nash's Woods, Posey County, Indiana. *Amer. Midl. Nat.*, **15**, 529–66.

CAIN, S. A. (1935). Ecological studies of the vegetation of the Great Smokey Mountains. II. *Amer. Midl. Nat.*, **16**, 566–84.

CAIN, S. A. (1939). The climax and its complexities. *Amer. Midl. Nat.*, **21**, 146–81.

CAIN, S. A. (1950). Life forms and phytoclimates. *Bot. Rev.*, **16**, 1–32.

CAIN, S. A., and CASTRO, G. M. de O. (1959). *Manual of Vegetation Analysis.* Harper, New York.

CAIN, S. A., CASTRO, G. M. de O., PIRES, J. M., and DA SILVA, N. T. (1956). Application of some phytosociological techniques to Brazilian rain forest. *Am. J. Bot.*, **43**, 911–41.

CAJANDER, A. K. (1903). Beiträge zur Kenntniss der Vegetation der Alluvionen des nordlichen Eurasiens. *Acta. Soc. Sci. Fenn.*, **37**, 1–182.

CAJANDER, A. K. (1909). Über Waldtypen. *Acta Forestalia Fenn.*, **1**, 1–175.

CANNON, W. A. (1911). The root habits of desert plants. *Carnegie Inst. Wash. Publ.*, *No. 131.*

CHIPP, T. F. (1927). The Gold Coast forest. A study in synecology. *Oxf. For. Mem.*, **7**, 1–94.

CHRISTIAN, C. S., and PERRY, R. A. (1953). The systematic description of plant communities by the use of symbols. *J. Ecol.*, **41**, 100–5.

CLAPHAM, A. R. (1932). The form of the observational unit in quantitative ecology, *J. Ecol.*, **20**, 192–7.

CLAPHAM, A. R., TUTIN, T. G., and WARBURG, E. F. (1962). *Flora of the British Isles*. 2nd Ed. Cambridge.

CLEMENTS, F. E. (1904). Development and structure of vegetation. *Rep. Bot. Surv. Nebr.*, **7**.

CLEMENTS, F. E. (1905). *Research Methods in Ecology*. Univ. Publ. Lincoln.

CLEMENTS, F. E. (1916). Plant succession: an analysis of the development of vegetation. *Carnegie Inst. Wash. Publ.*, **242**.

CLEMENTS, F. E. (1928). *Plant Succession and Indicators*. New York.

CLYMO, R. S. (1970). The growth of *Sphagnum:* methods of measurement. *J. Ecol.*, **58**, 13–50.

COOPER, W. S. (1916). Plant successions in the Mount Robson region, British Columbia. *Plant World*, **19**, 211–38.

COOPER, W. S. (1926). The fundamentals of vegetational change. *Ecology*, **7**, 391–413.

COPPINS, B. J., and SHIMWELL, D. W. (1971). Variations in cryptogam complement and biomass in dry *Calluna* heath of different ages. *Oikos*, **22** (2).

COTTAM, G., and CURTIS, J. T. (1949). A method for making rapid surveys of woodlands by means of pairs of randomly selected trees. *Ecology*, **30**, 101–4.

COTTAM, G., and CURTIS, J. T. (1955). Correction for various exclusion angles in the random pairs method. *Ecology*, **36**, 767.

COTTAM, G., and CURTIS, J. T. (1956). The use of distance measures in phytosociological sampling. *Ecology*, **37**, 451–60.

COTTAM, G., CURTIS, J. T., and HALE, B. W. (1953). Some sampling characteristics of a population of randomly dispersed individuals. *Ecology*, **34**, 741–57.

COWLES, H. C. (1899). The ecological relations of the vegetation on the sand dunes of Lake Michigan. I. *Bot. Gaz.*, **27**, 95–117 *et seq.*

CROCKER, R. L. (1952). Soil genesis and the pedogenic factors. *Quart. Rev. Biol.*, **27**, 139.

CURTIS, J. T. (1959). *The Vegetation of Wisconsin: an ordination of plant communities*. Madison, Wisconsin.

CURTIS, J. T., and McINTOSH, R. P. (1951). An upland forest continuum in the prairie-forest border region of Wisconsin. *Ecology*, **32**, 476–96.

CZEKANOWSKI, J. (1913). *Zarys Metod Statystycznyck*. Warsaw.

DAHL, E. (1956). Rondane: Mountain vegetation in south Norway and its relation to the environment. *Skr. norske Vidensk-Akad. Mat. Naturv. Kl. No. 3*, 374 pp.

DAHL, E., and HADAČ, E. (1941). Strandgesellschaften der Insel Ostøy im Oslofjord. *Nyt. Mag. Naturv.*, **82**, 251–312.

DAHL, E., and HADAČ, E. (1949). Homogeneity of plant communities. *Studia bot. Čechosl.*, **10**, 159–76.

DANSEREAU, P. (1951). Description and recording of vegetation upon a structural basis. *Ecology*, **32**, 172–229.

DANSEREAU, P. (1957). *Biogeography: an ecological perspective*. Ronald Press, New York.

DANSEREAU, P. (1961). Essai de représentation cartographique des elements structuraux de la végétation. In: GAUSSEN, H. (ed.) *Methodes de la Cartographie de la Végétation*, 233–55. Paris.

DANSEREAU, P., and ARROS, J. (1959). Essais d'application de la dimension structurale en phytosociologie. I. Quelques exemples européens. *Vegetatio*, **9**, 48–99.

DANSEREAU, P., and LEMS, K. (1957). The grading of dispersal types in plant communities and their ecological significance. *Cont. Inst. Bot. Univ. Montréal*, *No. 71*.

DAVIS, T. A. W., and RICHARDS, P. W. (1933). The vegetation of Moraballi Creek, British Guiana; an ecological study of a limited area of tropical rain forest. *J. Ecol.*, **21**, 350–85.

DEN HARTOG, C., and SEGAL, S. (1964). A new classification of the water plant communities. *Acta Bot. Neerl.*, **13**, 367–93.

DE VRIES, D. M. (1953). Objective combinations of species. *Acta Bot. Neerl.*, **4**, 497–9.

DOMIN, K. (1905). Das böhmische Mittelgebirge. Ein phytogeographische Studie. *Bot. Jahrb.*, **37**, 1–59.

DOMIN, K. (1923). Is the evolution of the earth's vegetation tending towards a small number of climatic formations. *Acta Bot. Bohemica*, **2**, 54–60.

DOMIN, K. (1928). The relations of the Tatra Mountain vegetation to the edaphic factors of the habitat; a synecological study. *Acta Bot. Bohemica*, **6/7**, 133–64.

DRUDE, O. (1896). *Deutschlands Pflanzengeographie*. Englehorn, Stuttgart.

DRUDE, O. (1905). Die Methoden der speziellen pflanzengeographischen Kartographie. *Wissensch. Ergebn. 2nd Int. Bot. Cong.*

DRUDE, O. (1928). *Pflanzengeographische Ökologie.*—Abderhalden's *Handbuch der biologischen Arbeitsmetoden* 11 (4). Berlin and Vienna.

DU RIETZ, G. E. (1921). *Zur methodologischen Grundlage der modernen Pflanzensoziologie*. Holzhausen, Wien.

DU RIETZ, G. E. (1930). Classification and nomenclature of vegetation. *Svensk Bot. Tidskr.*, **24**, 489–503.

DU RIETZ, G. E. (1931). Life forms of terrestrial flowering plants. *Acta Phytogeogr. Suecica*, **3**, 1–95.

DU RIETZ, G. E. (1936). Classification and nomenclature of vegetation units, 1930–1935. *Svensk. Bot. Tidskr.*, **30**, 580–9.

DU RIETZ, G. E. (1942). Rishedsförband i Torneträskskomradets lagfjällbalte. *Svensk Bot. Tidskr.*, **36**, 124–46.

DU RIETZ, G. E. (1954). Die Mineralbodenwasserzeigergrenze als Grundlage einer natürlichen Zweigliederung der nord- und mitteleuropäischen Moore. *Vegetatio*, **5/6**, 571–85.

DU RIETZ, G. E. (1957). Linne som myrforskare. *Uppsala Univ. Arsskr.*, No. 5.

DU RIETZ, G. E., FRIES, T. C. E., and TENGWALL, T. A. (1918). Vorschlag zur Nomenklatur der soziologischen Pflanzengeographie. *Svensk Bot. Tidskr.*, **12**, 145–70.

ELLENBERG, H. (1939). Über Zusammensetzung, Standort und Stoffproduktion bodenfeuchter Eichen- und Buchen-Mischwaldgesellschaften Nordwestduetschland, *Mitt. flor.-soz. Arbeitsgem.*, **5**, 135 pp.

ELLENBERG, H. (1950). *Landwirtschaftliche Pflanzensoziologie. I. Unkrautgemeinschaften als Zeiger für Klima und Boden*. Stuttgart.

ELLENBERG, H. (1952). *Landwirtschaftliche Pflanzensoziologie. II. Wiesen und Weiden und ihre standörtliche Bewertung*. Stuttgart.

ELLENBERG, H. (1954). Zur Entwicklung der Vegetationssystematik in Mitteleuropa. *Angew. Pflanzensoz.*, **1**, 134–43.

ELLENBERG, H. (1956). Aufgaben und Methoden der Vegetationskunde. *In* WALTER, H. (ed.) *Einführung in die Phytologie*. Vol. IV, pt. 1.

ELLENBERG, H. (1963). *Die Vegetation Mitteleuropas mit den Alpen in kausaler dynamischer und historischer Sicht*. Stuttgart.

ELLENBERG, H., and MÜLLER-DOMBOIS, D. (1966). A tentative physiognomic-ecological classification of the formations of the earth. *Ber. geobot. Inst. ETH, Stiftg. Rübel*, **37**, 21–55.

EVANS, F. C. (1952). The influence of size of quadrat on the distributional patterns of plant populations. *Contr. Lab. Vert. Biol. Univ. Mich.*, **54**, 1–15.

FAEGRI, K. (1933). Über die Längenvariationen einiger Gletscher des Jostedalsbre und die dadurch bedingten Pflanzensukzessionen. *Bergens. Mus. Aarbok Naturv. Rekke*, 1933 (7), 1–255.

FISHER, R. A., and YATES, F. (1943). *Statistical tables for biological, agricultural and medical research.* 2nd Ed. London.

FLAHAULT, C. (1893). Les zones botaniques dans le Bas-Languedoc et les pays voisins. *Bull. Soc. Bot. France*, **40**, 36–62.

FLAHAULT, C. (1901). A project for phytogeographic nomenclature. *Bull. Torrey. Bot. Club*, **28**, 391–409.

FLAHAULT, C., and SCHRÖTER, C. (1910), Rapport sur la nomenclature phytogéographique. *Proc. 3rd Int. Bot. Cong.*, *Brussels 1910*, **1**, 131–64.

FOSBERG, F. R. (1961). A classification of vegetation for general purposes. *Trop. Ecol.*, **2**, 1–28.

FOSBERG, F. R. (1967). A classification of vegetation for general purposes. In: PETERKEN, G. F. (ed.) Guide to the checklist for I.B.P. areas. *I.B.P. Handbook No. 4.* Cambridge.

FRIES, T. C. E. (1913). Botanische Untersuchungen im nördlichsten Schweden: ein Beitrag zur Kenntnis der alpinen und subalpinen Vegetation in Torne Lappmark. *Flora & Fauna*, **2**, 1–361.

FÜLLEKRUG, E. (1967). Phänologische Diagramme aus einem Melico-Fagetum. *Mitt. flor-soz. Arbeitsgem.*, N.F. **11/12**, 142–58.

GAMS, H. (1918). Prinzipienfragen der Vegetationsforschung. Ein Beitrag zur Begriffsklärung und Methodik der Biocoenologie. *Naturf. Gesell. Zurich, Vierteljahrschr.*, **63**, 293–493.

GILBERT, M. L., and CURTIS, J. T. (1953). Relation of the understorey of the upland forest in the prairie-forest border region of Wisconsin. *Trans. Wis. Acad. Sci. Arts Lett.*, **42**, 183–95.

GILLMAN, C. (1936). East African vegetation types. *J. Ecol.*, **24**, 502–5.

GITTINS, R. (1965a). Multivariate approaches to a limestone grassland community. I. A stand ordination. *J. Ecol.*, **53**, 385–401.

GITTINS, R. (1965b). Multivariate approaches to a limestone grassland community. II. A direct species ordination. *J. Ecol.*, **53**, 403–9.

GITTINS, R. (1965c). Multivariate approaches to a limestone grassland community. III. A comparative study of ordination and association analysis. *J. Ecol.*, **53**, 411–25.

GJAEREVOLL, O. (1956). *The Plant Communities of the Scandinavian Alpine Snow Beds.* Trondheim.

GLEASON, H. A. (1917). The structure and development of the plant association. *Bull. Torrey Bot. Club*, **44**, 463–81.

GLEASON, H. A. (1920). Some applications of the quadrat method. *Bull. Torrey Bot. Club*, **47**, 21–33.

GLEASON, H. A. (1926). The individualistic concept of the plant association. *Bull. Torrey Bot. Club*, **53**, 7–26.

GLEASON, H. A. (1929). The significance of Raunkiaer's law of frequence. *Ecology*, **10**, 406–8.

GODWIN, H. (1929). The subclimax and deflected succession. *J. Ecol.*, **17**, 144–7.

GODWIN, H., and CONWAY, V. M. (1939). The ecology of a raised bog near Tregaron, Cardiganshire, *J. Ecol.*, **27**, 313–63.

GOOD, R. D'O (1947). *The Geography of Flowering Plants.* Longmans, London.

GOODALL, D. W. (1952). Some considerations in the use of point quadrats for the analysis of vegetation. *Aust. J. Sci. Res. Ser. B*, **5**, 1–41.

GOODALL, D. W. (1953a). Objective methods for the classification of vegetation. I. The use of positive interspecific correlation. *Aust. J. Bot.*, **1**, 39–63.

GOODALL, D. W. (1954a). Vegetational classification and vegetational continuua. *Angew. Pflanzensoz.*, **1**, 168–82.

GOODALL, D. W. (1954b). Minimal area: a new approach. *Int. Bot. Cong.*, **8**, Sect. 7 and 8, 19–21.

GOODE, D. A. (1970). *Ecological studies on the Silver Flowe Nature Reserve.* Ph.D Thesis, Univ. of Hull.

GOODMAN, R. (1957). *Teach Yourself Statistics.* London.

GRADMANN, R. (1909). Über Begriffsbildung in der Lehre von den Pflanzenformationen. *Bot. Jahrb.*, **43**, 91–103.

GRAHAM, S. A. (1941). Climax forests of the upper peninsula of Michigan. *Ecology*, **22**, 355–62.

GRANT, S. A., and HUNTER, R. F. (1962). Ecotypic differentiation of *Calluna vulgaris* (L.) Hull in relation to altitude. *New Phytol.*, **61**, 44–55.

GREIG-SMITH, P. (1952). The use of random and contiguous quadrats in the study of the structure of plant communities. *Ann Bot. Lond.*, N.S. **16**, 293–316.

GREIG-SMITH, P. (1957) (1964). *Quantitative Plant Ecology.* Butterworth, London. 1st and 2nd Eds.

GREIG-SMITH, P. (1961a). The use of pattern analysis in ecological investigations. *Recent Advances in Botany*, **2**, 1354–8.

GREIG-SMITH, P. (1961b). Data on pattern within plant communities. I. The analysis of pattern. *J. Ecol.*, **49**, 695–702.

GREIG-SMITH, P., and CHADWICK, M. J. (1965). Data on pattern within plant communities. III. *Acacia-Capparis* semi-desert scrub in the Sudan. *J. Ecol.*, **53**, 465–74.

GRISEBACH, A. (1838). Über den Einfluss des Climas auf die Begränzung der natürlichen Floren. *Linnaea*, **12**, 159–200.

GRISEBACH, A. (1872). *Die Vegetation der Erde.* Engelmann, Leipzig.

GROENEWOUD, H. van. (1965). Ordination and classification of Swiss and Canadian forests by various biometric and other methods. *Ber. geobot. Inst. E.T.H., Stiftg. Rübel*, **35**, 28–102.

GRUBB, P. J., *et al.* (1963). A comparison of montane and lowland rain forest in Ecuador. I. The forest structure, physiognomy and floristics. *J. Ecol.*, **51**, 576–601.

HALE, M. E. (1955). Phytosociology of corticolous cryptogams in the upland forests of southern Wisconsin. *Ecology*, **36**, 45–63.

HANSEN, H. M. (1932). Nørholm Hede, en formationsstatistisk Vegetationsmonografi. *K. danske Vidensk. Selsk. Skr., Naturv, Ser. 9*, **3**, 99–106.

HANSON, H. C. (1934). A comparison of methods of botanical analysis of the native prairie in western North Dakota. *J. Agric. Res.*, **49**, 815–42.

HARPER, J. L. (1964). The individual in the population. *J. Ecol.*, **52**, 149–58.

HAWKSWORTH, D. L., and ROSE, F. (1970). Qualitative scale for estimating sulphur dioxide air pollution in England and Wales using epiphytic lichens. *Nature, Lond.*, **227**, 145–8.

HESLOP-HARRISON, J., and RICHARDSON, J. A. (1953). The Magnesian Limestone of Durham and its vegetation. *Trans. North. Nat. Union*, **2**, 1–28.

HEYLIGERS, P. C. (1965). Structure formulae in vegetation analysis on aerial photographs and in the field. *Symp. Ecol. Res. in Humid Tropics Veg., Sarawak, 1963*, 249–54.

HOPE-SIMPSON, J. F. (1940). On the errors in the ordinary use of subjective frequency estimations in grassland. *J. Ecol.*, **28**, 193–209.

HOPKINS, B. (1955). The species-area relations of plant communities. *J. Ecol.*, **43**, 409–26.

HOPKINS, B. (1957). Pattern in the plant community. *J. Ecol.*, **45**, 451–63.

HUGHES, R. E., and LINDLEY, D. V. (1955). Application of biometric methods to problems of classification in ecology. *Nature, Lond.*, **175**, 806–7.

HULT, R. (1885). Blekinges vegetation. Ett bidrag till vaxtformationernas utvecklingshistorie. *Medd. Soc. Fenn.*, **12**, 161.

HUMBOLDT, A. von (1805). *Essai sur la Géographie des Plantes: accompagne d'un tableau physique des régions equinoxiales.* Levrault, Paris.

HUMBOLDT, A. von (1808). Ansichten der Natur mit wissenschaftlichen Erläuterungen. Tübingen.

HUXLEY, J. S. (ed.) (1940). *The New Systematics.* Clarendon, Oxford.

IVERSEN, J. (1936). *Biologische Pflanzentypen als Hilfsmittel in der Vegetationsforschung.* Copenhagen.

IVIMEY-COOK, R. B., and PROCTOR, M. C. F. (1966). The application of association analysis to phytosociology. *J. Ecol.*, **54**, 179–92.

JACCARD, P. (1902). Lois de distribution florale dans la zone alpine. *Bull. Soc. vaud. Sci. nat.*, **38**, 69–130.

JACKSON, G., and SHELDON, J. (1949). The vegetation of Magnesian limestone cliffs at Markland Grips near Sheffield. *J. Ecol.*, **37**, 38–50.

JASNOWSKI, M. (1966). The peat resources of Poland, *Contr. Proj. TELMA I.B.P.* Mscr.

JENNY, H. (1941). *Factors of Soil Formation.* New York and London.

KENOYER, L. A. (1927). A study of Raunkiaer's law of frequence. *Ecology*, **8**, 341–9.

KERSHAW, K. A. (1957). The use of cover and frequency in the detection of pattern in plant communities. *Ecology*, **38**, 291–9.

KERSHAW, K. A. (1958). An investigation of the structure of a grassland community. I. The pattern of *Agrostis tenuis. J. Ecol.*, **46**, 571–92.

KERSHAW, K. A. (1960). The detection of pattern and association. *J. Ecol.*, **48**, 233–42.

KERSHAW, K. A. (1963). Pattern in vegetation and its causality. *Ecology*, **44**, 377–88.

KERSHAW, K. A. (1964). *Quantitative and Dynamic Ecology.* Arnold, London.

KÖPPEN, W. (1923). *Die Klimate der Erde.* De Gruyte.

KORSTIAN, C. F., and STICKEL, P. W. (1927). Natural replacement of blight-killed chestnut in the hardwood forests of the north-east. *J. Agric. Res.*, **34**, 631–48.

KÜCHLER, A. W. (1949). A physiognomic classification of vegetation. *Ann. Ass. Amer. Geog.*, **39**, 201–10.

KÜCHLER, A. W. (1966). Analyzing the physiognomy and structure of vegetation. *Ann. Ass. Amer. Geogr.*, **56**, 112–27.

KÜCHLER, A. W. (1967). *Vegetation Mapping.* Ronald Press, New York.

KYLIN, H. (1926). Über Begriffsbildung und Statistik in der Pflanzensoziologie. *Bot. Not.*, (1926), 81–180.

LAMBERT, J. M., and DALE, M. B. (1964). The use of statistics in phytosociology. *Adv. Ecol. Res.*, **2**, 59–99.

LAMBERT, J. M., and WILLIAMS, W. T. (1962). Multivariate methods in plant ecology. IV. Nodal analysis. *J. Ecol.*, **50**, 775–802.

LAWLEY, D. N., and MAXWELL, A. E. (1963). *Factor Analysis as a Statistical Method.* London.

LEEUWEN, C. G. van (1965). Het verband tussen natuurlijke en anthropogene landschapsvormen bezien vanuit de betrekkingen in grensmillieu's. *Gorteria*, **2 (8)**, 93–105.

LEEUWEN, C. G. van (1966). A relation theoretical approach to pattern and process in vegetation. *Wentia*, **15**, 25–46.

LEEUWEN, W. M. D. van (1936). Krakatau, 1883–1933. A. Botany. *Ann. Jard. bot. Buitenz.*, **46/47**, 1–506.

LENOBLE, F. (1927). À propos des associations végétales. *Bull. Soc. bot. Fr.*, **73**, 873–93.

LEVY, E. G., and MADDEN, E. A. (1933). The point method of pasture analysis. *N.Z.J. Agric.*, **46**, 267–79.

LIPMAA, T. (1931). Pflanzensoziologische Betractungen. *Acta Inst. Hort. Bot. Univ. Tartuensis*, **2**, 1–32.

LIPMAA, T. (1935). La méthode des associations unistrates et le système écologiques des associations. *Acta Inst. Hort. Bot. Univ. Tartuensis*, **4**, 1–7.

LOHMEYER, W., *et al.* (1962). Contribution a l'unification du système phytosociologique pour l'Europe moyénne et nordoccidentale. *Melhoramento*, **15**, 137–51.

LOUCKS, O. L. (1962). Ordinating forest communities by means of environmental scalars and phytosociological indices. *Ecol. Monogr.*, **32**, 137–66.

LOVELESS, A. R. (1961). A nutritional interpretation of sclerophylly based on differences in the chemical composition of sclerophyllous and mesophytic leaves. *Ann. Bot. Lond.*, N.S. **25**, 168–84.

LOVELESS, A. R., and ASPREY, G. F. (1957). The dry evergreen formations of Jamaica. I. The limestone hills of the south coast. *J. Ecol.*, **45**, 799–822.

LOVIS, J. D. (1964). The taxonomy of *Asplenium trichomanes* in Europe. *Brit. Fern. Gaz.*, **9**, 147–60.

LÜDI, W. (1920). Die Suksession der Pflanzenvereine. Allgemeine Betrachtungen über die dynamisch-genetischen Verhältnisse der Vegetation in einem Gebiet des Berner Oberlanders. *Naturf. Gesell. Bern Mitt.*, 1919, 9–87.

LÜDI, W. (1928). Der Assoziationsbegriff in der Pflanzensoziologie. *Biblioth. Bot.*, **96**, 1–93.

LÜDI, W. (1932). Die Methoden der Suksessionsforschung in der Pflanzensoziologie. In: *Abderhalden, Handb. Biol. Arbeitsmeth.*, Sect. 11, 5, 527–728.

LUTZ, H. J. (1930). The vegetation of Heart's Content, a virgin forest in northwestern Pennsylvania. *Ecology*, **11**, 1–29.

MAAREL, E. van der (1966). Dutch studies on coastal sand dune vegetation, especially in the Delta region. *Wentia*, **15**, 47–82.

MAAREL, E. van der (1969). On the use of ordination models in phytosociology. *Vegetatio*, **19**, 21–46.

MAJOR, J. (1951). A functional, factorial approach to plant ecology. *Ecology*, **32**, 392.

MALMER, N. (1957). *Myrvegetationsundersökningar; SV Götaland*. Ph.D. Thesis, Univ. Lund.

MALMER, N. (1962). Studies on mire vegetation in the Archaean area of southwestern Götaland (South Sweden). *Opera Bot.*, **7**, 1–322.

MALMER, N. (1964) (1968). Über die Gliederung der Oxycocco-Sphagnetea und der Scheuchzerio-Caricetea fuscae in Südschweden. In: Tüxen, R. (ed.) *Pflanzensoziologische Systematik*. Ber. Int. Symp. Stolzenau/Weser, 1964.

MARGALEF, R. (1957). La téoria de la infomacion en Ecologia *Mem. Resl. Acad. Ciencias. Artes Barcelona*, **32**.

MARGALEF, R. (1968). *Perspectives in Ecological Theory*. Chicago.

McINTOSH, R. P. (1967). The continuum concept of vegetation *Bot. Rev.*, **33**, 130–87.

McKEE, R. F. (1965). *An investigation in the East Durham plateau into problems of soil survey in relation to agricultural productivity*. Ph.D. Thesis, Univ. of Durham.

McVEAN, D. N., and RATCLIFFE, D. A. (1962). *Plant Communities of the Scottish Highlands*. H.M.S.O., London.

MEIJER DREES, E. (1951). Capita selecta from modern plant sociology and a design for rules of phytosociological nomenclature. *Bosbow. Bogor Ind. Stat. Rapp.*, **52**, 1–58.

MEIJER DREES, E. (1953). A tentative design for rules of phytosociological nomenclature. *Vegetatio*, **4**, 205–14.

MOLINIER, R., and MÜLLER, P. (1939). La dissemination des espèces végétales. *S.I.G.M.A. Comm.*, **64**.

MOORE, J. J. (1962). The Braun-Blanquet system: a reassessment. *J. Ecol.*, **50**, 761–9.

MOORE, J. J., *et al.* (1970). A comparison and evaluation of some phytosociological techniques. *Vegetatio*, **20**, 1–20.

MORAVEC, J. (1964) (1968). Zu den Problemen der pflanzensoziologischen Nomenklatur. In: Tüxen R. (ed.) *Pflanzensoziologische Systematik.* Ber. Int. Symp. Stolzenau/Weser, 1964.

MORONEY, M. J. (1951). *Facts from Figures.* Penguin.

MOSS, C. E. (1907). Succession of plant formations in Britain. *Brit. Assoc. Adv. Sci. Rept.*, 1906, 724–43.

MOSS, C. E. (1910). The fundamental units of vegetation. *New Phytol.*, **9**, 18–53.

MOSS, C. E. (1913). *Vegetation of the Peak District.* Cambridge Univ. Press.

NICHOLS, G. E. (1917). The interpretation and application of certain terms and concepts in the ecological classification of plant communities. *Plant World*, 305–19, 341–53.

NICHOLS, G. E. (1923). A working basis for the ecological classification of plant communities. *Ecology*, **4**, 11–23, 154–79.

NORDHAGEN, R. (1920). Om nomenklatur og begrepsdannelse i plantesociologien. *Nyt. Mag. Naturv.*, **57**, 17–128.

NORDHAGEN, R. (1936). Versuch einer neuen Einteilung der subalpinen-alpinen Vegetation Norwegens. *Bergens Mus. Aarbok, Naturv. Rekke*, 1936, 7.

NORDHAGEN, R. (1943). *Sikilsdalen og Norges Fjellbeiter.* Bergen.

NORRLIN, J. P. (1870). Bidrag till sydöstra Tavastlands flora. *Sällsk Fauna och Flora Fennica Not.*, **11**, 73–196.

ODUM, E. P. (1950). Bird populations of the Highlands (North Carolina) Plateau in relation to plant succession and avian invasion. *Ecology*, **31**, 587–605.

OLSON, J. S. (1958). Rates of succession and soil changes on Southern Lake, Michigan sand dunes. *Bot. Gaz.*, **119**, 125–70.

OOSTING, H. J. (1948). *The Study of Plant Communities: An introduction to plant ecology.* Freeman, San Francisco.

ORLOCI, L. (1966). Geometric models in ecology. I. The theory and application of some ordination methods. *J. Ecol.*, **54**, 193–215.

OSTENFELD, C. H. (1905). Skildringer af vegetationen i Island. *Bot. Tidsskr.*, **27**, 111–22.

OSTENFELD, C. H. (1908). The land vegetation of the Faeroes. In: WARMING, E. (ed.) (1908). *Botany of the Faeroes*, **3**, 867–1026.

O'SULLIVAN, A. M. (1965). *A phytosociological survey of Irish lowland pastures.* Ph.D. Thesis, University College, Dublin.

O'SULLIVAN, A. M. (1968). The lowland grasslands of Co. Limerick. *An Foras Taluntais, Irish, Veg. Stud.*, **2**, Dublin.

OSVALD, H. (1923). Die vegetation des Hochmoores Komosse. *Svenska Växtsociol. Sällsk, Handl.*, **1**, 1–436.

PAVILLARD, J. (1920). *Espèces et associations.* Montpellier.

PEARSALL, W. H. (1956). Two blanket bogs in Sutherland. *J. Ecol.*, **44**, 493–516.

PERRING, F. H. (1958). A theoretical approach to the study of chalk grassland. *J. Ecol.*, **46**, 665–79.

PERRING, F. H. (1959). Topographical gradients of chalk grassland. *J. Ecol.*, **47**, 447–82.

PERRING, F. H. (1960). Climatic gradients of chalk grassland. *J. Ecol.*, **48**, 415–42.

PERRING, F. H., and WALTERS, S. M. (1962). *Atlas of the British Flora.* Nelson & Sons, London.

PETERKEN, G. F. (1967). Guide to the checklist for I.B.P. areas. *I.B.P. Handbook No. 4.* Cambridge.

PHILLIPS, J. F. V. (1930). Some important vegetation communities in the central province of Tanganyika Territory. *J. Ecol.*, **18**, 193–234.

PIGOTT, C. D. (1956). The vegetation of Upper Teesdale in the North Pennines. *J. Ecol.*, **44**, 545–86.

PIGOTT, C. D. (1968). Biological flora of the British Isles: *Cirsium acaulon* (L.) Scop. *J. Ecol.*, **56**, 597–612.

POORE, M. E. D. (1955a, b, c). The use of phytosociological methods in ecological investigations. Parts I, II, III. *J. Ecol.*, **43**, 226–44, 245–69, 606–51.

POORE, M. E. D. (1956). The use of phytosociological methods in ecological investigations. Part IV. *J. Ecol.*, **44**, 28–50.

POORE, M. E. D., and McVEAN, D. N. (1957). A new approach to Scottish mountain vegetation. *J. Ecol.*, **45**, 401–39.

POUND, R., and CLEMENTS, F. E. (1898). The vegetation regions of the prairie province. *Bot. Gaz.*, **25**, 381–94.

PREIS, K. (1939). Die *Festuca vallesiaca-Erysimum crepidifolium* Assoziation auf Basalt, Glimmerschiefer und Granitgneis. *Beih bot. Cbl.*, *59B*, 478–530.

PRESTON, F. W. (1948). The commonness and rarity of species. *Ecology*, **29**, 254–83.

PUTWAIN, P. D., and HARPER, J. L. (1970). Studies on the dynamics of plant populations. III. The influence of associated species on populations of *Rumex acetosa* L. and *R. acetosella* L. in grassland. *J. Ecol.*, **58**, 251–64.

QUANTIN, A. (1935). L'évolution de la végétation a l'étage de la chênaie dans le Jura meridional. *S.I.G.M.A.*, **37**, 382 pp.

RAMENSKY, L. G. (1924). Die Grundgesetzmässigkeiten im Aufbau der Vegetationsdecke (German summary to Russian paper). *Bot. Centralbl.*, N.F. **7**, 453–5, 1926.

RAMENSKY, L. G. (1930). Zur Methodik der vergleichenden Bearbeitung und Ordnung von Pflanzenlisten und anderen Objekten, die durch mehrere, verschiedenartig wirkende Faktoren bestimmt werden. *Beitr. Biol. Pfl.*, **18**, 269–304.

RATCLIFFE, D. A., and WALKER, D. (1958). The Silver Flowe, Galloway, Scotland. *J. Ecol.*, **46**, 407–45.

RAUNKIAER, C. (1905). Types biologiques pour la géographie botanique. *K. Danske Vidensk. Selsk.*, *1905*, 347–438.

RAUNKIAER, C. (1907). *Planterigets Livsformer og deres Betydning for Geografien.* Copenhagen.

RAUNKIAER, C. (1908). Livsformernes Statistik som Grundlag for biologisk Plantegeografi. *Bot. Tidsskr.*, **29**.

RAUNKIAER, C. (1910). Formationsundersøgelse og Formation sstatistik. *Bot. Tidsskr.*, **30**, 20–132.

RAUNKIAER, C. (1918). Récherches statistiques sur les formations végétales. *K. Danske Vidensk Selsk. Biol. Meddel.*, **1**, 1–47.

RAUNKIAER, C. (1928). Dominansareal, Artstaethed og Formationsdominanter. *Kgl. Danske. Vidensk Selsk. Biol. Med.*, **7**, 1.

RAUNKIAER, C. (1934). *The Life Forms of Plants and Statistical Plant Geography.* Oxford, 632 pp.

RESVOLL-HOLMSEN, H. (1932). Om planteveksten i grensetrakter mellem Hallingdal og Valdres. *Skr. Vidensk Akad. Mat-naturv Kl.*, No. 9.

RICHARDS, P. W. (1936). Ecological observations on the rain forest of Mount Dulit, Sarawak. *J. Ecol.*, **24**, 1–37, 340–60.

RICHARDS, P. W. (1939). Ecological studies on the rain forest of Southern Nigeria. I. The structure and floristic composition of the primary forest. *J. Ecol.*, **27**, 1–61.

RICHARDS, P. W. (1952). *The Tropical Rain Forest.* Cambridge.

RICHARDS, P. W., TANSLEY, A. G., and WATT, A. S. (1940). The recording of structure, life form and flora of tropical forest communities as a basis for their classification. *J. Ecol.*, **28**, 224–39.

ROSS ASHBY, W. (1956). *An Introduction to Cybernetics.*

RÜBEL, E. (1912a). Pflanzegeographische Monographie des Bernina-Gebietes. *Bot. Jahrb.*, **47**, 1–616.

RÜBEL, E. (1912b). The International Phytogeographical Excursion in the British Isles. V. The Killarney woods. *New Phytol.*, **11**, 54–7.

RÜBEL, E. (1913). Geographie der Pflanzen: Oekologische Pflanzengeographie. *Handwörterbuch der Naturwiss.*, **4**, 858–907.

RÜBEL, E. (1925). Betrachtung über einige pflanzensoziologische Auffassungsdifferenzen. Verständigungsbeitrag Schweden-Schweiz. *Veröff. Geobot. Inst. Rübel*, **2**, 1–12.

RÜBEL, E. (1927). Einige skandinavische Vegetationsprobleme. *Veröff. Geobot. Inst. Rübel.*, **4**, 19–41.

RÜBEL, E. (1930). *Pflanzengesellschaften der Erde.* Bern–Berlin.

RÜBEL, E. (1931). A standard description of a plant community. *Proc. Fifth Int. Bot. Cong.*, *1930*, Sect. E, 176–7.

SAGAR, G. R., and HARPER, J. L. (1961). Controlled interference with natural populations of *Plantago lanceolata*, *P. major* and *P. media. Weed Res.*, **1**, 163–76.

SCHENNIKOW, A. P. (1932). Phänologische Spektra der Pflanzengesellschaften. *Handb. der biol. Arbeitsmethoden*, 11 (6). Berlin.

SCHIMPER, A. F. W. (1898). *Pflanzengeographie auf physiologischer Grundlage.* Jena.

SCHIMPER, A. F. W., and von FABER, F. C. (1935). *Pflanzengeografie auf physiologischer Grundlage.* Fischer, Jena.

SCHOUW, J. F. (1823). *Grundzüge einer allgemeinen Pflanzengeografie.* Reimer, Berlin.

SCHRÖTER, C. (1894). Notes sur quelques associations de plantes rencontrées pendant les excursions dans la Valais. *Bull. Soc. Bot. France*, **41**, 222–325.

SCHRÖTER, C., and KIRCHNER, O. (1902). Die Vegetation des Bodensees. *Schr. des Ver d. Gesch. des Bodensees*, **9**, 1–86.

SEAL, H. L. (1964). *Multivariate Statistical Analysis for Biologists.* London.

SELLECK, G. W. (1960). The climax concept. *Bot. Rev.*, **26**, 534–45.

SERNANDER, R. (1898). Studier öfver vegetationen; mellersta Skandinaviens fjälltrakter. *K. Svensk. Vetensk.-Akad.*, **6**, 325–67.

SERNANDER, R. (1901). *Den Skandinaviska vegetations spridungsbiologie.* Lundequistska, Uppsala.

SERNANDER, R. (1925). Exkursionsführer für Skane. *4th Int. Pflanzengeog. Exkurs.*, *1925*, 16 pp.

SHIMWELL, D. W. (1968a). *The Vegetation of the Derbyshire Dales.* Nat. Cons. Publ., Shrewsbury.

SHIMWELL, D. W. (1968b) *The Phytosociology of calcareous grasslands in the British Isles.* Ph.D. Thesis, Univ. of Durham.

SHIMWELL, D. W. (1971a, b) Festuco-Brometea Br.-Bl. & R. Tx. 1943 in the British Isles: the phytogeography and phytosociology of limestone grasslands, Parts I & II. *Vegetatio*, **22**.

SJÖRS, H. (1948). Myrvegetation i Bergslagen. *Acta Phytogeogr. Suecica*, **21**, 1–299.

SJÖRS, H. (1954). Slatterängar i Grangärde finnmark. *Acta Phytogeogr. Suecica*, **34**, 1–135.

SJÖRS, H. (1965). Regional ecology of mire sites and vegetation. *Acta Phytogeogr Suecica.*, **50**, 180–8.

SMITH, R. (1898). The plant associations of the Tay Basin. *Proc. Perthshire Nat. Sci. Soc.*, **2**, 200–17.

SOKAL, R. R., and SNEATH, P. H. A. (1963). *Principles of Numerical Taxonomy*. San Francisco and London.

SØRENSEN, T. A. (1948). A method of establishing groups of equal amplitude in plant sociology based on similarity of species content. *K. Danske Vidensk Selsk. Biol. Skr.*, 5 (4), 1–34.

STAMP, L. D. (1931). Vegetation Formulae. *Proc. 5th Inst. Bot. Cong. 1930, Sect. E*, 118–22.

STAMP, L. D. (1934). Vegetation formulae. *J. Ecol.*, 22, 299–303.

SVEDBERG, T. (1922). Ettbidrag till de statistika metodernas användninginom växtbiologien. *Svensk. Bot. Tidskr.*, 16, 1–8.

TANSLEY, A. G. (1904). The problems of ecology. *New Phytol.*, 3, 191–200.

TANSLEY, A. G. (ed.) (1911). *Types of British Vegetation*. Cambridge Univ. Press.

TANSLEY, A. G. (1920). The classification of vegetation and the concept of development. *J. Ecol.*, 8, 118–49.

TANSLEY, A. G. (1935). The use and abuse of vegetational concepts and terms. *Ecology*, 16, 284–307.

TANSLEY, A. G. (1939) (1953). *The British Islands and their Vegetation*. Cambridge Univ. Press.

TANSLEY, A. G., and ADAMSON, R. S. (1926). Studies on the vegetation of the English chalk. IV. A preliminary survey of the chalk grasslands of the Sussex Downs. *J. Ecol.*, 14, 1–32.

TANSLEY, A. G., and CHIPP, T. F. (1926). *Aims and Methods in the Study of Vegetation*. London.

THOMSON, G. W. (1952). Measures of plant aggregation based on contagious distributions. *Contr. Lab. Vertebr. Biol. Univ. Mich.*, 53, 1–16.

TOMASELLI, R. (1956). *Introduzione allo Studio Della Fitosociologia*. Ind. Poli. Lombarda, Milan.

TURNER, J. S., and WATT, A. S. (1939). The Oakwoods (Quercetum sessiliflorae) of Killarney, Ireland, *J. Ecol.*, 27, 202–33.

TÜXEN, R. (1937). Die Pflanzengesellschaften Nordwestdeutschlands. *Mitt. Flor.-soz. Arbeitsgem.*, 3.

TÜXEN, R. (1962). Das phänologische Gesellschaftsdiagramm. *Mitt. Flor.-soz. Arbeitsgem*, N.F. 9, 51–2.

TÜXEN, R. (1967). Pflanzensoziologische Beobachtung südwestnorwegischer Küstendünengebiete. *Aquilo Ser. Bot.*, 6, 241–72.

TÜXEN, R. (1969). Réflexions sur l'importance de la sociologie végétale pour l'économie de l'herbage europeen. *Melhoramento*, 21, 187–99.

TÜXEN, R., and ELLENBERG, H. (1937). Der systematische und der ökologische Gruppenwert. *Mitt. flor.-soz. Arbeitsgem*, 3, 171–84.

TÜXEN, R., and OBERDORFER, E. (1958). Eurosibirische Phanerogamen-Gesellschaften Spaniens. *Veröff. Geobot. Inst. Rübel*, 32.

USHER, M. B. (1969). The relation between mean square and block size in the analysis of similar patterns. *J. Ecol.*, 57, 505–14.

VAUPELL, C. (1857). Bögens invandring i de Danske skove. *Ann. Sci. Nat.*, 4 (7), 55.

VON POST, H. (1862). *Forsök till en systematisk uppstallning af vextstallena i mellersta Sverige*. Bonnier, Stockholm.

VON POST, H., and SERNANDER, R. (1910). Pflanzen-physiognomischen Studien auf Torfmooren in Närke. *Inst. Cong. Geol., Guide excurs. en Suede, Sess., 11*, 14, 1–48.

WALTER, H., and LIETH, H. (1967). *Klimadiagramm Weltatlas*. Jena.

WARMING, E. (1891). *De psammofile vegetationer i Denmark*. Fest Nat. For. Kjöbenhaven.

WARMING, E. (1895). *Plantesamfund. Grundträk af den ökologiska Plantegeografi*.

WARMING, E. (1909). *Oecology of Plants*. Oxford, 422 pp.

WARMING, E. (1923). Økologiens Grundformer. Udkast til en systematisk Ordning. *K. Dansk Vidensk. Selsk. Naturv. Math. Skr.*, **8**, R4, 119–87.

WATT, A. S. (1924). On the ecology of British beechwoods with special reference to their regeneration. Part II, Section I. *J. Ecol.*, **12**, 145–204.

WATT, A. S. (1925). On the ecology of British beechwoods with special reference to their regeneration. Part II. Sections II and III. *J. Ecol.*, **13**, 27–73.

WATT, A. S. (1934a, b). The vegetation of the Chiltern Hills with special reference to the beechwoods and their seral relationships. *J. Ecol.*, **22**, 230–70.

WATT, A. S. (1945). Contributions to the ecology of bracken (*Pteridium aquilinum*). III. Frond types and the make up of the population. *New Phytol.*, **44**, 156–78.

WATT, A. S. (1947). Pattern and process in the plant community. *J. Ecol.*, **35**, 1–22.

WATT, A. S. (1955). Bracken versus heather, a study in plant sociology. *J. Ecol.*, **43**, 490–506.

WATT, A. S. (1957). The effects of excluding rabbits from grassland B (Meso-brometum) in Breckland. *J. Ecol.*, **45**, 861–78.

WATT, A. S. (1962). The effect of excluding rabbits from grassland A (Xero-brometum) in Breckland, 1936–60. *J. Ecol.*, **50**, 181–98.

WEAVER, J. E. (1920). *Root Development in the Grassland Formation*. Carnegie Inst. Publ., Washington.

WEAVER, J. E., and CLEMENTS, F. E. (1938). *Plant Ecology*. McGraw-Hill, New York.

WEBB, D. A. (1954). Is the classification of plant communities either possible or desirable? *Bot. Tidsskr.*, **51**, 362–70.

WEBB, L. J. (1959). A physiognomic classification of Australian rain forests. *J. Ecol.*, **47**, 551–70.

WEBB, L. J., *et al.* (1967). Studies in the numerical analysis of complex rain-forest communities. I. A comparison of methods applicable to site/species data. *J. Ecol.*, **55**, 171–91.

WEBB, L. J., *et al.* (1970). Studies in the numerical analysis of complex rain-forest communities. V. A comparison of the properties of floristic and physiognomic-structural data. *J. Ecol.*, **58**, 203–32.

WELCH, J. R. (1960). Observations on deciduous woodland in the Eastern Province of Tanganyika. *J. Ecol.*, **48**, 557–73.

WELLS, T. C. E. (1967). Changes in a population of *Spiranthes spiralis* (L.) Chevall at Knocking Hoe National Nature Reserve, Bedfordshire, 1962–65. *J. Ecol.*, **55**, 83–99.

WESTHOFF, V. (1967). Problems and use of structure in the classification of vegetation. *Acta bot. Neerl.*, **15**, 495–511.

WHITMORE, T. C. (1962a). Studies in systematic bark morphology. II. General features of bark construction in Dipterocarpaceae. *New Phytol.*, **61**, 208–20.

WHITMORE, T. C. (1962b). Studies in systematic bark morphology. III. Bark taxonomy in Dipterocarpaceae. *Gdns. Bull.*, **19**, 321–71.

WHITTAKER, R. H. (1948). *A vegetation analysis of the Great Smokey Mountains*. Ph.D. Thesis, Univ. of Illinois, Urbana.

WHITTAKER, R. H. (1951). A criticism of the plant association and climatic climax concepts. *Northwest Sci.*, **25**, 17–31.

WHITTAKER, R. H. (1953). A consideration of climax theory: the climax as a population and pattern. *Ecol. Monogr.*, **23**, 41–78.

WHITTAKER, R. H. (1956). Vegetation of the Great Smoky Mountains. *Ecol. Monogr.*, **26**, 1–80.

WHITTAKER. R. H. (1960). Vegetation of the Siskiyou Mountains, Oregon and California. *Ecol. Monogr.*, **30**, 279–338.

WHITTAKER, R. H. (1962). Classification of natural communities. *Bot. Rev.*, **28**, 1–239.

WHITTAKER, R. H. (1965). Dominance and diversity in land plant communities. *Science (New York)*, **147**, 250–60.

WHITTAKER, R. H. (1967). Gradient analysis of vegetation. *Biol. Rev.*, **49**, 207–64.

WHITTAKER, R. H. (1970). *Communities and Ecosystems*. Macmillan, London.

WHITTAKER, R. H., and NIERING, W. A. (1965). Vegetation of the Santa Catalina Mountains, Arizona: a gradient analysis of the south slope. *Ecology*, **46**, 429–52.

WILLIAMS, J. T., and VARLEY, Y. W. (1967). Phytosociological studies of some British grasslands. *Vegetatio*, **15**, 169–89.

WILLIAMS, W. T., and LAMBERT, J. M. (1959). Multivariate methods in plant ecology. I. Association-analysis in plant communities. *J. Ecol.*, **47**, 83–101.

WILLIAMS, W. T., and LAMBERT, J. M. (1960). Multivariate methods in plant ecology. II. The use of an electronic digital computer for association-analysis. *J. Ecol.*, **48**, 689–710.

WILLIS, J. C., and BURKHILL, I. H. (1904). The phanerogamic flora of the Clova Mountains in special relation to flower-biology. *Trans. Bot. Soc. Edinb.*, **22**, 109–25.

WILSON, J. W. (1959a). Analysis of the spatial distribution of foliage by two-dimensional point quadrats. *New Phytol.*, **58**, 92–101.

WILSON, J. W. (1959b). Analysis of the distribution of foliage area in grassland. In: IVINS, J. D., *The measurement of grassland productivity*, pp. 51–61.

WINKWORTH, R. E., and GOODALL, D. W. (1962). A crosswire sighting tube for point quadrat analysis. *Ecology*, **43**, 342–3.

WINKWORTH, R. E. (1955). The use of point quadrats for the analysis of heath-land. *Aust. J. Bot.*, **3**, 68–81.

YARRANTON, G. A. (1966). A plotless method of sampling vegetation. *J. Ecol.*, **54**, 229–38.

ZONNEVELD, I. S. (1960). De Brabantse Biesbosch. A study of soil and vegetation of a fresh water tidal delta. *V.L.O.*, No. 65.20. PUDOC., Wageningen.

APPENDIX I

The Life-form System of Ellenberg (1956)

(N.B.—This system is preferred to the later elaborate system of Ellenberg and Müller-Dombois, 1967, because of its relative simplicity and its use of accepted terminology.)

A. RADIKANTE (rooted plants)
    I. MACRO PHANEROPHYTES (M or MP): trees or tree-like plants with buds more than 2 m above ground
        1. Eu-macro-phanerophytes: real trees (M)
            a. evergreen trees
                a.i.   evergreen rain forest trees: *ombro macro phan.* (oM)
                a.ii. evergreen soft leaves: *daphno macro phan.* (dM)
                a.iii. evergreen hard leaves: *sklero macro phan.* (sM)
                a.iv. evergreen needle-leaved trees; *belonido macro phan.* (bM)
            b. summer green trees
                b.i.   summer broad-leaved trees: *thero macro phan.* (tM)
                b.ii. summer needle-leaved trees: *thero-belonido macro phan.* (tbM)
            c. rainy green trees
                rainy broad-leaved trees: *cheimo macro phan.* (cM)
        2. Tree grasses: *macrophanerophyta graminidea*, e.g. bamboos (M gram)
        3. Rosette trees: *macrophanerophyta scaposa*, e.g. many palms (M scap)
        4. Herb-stem trees: *macrophanerophyta herbacea*, e.g. banana trees (M herb)
        5. High-stem succulents: *macrophanerophyta sukkulenta*, plants with water tissue in the stem (M sukk)
        6. Lianas: *macrophanerophyta scandentia* (M scand)

    II. NANO PHANEROPHYTES: shrubs; buds between 0·25 and 2 m above ground (NP or N)
                  same subdivision as I plus
        7. Leafless shrubs (not succulents): e.g. *Ruscus aculeatus*

    III. CHAMAEPHYTES: buds lower than 0·25 m, but not at ground level (Ch)
        1. Dwarf shrubs: *Chamaephyta frutescentia* (Ch frut)
            a. evergreen dwarf shrubs
                a.a. evergreen soft-leaved dwarf shrubs: *daphno cham. frut.* (dCH frut)
                a.b. evergreen hardwood dwarf shrubs: *skerlo cham. frut.* (sCH frut)
                a.c. leafless dwarf shrubs (1 CH frut)
            b. summer-green dwarf shrubs
                summer broad-leaved dwarf shrubs: *thero cham. frut.* (tCH frut)
        2. Semi shrubs: *Chamaephyta suffrutescentia* (CH suffr) same subdivision as 1

3. Trailing shrubs: *Chamaephyta velantia* (CH vel) same subdivision as 1 except a.iii
4. Cushion dwarf shrubs: *Chamaephyta frutescentia pulvinata* (CH frut pulv)
5. Cushion herbs: *Chamaephyta pulvinata* (CH pulv)
6. Creeping herbs: *Chamaephyta reptantia* (CH rept)
7. Low perennial climbers: *Chamaephyta scandentia* (CH scand)
8. Low succulents: *Chamaephyta sukkulenta* (CH sukk)
   a. leaf succulents
   b. stem succulents
9. Hard grasses: *Chamaephyta graminidea* (CH gram)

IV. HEMICRYPTOPHYTES: buds close to the ground surface and generally protected by a covering layer of dead leaves (H)
1. Tussock plants: *Hemicryptophyta caespitosa* (H caesp) grasses and grass-like plants forming tussocks
2. Creeping Hemicryptophytes: *Hemicryptophyta reptantia* (H rept)
3. Rosette plants: *Hemicryptophyta rosulata* (H ros)
4. Semi rosette plants: *Hemicryptophyta hemirosulata* (H hem)
5. Scapose plants: *Hemicryptophyta scaposa* (H scap)
6. Climbing plants: *Hemicryptophyta scandentia* (H scand)
7. Water-Hemicryptophytes: *Hydro-Hemicryptophyta* (Hyd H)

V. GEOPHYTES (or cryptophytes) (G)
1. Root-bud geophytes: *Geophyta radicigemma* (G rad)
2. Rhizome-geophytes: *Geophyta rhizomatosa* (G rhiz)
3. Bulb-geophytes: *Geophyta bulbosa* (G bulb)
4. Water-geophytes: *Hydro-Geophyta* (Hyd G)

VI. THEROPHYTES: annuals (T)
1. Tussock therophytes: *Therophyta caespitosa* (T caesp)
2. Creeping therophytes: *Therophyta reptantia* (T rept)
3. Rosette   "   *Therophyta rosulata* (T ros)
4. Semi-rosette   "   *Therophyta hemirosulata* (T hem)
5. Scapose   "   *Therophyta scaposa* (T scap)
6. Climbing   "   *Therophyta scandentia* (T scand)
7. Water   "   *Hydro-Therophyta* (Hyd T)

All groups 1–6 can be subdivided according to the season in which they germinate and are green.
   a. Winter annuals (*Theroph. hibernalis*)
      (germinate in autumn, green in winter)
   b. Summer annuals (*Theroph. aestivalis*)
      (germinate in spring-autumn, but die in winter)
   c. Rain annuals (*Theroph. pluvialis*)
      (germinate and live only in rainy season)
   d. In summerdress hibernating annuals (*Theroph. epigeios*)
      (germinate in spring and summer)

B. **ADNATE** (attached plants)
   I. VASCULAR EPIPHYTES: *Kormo-Epiphyta*
   II. THALLO-EPIPHYTES: mosses, lichens, fungi, algae living on other plants
   III. THALLO-CHAMAEPHYTES: mosses, lichens
   IV. THALLO-HEMICRYPTOPHYTES: mosses, lichens, algae
   V. THALLO-GEOPHYTES: fungi
   VI. THALLO-THEROPHYTES: short-living fungi and mosses

C. **ERRANTE** (motile plants)

    I. HIGHER FLOATING PLANTS: *Kormo-hydrophyta natantia*
    II. LOWER FLOATING PLANTS: *Thallo-hydrophyta natantia*
    III. HYDROPLANKTON: *Hydro-planktophyta*
    IV. CRYOPLANKTON: *Cryo-planktophyta*
    V. EDAPHOPHYTES: Microscopic soil flora

APPENDIX II

Classification and Description of Vegetation
(Rübel 1930–31)

(a) *The standard description of a community should include:*

I. Environmental factors
   1. Physiography: locality, exposure, slope, altitude
   2. Climatic factors
   3. Edaphic factors
   4. Biotic factors—including forestry and agricultural management

II. Morphology
   1. Ecological morphology
      Physiognomy, life forms
      Layers
      Synusiae
      Exclusiveness
      Vitality
      Periodicity and rhythm
      Aspects
   2. Floristic morphology
      List of species
      Abundance
      Frequency
      Constancy
      Extent. Sociability

III. Succession
      Early, transition and final stages
      Different phases of each stage
      Structural value of the species

(b) *Rübel's formations with examples from the regions of Figure 42*

(Examples are taken from as many of the floristic regions of Good as possible using readily available reference. The floristic regions are denoted according to their letter in the legend of Figure 42.)

1. Polar and alpine vegetation

   F.I. FRIGORIDESERTA—cold deserts
      (a) Late snow-bed moss vegetation, Gjaerevoll, 1956 (Norway)   A,B
      (b) *Rhacomitrium* tundra, Rübel, 1930 (Iceland, Greenland)   A,B
      (c) Moss heaths, McVean and Ratcliffe, 1963 (Scotland)   B
      (d) Arctic fjaeldmark, Rübel, 1930 (Iceland, Greenland)   B

(e) *Rhacomitrium-Andraea* cinder scree, Wace and Holdgate, 1958 (Tristan de Cunha) I
(f) Antarctic non-vascular cryptogam tundra, Gimingham and Smith, 1970 (Maritime Antarctic) —

2. Aciculilignosa—needle-leaf forests and scrub

F.II. ACICULISILVAE—needle-leaf forests
(a) Scandinavian pine-spruce forests, Rübel, 1930 (Norway) B
(b) *Pinus sylvestris* forest, McVean and Ratcliffe, 1962 (Scotland) B
(c) *Pinus cembra* woodland, Rivas-Goday, 1950 (Spain) C
(d) *Cedrus* forest, Rübel, 1930 (Atlas Mountains, N. Africa) C
(e) *Pinus canariensis* forest, Rübel, 1930 (Canary Isles) D

F.III. ACICULIFRUTICETA—needle-leaf scrub
(a) Sub-alpine juniper scrub, McV. and Rat. (Scotland) B

3. Aestilignosa—summer green deciduous temperate forests and scrub

F.IV. AESTILIGNOSAE—deciduous forests
(a) Beech, oak and ash woodlands, Tansley, 1939 (Gt. Britain) B
(b) *Quercus pubescens* forests, Rivas-Goday (Spain) C
(c) North African poplar-alder woodland, Rübel (N. Africa) C

F. V. AESTIFRUTICETA—DECIDUOUS scrub
(a) *Betula nana-B. pubescens* scrub, Rübel (Greenland, Iceland) A
(b) *Corylus* scrub Tan. (England and Iceland) B
(c) *Alnus-Salix* fen carr (England) B
(d) *Quercus pubescens* scrub, Rübel (Spain) C

4. Hiemilignosae—deciduous raingreen tropical and sub-tropical forests

F.VI. HIEMILIGNOSAE—raingreen forests
(a) Sudan Savanna F
(b) Southern Guinea Savanna F
(c) *Acacia-Commiphora* Sahel steppe, Nielson, 1965 F
(d) Types of S. African bushveld, e.g. *Sclerocarya-Acacia* lowveld, Acocks, 1953 K

F.VII. HIEMIFRUTICETA—raingreen scrub
(a) Doum palm-*Acacia* oases, Walter, 1962 E
(b) *Acacia-Panicum* wadi, Walter E
(c) *Acacia giraffae* thornveld, Acocks K

5. Laurilignosa—soft-leaved evergreen and temperate forest and scrub

F.VIII. LAURISILVAE—soft-leaved forests
(a) *Persea-Myrica* laurel forest, Tutin, 1953 (Azores) D
(b) *Laurus canariensis* forest, Rübel (Canaries) D
(c) *Cassine-Olea* Alexandria and Knysna forests, Ac. (S. Africa) K

F.IX. LAURIFRUTICETA—soft-leaved scrubs
(a) *Rhododendron ponticum* scrub, Rübel (Portugal, etc.) C

6. Durilignosa—sclerophyllous Mediterranean woodlands and scrub

F.X. DURISILVAE—sclerophyllous woodland
(a) *Quercus ilex* woodland, Rivas-Goday (Spain) C
(b) *Quercus suber* woodland, Rübel (Spain and Portugal) C
(c) *Phoenix* date-palm woodland, Rübel (N. Africa) C
(d) *Protea-Restionaceae* spp. sclerophyllous bush, Ac. (Cape) L

F.XI.　DURIFRUTICETA—sclerophyllous scrub
- (a) *Arbutus unedo* scrub, Tan. (S. Ireland)　　　　　　　　　　B
- (b) *Arbutus-Phillyrea* scrub, Rübel (Spain)　　　　　　　　　　C
- (c) *Cistus* Macchia, Rübel (S. France, Spain)　　　　　　　　　C
- (d) *Chamaerops* palm scrub (Spain, Morocco, Algeria)　　　　　C
- (e) S. African Fynbos, Ac. (Cape Province)　　　　　　　　　　K
- (f) *Phylica arborea* bush, Wace and Hold (Tristan da Cunha)　I

7. Deserta (Siccideserta). The deserts are divided by Rübel into six formations of which the Frigorideserta and Siccideserta are the only two climatically determined ones. Others which rely on a habitat definition are as follows:

Litorideserta—salt deserts and saltmarshes represented ± throughout the global transect

Mobilideserta—unstable dune systems (both coastal and inland) and screes

Rupideserta—alpine and other cliff faces

Saxideserta—cryptogamic vegetation of detached rocks, etc.

The latter groups are better regarded as sub-formations within each of the major climatic formations.

F.XI.　SICCIDESERTA—hot, dry sandy and stony deserts
- (a) *Stipa* semi-steppe, Rübel (N. Africa)　　　　　　　　　　C,E
- (b) *Zygophyllum-Zilla* shrubby desert, Walter (Sahara)　　　　E
- (c) *Fredolia* stone-polster desert, Walter (Algeria)　　　　　E
- (d) *Spartocytisus* alpine desert, Rübel (Canaries)　　　　　　D
- (e) *Crassula* succulent Karroo, Acocks (S. Africa)　　　　　　K
- (f) *Salsola* arid Karroo, Acocks (S. Africa)　　　　　　　　　K

8. Herbosa (Duriherbosa). As with the previous formation class, Rübel recognized a series of edaphically determined formations:

Emersiherbosa—vegetation subject to periodic inundation, e.g. salt marshes, flood plain marshes, etc.

Submersiherbosa—truly aquatic, floating vegetation

Sphagniherbosa—*Sphagnum* dominated mires

Sempervirentiherbosa—permanent low grasslands mainly above the present tree limit

Altherbosa—tall herb sub-alpine communities generally above the tree limit

Duriherbosa—climatically determined tall grass pampas, prairie and veld

F.XIII.　DURIHERBOSA
- (a) *Stipa-Aristida* steppe Rübel (Algeria)　　　　　　　　　　C
- (b) *Cymbopogon-Themeda* grassveld, Acocks (S. Africa)　　　　K

9. Pluviilignosa—tropical and sub-tropical evergreen rain forests

F.XIV.　PLUVIISILVAE—tropical rain forest
- (a) Mixed rain forest, Richards, 1952 (Nigeria)　　　　　　　　G
- (b) *Alstonia* swamp forest, Niel (Nigeria)　　　　　　　　　　G

F.XV.　PLUVIIFRUTICETA—tropical scrub
- (a) *Rhizophora*—mangrove swamp, Niel (W. Africa)　　　　　　G

One further major formation is recognized by Rübel—*Ericifruticeta*. This formation of ericaceous heaths is partly climatic in that it is ± restricted to the Atlantic regions and partly edaphic in that it frequently occurs above the tree zone. Examples are to be found in most northern and temperate regions:

(a) *Phyllodoce caerulea* heath, Rübel (Arctic Scandinavia)   A
(b) *Arctous-Calluna* alpine heath, McVean and Ratcliffe (Scotland)   B
(c) *Erica vagans* Cornish heath, Rübel (S.W. England)   B
(d) *Erica mediterranea* heath, Rübel (Portugal)   C
(e) *Calluna-Daboecia* heath, Tutin (Azores)   D
(f) *Erica arborea-E. scoparia* heath, Rübel (Canaries)   D
(g) *Erica-Blaeria* heath, Acocks (Cape Province)   L
(h) *Rhacomitrium-Empetrum rubrum* heath, Wace and Holdgate (Tristan da
    Cunha)   I

REFERENCES

ACOCKS, J. P. H. (1953). Veld types of South Africa. *S. Afr. Dept. Agric. Bot. Surv. Mem., No. 28.*

GIMINGHAM, C. H., and SMITH, R. I. L. (1970). Bryophyte and lichen communities in the maritime Atlantic. In: HOLDGATE, M. W. *Antarctic Ecology, 2,* pp. 752–85.

McVEAN, D. N., and RATCLIFFE, D. A. (1962). *Plant Communities of the Scottish Highlands,* H.M.S.O., London.

NIELSEN, M. S. (1965). *Introduction to the Flowering Plants of West Africa.* London.

RICHARDS, P. W. (1952). *The Tropical Rain Forest.* Cambridge.

RIVAS-GODAY, S. (1950). Essai sur les climax dans la Peninsule Ibérique. *Proc. 7th Int. Bot. Cong. Stockholm,* pp. 648–50.

RÜBEL, E. (1930). *Pflanzengesellschaften der Erde.* Berlin.

TANSLEY, A. G. (1939). *The British Islands and their Vegetation.* Cambridge.

TUTIN, T. G. (1953). The vegetation of the Azores. *J. Ecol.,* **41,** 53–61.

WACE, N. M., and HOLDGATE, M. W. (1958). The vegetation of Tristan da Cunha. *J. Ecol.,* **46,** 593–620.

WALTER, H. (1962). *Die vegetation der Erde,* Vol. 1.

APPENDIX III

The application of the Zurich–Montpellier classification to the vegetation of the British Isles: a synopsis of higher vegetation units which are present in Britain

The following system is based on that of Lohmeyer *et al.* (1962) for north-west Europe and the recent work on a Prodromus of European plant communities outlined in *Internationale Vereinigung für Vegetationskunde, Vegetatio*, 22, 251–83 (1971).

The units are recognized on the basis of the present author's own field experience and extensive surveys of the relevant literature. A more expanded form of the classification system is published elsewhere.

1. Class Lemnetea—free floating duckweed mats of eutrophic waters.
    O. Lemnetalia
    All. Lemnion minoris
    All. Lemnion trisulcae
2. Class Zosteretea—mud flat and estuarine communities of the lower inter-tidal zone exposed for a maximum of three hours.
    O. Zosteretalia
    All. Zosterion
3. Class Thero-Salicornietea—stands of annual *Salicornia* of saline mud flats in the eu- and supralittoral zones.
    O. Thero-Salicornietalia
    All. Thero-Salicornion
4. Class Spartinetea—*Spartina* salt marshes.
    O. Spartinetalia
    All. Spartinion
5. Class Ceratophylletea—floating *Ceratophyllum* mats of slow moving eutrophic waters.
    O. Ceratophylletalia
    All. Ceratophyllion
6. Class Utricularietea—floating and submerged communities of ombrogenous and other peaty pools.
    O. Utricularietalia
    All. Utricularion
7. Class Corynephoretea—pioneer communities of dry, continental acidic, sandy soils.
    O. Corynephoretalia
    All. Corynephorion
8. Class Ammophiletea—vegetation of embryo and mobile dunes.
    O. Elymo-Ammophiletalia
    All. Agropyrion boreoatlanticum

312

All. Ammophilion borealis
O. Euphorbio-Ammophiletalia

9. Class Cakiletea maritimae—open communities of annual halonitrophile species of strand lines and coastal jetsam.
   O. Cakiletalia maritimae
   All. Salsolo-Minuartion peploidis
   O. Euphorbietalia peplis
   All. Euphorbion peplis
   O. Thero Suaedetalia
   All. Thero-Suaedion

10. Class Honckenyo-Agropyretea pungentis—perennial communities of the halonitrophile strand boundary zone, mainly in exposed situations.
   O. Honckenyo-Crambetalia
   All. Honckenyo-Crambion
   O. Agropyretalia pungentis
   All. Agropyrion pungentis

11. Class Asplenietea rupestris—small fern-dominated communities of rocks and walls.
   O. Tortulo-Cymbalarietalia
   All. Parietarion judaicae
   All. Cymbalario-Asplenion
      Sub. All. Tortulo-Asplenion
      Sub. All. Tortulo-Saginion
   O. Asplenio-Cystopteridetalia
   All. Asplenio-Cystopteridion
   All. Oxyrio-Sedion roseae

12. Class Adiantetea—fern and moss-dominated vegetation of calcareous tufa.
   O. Adiantetalia
   All. Adiantion
   All. Cratoneurion commutati

13. Class Crithmo-Limonietea—maritime rock crevice communities.
   O. Crithmo-Limonietalia
   All. Crithmion maritimi

14. Class Thlaspeetea rotundifolii—alpine and sub-alpine scree vegetation.
   O. Thlaspeetalia rotundifolii
   (All. Stipion calamagrostis)
   All. Arenarion norvegicae
   O. Androsacetalia
   (All. Androsacion alpinae)

15. Class Isoeto-Nanojuncetea—short-lived small rush—annual dominated communities of sandy habitats which are inundated throughout winter.
   O. Cyperetalia fuscae
   All. Cicendion
   All. Nanocyperion

16. Class Bidentetea tripartiti—nitrophilous weed communities in habitats subject to periodic inundation.
   O. Bidentetalia tripartiti
   All. Bidention tripartiti

17. Class Secalinetea—thermophilous weed communities of grain and flax fields.
   O. Aperetalia
   All. Arnoserion
   All. Aphanion

18. Class Chenopodietea—nitrophilous weed communities of rootcrop fields and waste places.

O. Polygono-Chenopodietalia
  All. Eu-Polygono-Chenopodion
O. Sisymbrietalia
  All. Sisymbrion
  All. Polygono-Coronopion
19. Class Artemisietea—nitrophilous tall herb weed communities of border zones, e.g. hedgerows and around lakes.
  O. Artemisietalia
    All. Eu-Arction
  O. Galio-Alliarietalia
    All. Geo-Alliarion
    All. Convolvulion sepii
20. Class Epilobietea angustifolii—nitrophilous woodland edge and clearing communities.
  O. Epilobietalia angustifolii
    All. Epilobion angustifolii
    All. Fragarion vescae
21. Class Potametea—perennial rooted aquatic vegetation composed mainly of species of *Potamogeton, Ranunculus* and *Callitriche.*
  O. Magnopotametalia
    All. Magnopotamion
    All. Nymphaeion
  O. Parvopotametalia
    All. Parvopotamion
    All. Callitricho-Batrachion
  O. Luronio-Potametalia
    All. Potamion graminei
22. Class Charetea—submerged stonewort dominated vegetation.
  O. Charetalia
    All. Charion fragilis
    All. Charion canescentis
23. Class Stratiotetea—floating communities composed of members of the family Hydrocharitaceae with a marked periodicity.
  O. Stratiotetalia
    All. Stratiotion
24. Class Ruppietea—submerged meadows of brackish water.
  O. Ruppietalia
    All. Ruppion maritimae
25. Class Littorelletea—rooted aquatic vegetation composed of rosette species in oligotrophic and dystrophic waters.
  O. Littorelletalia
    All. Littorellion
26. Class Montio-Cardaminetea—vegetation of oligotrophic springs and flushes mainly in the montane zone.
  O. Montio-Cardaminetalia
    All. Cardamino-Montion
    All. Mniobryo-Epilobion
27. Class Scheuchzerietea—*Sphagnum*-dominated communities of ombrogeneous bog pools and flats.
  O. Scheuchzerietalia palustris
    All. Rhynchosperion albae
28. Class Phragmitetea—emergent, aquatic tall sedge, grass and herb vegetation at the margins of lakes, canals, etc.
  O. Phragmitetalia

All. Phragmition
O. Nasturtio-Glycerietalia
    All. Glycerio-Sparganion
    All. Apion nodiflori
O. Magnocaricetalia
    All. Magnocaricion
29. Class Bulboschoenetea maritimi—tall sedge vegetation of brackish waters.
    O. Bulboschoenetalia
        All. Bulboschoenion maritimi
        All. Magnocaricion paleaceae
30. Class Juncetea maritimi—halophyte communities dominated by tall *Juncus* species.
    O. Juncetalia maritimi
        All. Juncion maritimi
31. Class Asteretea tripolium—grass and herb-rich salt marsh vegetation.
    O. Glauco-Puccinellietalia
        All. Puccinellion maritimae
        All. Armerion maritimae
        All. Puccinellio-Spergularion salinae
32. Class Saginetea maritimae—small winter-annual and rosette communities of slightly saline dune slacks.
    O. Saginetalia maritimae
        All. Saginion maritimae
33. Class Salicornietea fruticosae—*Salicornia fruticosa* dominated maritime strand vegetation.
    O. Salicornietalia fruticosae
        All. Salicornion fruticosae
    O. Limonietalia
34. Class Violetea calaminariae—open communities of heavy metal mine spoil heaps.
    O. Violetalia calaminariae
        All. Thlaspeion calaminariae
35. Class Molinio-Arrhenatheretea—vegetation of hay meadows, permanent pastures and adjacent footpaths.
    O. Trifolio-Agrostietalia
        All. Agropyro-Rumicion crispi
        All. Trifolio-Cynodontion
    O. Plantaginetalia maioris
        All. Lolio-Plantaginion
    O. Arrhenatheretalia
        All. Arrhenatherion elatioris
        All. Cynosurion cristati
        All. Ranunculo-Anthoxantion
    O. Molinietalia
        All. Calthion palustris
        All. Filipendulion
        All. Junco-Molinion
36. Class Sedo-Scleranthetea—± closed herbaceous vegetation of sand dunes and other acidic sandy soils.
    O. Festuco-Sedetalia
        All. Thero-Airion
        All. Sedo-Cerastion
        All. Galio-Koelerion
        All. Armerion elongatae

37. Class Festuco-Brometea—vegetation of calcareous grasslands.
    O. Brometalia erecti
        All. Bromion
        All. Mesobromion
    Class Elyno-Seslerietea—alpine and sub-alpine dwarf shrub heaths and grasslands mainly on base-rich soils.
    O. Elyno-Dryadetalia
        All. Kobresio-Dryadion
        All. Potentillo-Polygonion vivipari
39. Class Caricetea nigrae (Parvocaricetea)—small sedge-dominated vegetation of mires.
    O. Caricetalia nigrae
        All. Caricion curto-nigrae
    O. Tofieldietalia
        All. Eriopherion latifolii
        All. Caricion bicoloris-atrofuscae
40. Class Oxycocco-Sphagnetea—vegetation of ombrogeneous and topogeneous peats.
    O. Ericetalia tetralicis
        All. Ericion tetralicis
    O. Sphagnetalia magellanici
        All. Erico-Sphagnion
        All. Sphagnion fusci
41. Class Nardo-Callunetea—species-poor acidophilous grasslands and grass-heaths.
    O. Nardetalia
        All. Eu-Nardion
        All. Nardo-Galion saxatilis
    O. Calluno-Ulicetalia
        All. Calluno-Genistion
        All. Sarothamnion
        All. Empetrion nigri
42. Class Trifolio-Geranietea—species rich, often calcicolous communities of scrub-grassland boundary zones.
    O. Origanetalia
        All. Trifolion medii
        All. Geranion sanguinei
43. Class Betulo-Adensotyletea—Betula—tall herb sub-alpine and alpine communities often lacking a tree layer.
    O. Adenostyletalia
        All. Lactucion alpinae
44. Class Salicetea purpureae—mainly montane and pre-alpine willow scrubs.
    O. Salicetalia purpureae
        All. Salicion albae
        All. Salicion cinereae
45. Class Rhamno-Prunetea—woodland edge scrub communities.
    O. Prunetalia spinosae
        All. Rubion sub-atlanticum
        All. Berberidion
        All. Salicion arenariae
        All. Potentillion fruticosae
    O. Sambucetalia
        All. Sambuco-Salicion capreae
        All. Lonicero-Rubion sylvatici

46. Class Alnetea glutinosae—species-poor alder woodlands of oligotrophic marshes.
    O. Alnetalia glutinosae
    All. Alnion glutinosae
47. Class Vaccinio-Piceetea—pine forest and associated juniper scrub vegetation.
    O. Vaccinio-Piceetalia
    All. Vaccinio-Piceion
    All. Juniperion nanae
48. Class Loiseleurio-Vaccinietea—alpine heaths and grass heaths above the tree
    limit.
    O. Caricetalia curvulae
    All. Loiseleurieto-Arctostaphylion
49. Class Salicetea herbaceae—dwarf willow—moss-dominated communities of late
    snow-lie beds.
    O. Salicetalia herbaceae
    All. Cassiopeto-Salicion herbaceae
50. Class Quercetea robori-petraeae—acidophilous oak woodlands.
    O. Quercetalia robori-petraeae
    All. Quercion robori-petraeae
51. Class Querco-Fagetea—species-rich oak, ash and beech woodland of calcareous
    soils and eutrophic river plains.
    O. Fagetalia sylvaticae
    All. Alno-Padion
    All. Fagion sylvaticeae
    All. Fraxino-Brachypodion

AUTHOR INDEX

SUBJECT INDEX